Advanced Digital Design and Intelligent Manufacturing

Advanced Digital Design and Intelligent Manufacturing

Guest Editors

Jozef Svetlík
Rudolf Jánoš
Ján Semjon

MDPI

Basel • Beijing • Wuhan • Barcelona • Belgrade • Novi Sad • Cluj • Manchester

Guest Editors

Jozef Svetlík
Department of
Manufacturing Machinery
and Robotics
Technical University of Košice
Košice
Slovakia

Rudolf Jánoš
Department of
Manufacturing Machinery
and Robotics
Technical University of Košice
Košice
Slovakia

Ján Semjon
Department of
Manufacturing Machinery
and Robotics
Technical University of Košice
Košice
Slovakia

Editorial Office
MDPI AG
Grosspeteranlage 5
4052 Basel, Switzerland

This is a reprint of the Special Issue, published open access by the journal *Applied Sciences* (ISSN 2076-3417), freely accessible at: https://www.mdpi.com/journal/applsci/special_issues/I1N70N2639.

For citation purposes, cite each article independently as indicated on the article page online and as indicated below:

Lastname, A.A.; Lastname, B.B. Article Title. *Journal Name* **Year**, *Volume Number*, Page Range.

ISBN 978-3-7258-7807-9 (Hbk)
ISBN 978-3-7258-7808-6 (PDF)
https://doi.org/10.3390/books978-3-7258-7808-6

Cover image courtesy of Rudolf Jánoš

Contents

Preface . . . . **vii**

Jozef Svetlík, Rudolf Jánoš and Ján Semjon
Advanced Digital Design and Intelligent Manufacturing
Reprinted from: *Appl. Sci.* **2026**, *16*, 2754, https://doi.org/10.3390/app16062754 . . . . **1**

Štefan Ondočko, Jozef Svetlík, Rudolf Jánoš, Ján Semjon and Miroslav Dovica
Calculation of Trusses System in MATLAB—Multibody
Reprinted from: *Appl. Sci.* **2024**, *14*, 9547, https://doi.org/10.3390/app14209547 . . . . **5**

Daria Fedorova, Vladimír Tlach, Ivan Kuric, Tomáš Dodok, Ivan Zajačko and Karol Tucki
Technical Diagnostics of Industrial Robots Using Vibration Signals: Case Study on Detecting Base Unfastening
Reprinted from: *Appl. Sci.* **2024**, *15*, 270, https://doi.org/10.3390/app15010270 . . . . **25**

Patrik Fejko, Damián Peti, Jozef Dobránsky, Miroslav Gombár and Peter Michalík
Comprehensive Analysis of Deposition Parameters and Energy-Dispersive X-Ray Spectroscopy Characterization in Cataphoretic Coating Processes
Reprinted from: *Appl. Sci.* **2025**, *15*, 3760, https://doi.org/10.3390/app15073760 . . . . **47**

Peter Gabštur, Marek Kočiško, Jakub Kaščak and Martin Pollák
Methodology for Verification of Geometrically Complex Components Through Reverse Engineering
Reprinted from: *Appl. Sci.* **2025**, *15*, 3963, https://doi.org/10.3390/app15073963 . . . . **71**

Ciarán-Victor Veitenheimer, Dirk Alexander Molitor, Viktor Arne and Peter Groche
Controlling Product Properties in Forming Processes Using Reinforcement Learning—An Application to V Die Bending
Reprinted from: *Appl. Sci.* **2025**, *15*, 5483, https://doi.org/10.3390/app15105483 . . . . **93**

Claudiu-Damian Petru, Fineas Morariu, Radu-Eugen Breaz, Mihai Crenganiș, Sever-Gabriel Racz, Claudia-Emilia Gîrjob, et al.
Development of a Small CNC Machining Center for Physical Implementation and a Digital Twin
Reprinted from: *Appl. Sci.* **2025**, *15*, 5549, https://doi.org/10.3390/app15105549 . . . . **111**

Fedor Burčiar, Monika Herchlová, Bohuslava Juhásová, Martin Juhás and Pavel Važan
Proactive Approach to Production Control Utilizing Heterogeneous Shop-Level Production Data
Reprinted from: *Appl. Sci.* **2025**, *15*, 7570, https://doi.org/10.3390/app15137570 . . . . **144**

Štefan Ondočko, Jozef Svetlík, Rudolf Jánoš, Ján Semjon, Marek Sukop, Tomáš Stejskal and Peter Marcinko
Conceptually Simple Method for Optimizing Model Computations in MATLAB Simulink
Reprinted from: *Appl. Sci.* **2025**, *15*, 11312, https://doi.org/10.3390/app152111312 . . . . **163**

Carlos Schmidt, Friedrich Volz, Ljiljana Stojanovic and Holger Kett
Integration Approaches for Digital Twins in Dataspaces
Reprinted from: *Appl. Sci.* **2025**, *15*, 11623, https://doi.org/10.3390/app152111623 . . . . **194**

Lukáš Hanko, Ondrej Chlebo, Peter Križan, Stanislav Strigáč, Miloš Matúš and Lenka Matejáková
Investigation of Key Process Parameters Affecting Product Quality in Robotic Milling: A Comprehensive Analysis
Reprinted from: *Appl. Sci.* **2026**, *16*, 1832, https://doi.org/10.3390/app16041832 . . . . **215**

Preface

The rapid evolution of industrial technologies has initiated a fundamental transformation in how products are designed and manufactured. This Reprint captures a pivotal moment in mechanical engineering where digital tools and intelligent systems converge to redefine production efficiency. Our primary motivation for compiling this work was to bridge the gap between advanced theoretical simulations and the practical challenges faced in modern factory environments.

The scope of this Reprint encompasses a wide range of topics including industrial robotics, advanced CNC machining and the implementation of digital twins. A significant portion of the contributed research focuses on the technical nuances of robotic milling addressing the critical issues of system rigidity and geometric precision. By analyzing the impact of key technological parameters such as feed rates and depth of cut, the authors provide actionable data for optimizing complex manufacturing workflows.

The aim and purpose of this collection are to provide a robust scientific foundation for researchers and engineers who are navigating the transition toward Industry 4.0. It is addressed to both academic scholars and industry professionals seeking to enhance operational performance through automation and artificial intelligence. We believe that the insights presented here will foster further innovation in creating more flexible and sustainable manufacturing systems. As Guest Editors, we are honored to present these contributions and hope that this Reprint serves as a valuable resource for the global engineering community.

Jozef Svetlík, Rudolf Jánoš, and Ján Semjon
Guest Editors

applied sciences

Editorial

Advanced Digital Design and Intelligent Manufacturing

Jozef Svetlík, Rudolf Jánoš and Ján Semjon *

Department of Production Technology and Robotics, Faculty of Mechanical Engineering, Technical University of Kosice, Park Komenskeho 8, 040 01 Kosice, Slovakia; jozef.svetlik@tuke.sk (J.S.); rudolf.janos@tuke.sk (R.J.)
* Correspondence: jan.semjon@tuke.sk; Tel.: +421-55-602-2625

1. Introduction

This Special Issue was initiated to reflect how digital technologies, data, and intelligence are reshaping mechanical engineering and production systems across the entire product life cycle [1–3]. Advanced computer aided design (CAD), high-fidelity simulations, and automation are now central to achieving higher precision, efficiency, and innovation in products and processes, while artificial intelligence (AI), machine learning (ML), and the Industrial Internet of Things (IIoT) enable real-time monitoring, predictive maintenance, and adaptive optimization in smart factories [4–7]. This Closing Editorial situates the contributions of the Special Issue within current research on Industry 4.0, digital twins, and intelligent manufacturing, highlighting their impact and outlining future directions for both research and industrial practice.

2. Digital Design, Modeling, and Simulation

Digital design is increasingly grounded in multi-domain modeling and simulation, which together form the backbone of "design for intelligent manufacturing". This Special Issue shows how general purpose environments such as MATLAB/Simulink R2025b and Simscape can be exploited beyond classical control design to support structural analysis and evaluation of mechatronic systems, thereby reducing dependence on labor-intensive analytical calculations [8–11]. This trend is aligned with the broader shift toward simulation-driven design, where virtual models are iteratively refined before physical prototyping, effectively establishing a digital thread from conception to realization.

A key contribution in this context is the conceptually simple method for optimizing model computations in MATLAB Simulink, which focuses on solver selection, time step control, and numerical settings to improve accuracy without altering the underlying physical model. Similar numerical concerns have been addressed in hybrid simulation frameworks for cyber physical systems, where inappropriate integration schemes or step sizes can compromise stability and reliability of simulation-based decisions. By providing a practical, well-structured procedure, the article supports wider industrial adoption of advanced simulation, and is compatible with model-based systems engineering (MBSE) approaches that integrate models across disciplines and life-cycle stages [12–14].

More generally, robust simulation is a prerequisite for trustworthy digital twins and virtual commissioning, where the virtual representation must faithfully capture dynamic behavior, constraints, and interactions among system components. The works in this Special Issue demonstrate that when numerical configuration is treated as a first-class design variable, multi-body and system-level simulations can provide reliable guidance for later stages of manufacturing system design and operation [15,16].

3. Digital Twins and Data-Driven Integration

Digital twins have emerged as a cornerstone of intelligent manufacturing, providing dynamic, data-enriched virtual representations of physical assets, processes, and systems. This Special Issue presents implementations of digital twins for CNC machining centers and entire production systems, where simulation models are coupled with real-time and historical data from sensors, controllers, and manufacturing execution systems (MES). Such implementations resonate with established frameworks for digital-twin-based sustainable intelligent manufacturing, which emphasize closed-loop interactions between design, production, and service to support continuous performance improvement and life-cycle optimization.

A distinctive feature of the reported work is the integration of heterogeneous data sources—including CSV logs, database connections, OPC-based communication, and scenario files—into coherent digital twin environments that support monitoring, planning, and decision support. Similar architectures have been proposed in the literature to enable flexible and scalable data pipelines for Industry 4.0 applications, particularly in small and medium-sized enterprises (SMEs) that require cost effective solutions [14,17]. By validating their approaches on different manufacturing systems, the contributions in this Special Issue illustrate concrete pathways for SMEs to adopt digital twin technologies, complementing conceptual surveys and reference frameworks reported elsewhere [17–19].

Beyond individual factories, there is growing attention to industrial data spaces and standardized reference architectures, such as the Asset Administration Shell (AAS), which aim to ensure secure, interoperable data exchange across organizational boundaries. Contributions to this Special Issue that demonstrate automatic registration and synchronization of AAS-based digital twins within data space infrastructures align with ongoing research on federated industrial data management and cross company value networks [19]. These developments are tightly coupled with research on AI-enabled digital twins, where generative and predictive models are orchestrated across cloud and edge resources to support diagnostics, optimization, and decision making in distributed manufacturing ecosystems.

4. Intelligent Monitoring, Control, and Process Optimization

Intelligent manufacturing is closely associated with advanced monitoring and control strategies that leverage ML and data-driven models to enhance process performance, quality, and reliability. In this Special Issue, several contributions examine condition monitoring of robotic systems and intelligent control of forming and machining processes, thereby extending the reach of AI-based analytics into domains that have historically relied on heuristic or rule-based approaches. These efforts echo a broader body of work demonstrating the effectiveness of deep learning and statistical learning for fault diagnosis and predictive maintenance in complex industrial systems [20].

Vibration-based monitoring of industrial robots, for instance, combines statistical feature selection methods (such as ANOVA and *t* tests) with classification algorithms to detect loosening of robot bases during operation, providing an example of data-driven condition monitoring tailored to sequential and multi-axis machinery. Similar strategies have been widely applied to rotating machines, where carefully engineered features and supervised learning have yielded significant improvements over traditional threshold-based approaches. Extending these methods to robots addresses a recognized gap in the diagnostics literature and supports the development of predictive maintenance programs for highly automated production lines [20,21].

Process optimization is further illustrated by studies on robotic milling and metal forming, where the influence of parameters like feed rate, depth of cut, and forming angles on product quality and springback is systematically analyzed. These works are in

https://doi.org/10.3390/app16062754

line with research on data-driven process optimization and virtual machining, which use simulation and experimental data to build models that predict surface quality, dimensional accuracy, and energy consumption for different parameter combinations. In particular, the application of reinforcement learning (RL) for springback compensation in V bending demonstrates how "learning in the loop" strategies can adapt to uncertainties in material properties and tooling, reflecting a wider trend toward integrating RL into industrial process control [22–24].

Collectively, these contributions show that intelligent monitoring and control can be integrated into existing manufacturing systems incrementally, augmenting rather than replacing established automation and control architectures.

5. Conclusions

The contributions gathered in this Special Issue collectively demonstrate that advanced digital design and intelligent manufacturing are deeply intertwined dimensions of the emerging cyber physical production paradigm [1–3]. Multi-domain modeling, numerically robust simulation, and digital twins form the foundation upon which intelligent monitoring, AI-based control, and data-driven optimization are built. At the same time, developments in industrial data spaces, standardized architectures, and MBSE provide the structural and organizational scaffolding needed to integrate these technologies across life-cycle stages and organizational boundaries.

Looking ahead, several research challenges and opportunities can be identified. First, there is a need for interoperable reference architectures that support modular digital twin components, hybrid physics-based and data-driven models, and standardized interfaces for analytics and control services. Second, robust AI methods that can cope with non-stationary conditions, limited labeled data, and strict safety requirements are essential to move from prototypes to large-scale, mission-critical deployments. Third, specific attention should be paid to SMEs, ensuring that digital twin and AI solutions are economically viable, scalable, and compatible with existing equipment and skills. Finally, education and training in digital design, data science, and AI for manufacturing will be crucial to fully exploit the potential of these technologies and to foster a workforce capable of working effectively in data rich, highly automated environments.

As Guest Editors, we believe that the contributions collected in this Special Issue provide a timely and representative snapshot of ongoing developments in advanced digital design and intelligent manufacturing [1–3]. We hope that the presented concepts, methods, and case studies will stimulate further interdisciplinary collaboration between academia and industry, and will support the transition towards more intelligent, sustainable, and resilient manufacturing systems in line with the vision of Industry 4.0 and beyond.

Author Contributions: J.S. (Jozef Svetlík): writing—original draft preparation; R.J. and J.S. (Ján Semjon): writing—review and editing. All authors have read and agreed to the published version of the manuscript.

Funding: This research received no external funding.

Conflicts of Interest: The authors declare no conflicts of interest.

References

1. He, B.; Bai, K.-J. Digital twin-based sustainable intelligent manufacturing: A review. *Adv. Manuf.* **2020**, *9*, 1–21. [CrossRef]
2. Huang, Z.; Shen, Y.; Li, J.; Fey, M.; Brecher, C. A survey on AI-driven digital twins in industry 4.0: Smart manufacturing and advanced robotics. *Sensors* **2021**, *21*, 6340. [CrossRef] [PubMed]
3. Singh, A.P.; Gameti, N. Digital Twins in Manufacturing: A Survey of Current Practices and Future Trends. *Int. J. Sci. Res. Arch.* **2024**, *13*, 1240–1250. [CrossRef]

https://doi.org/10.3390/app16062754

4. Mon, B.F.; Hayajneh, M.; Ali, N.A.; Ullah, F.; Ullah, H.; Alkobaisi, S. Digital Twins in the IIoT: Current Practices and Future Directions Toward Industry 5.0. *Comput. Mater. Contin.* **2025**, *83*, 3675–3712. [CrossRef]
5. Rosen, D.W. Thoughtson design for intelligent manufacturing. *Engineering* **2019**, *5*, 609–614. [CrossRef]
6. Rahman, M.A.; Shahrior, M.F.; Iqbal, K.; Abushaiba, A.A. Enabling Intelligent Industrial Automation: A Review of Machine Learning Applications with Digital Twin and Edge AI Integration. *Automation* **2025**, *6*, 37. [CrossRef]
7. Malik, A.A. Simulation Based High Fidelity Digital Twins of Manufacturing Systems: An application model and industrial use case. In *Proceedings of the 2023 Winter Simulation Conference (WSC), San Antonio, TX, USA, 10–13 December 2023*; IEEE: New York, NY, USA, 2023; pp. 3262–3271. [CrossRef]
8. Feng, Y.; Zhao, Y.; Zheng, H.; Li, Z.; Tan, J. Data-driven product design toward intelligent manufacturing: A review. *Int. J. Adv. Robot. Syst.* **2020**, *17*, 1729881420911257. [CrossRef]
9. Ondocko, S.; Svetlik, J.; Janos, R.; Semjon, J.; Sukop, M.; Stejskal, T.; Marcinko, P. Conceptually Simple Method for Optimizing Model Computations in MATLAB Simulink. *Appl. Sci.* **2025**, *15*, 11312. [CrossRef]
10. Zhang, L.; Zhou, L.; Ren, L.; Laili, Y. Modeling and simulation in intelligent manufacturing. *Comput. Ind.* **2019**, *112*, 103123. [CrossRef]
11. Das, S. *Modeling and Simulation of Mechatronic Systems Using Simscape*; Morgan & Claypool Publishers: San Rafael, CA, USA, 2020. Available online: https://www.scribd.com/document/574937143/9781681737355-sample (accessed on 10 March 2026).
12. Nguyen, P.H.; Ali, S.; Yue, T. Model-based security engineering for cyber-physical systems: A systematic mapping study. *Inf. Softw. Technol.* **2017**, *83*, 116–135. [CrossRef]
13. Tao, F.; Qi, Q.; Liu, A.; Kusiak, A. Data-driven smart manufacturing. *J. Manuf. Syst.* **2018**, *48*, 157–169. [CrossRef]
14. Krajcovic, M.; Bulej, V.; Kuric, I.; Sapietova, A. Intelligent manufacturing systems in concept of digital factory. *Commun. Sci. Lett. Univ. Žilina* **2013**, *15*, 77–87. [CrossRef]
15. Wang, J.; Niu, X.; Gao, R.X.; Huang, Z.; Xue, R. Digital twin-driven virtual commissioning of machine tool. *Robot. Comput.-Integr. Manuf.* **2023**, *81*, 102499. [CrossRef]
16. Ugarte, M.; Etxeberria, L.; Unamuno, G.; Bellanco, J.L.; Ugalde, E. Implementation of digital twin-based virtual commissioning in machine tool manufacturing. *Procedia Comput. Sci.* **2022**, *200*, 527–536. [CrossRef]
17. Schnicke, F.; Kuhn, T. Reference Architectures for Industry 4.0. In *Reference Architectures for Critical Domains*; Nakagawa, E.Y., Oliveira Antonino, P., Eds.; Springer: Cham, Switzerland, 2023. [CrossRef]
18. Qi, Q.; Tao, F. Digital Twin and Big Data Towards Smart Manufacturing and Industry 4.0: 360 Degree Comparison. *IEEE Access* **2018**, *6*, 3585–3593. [CrossRef]
19. Dai, D.; Zhao, B.; Yu, Z.; Franciosa, P.; Ceglarek, D. Generative and Predictive AI for digital twin systems in manufacturing. *Front. Artif. Intell.* **2025**, *8*, 1655470. [CrossRef] [PubMed]
20. Alkhatib, F.; Allam, M.; Swarnakar, V.; Alsadi, J.; Maalouf, M. Implementing and evaluating the quality 4.0 PMQ framework for process monitoring in automotive manufacturing. *Sci. Rep.* **2025**, *15*, 24742. [CrossRef] [PubMed]
21. Tolk, A.; Page, E.H.; Mittal, S. Hybrid Simulation for Cyber Physical Systems: State of the Art and a Literature Review. In *Proceedings of the Symposium on M&S of Complexity in Intelligent, Adaptive and Autonomous Systems, Spring Simulation Conference*; Society for Computer Simulation International: San Diego, CA, USA, 2018; pp. 10–11. [CrossRef]
22. Cevik, Z.A.; Ozsoy, K.; Ercetin, A.; Sariisik, G. Machine Learning-Driven Optimization of Machining Parameters Optimization for Cutting Forces and Surface Roughness in Micro-Milling of AlSi10Mg Produced by Powder Bed Fusion Additive Manufacturing. *Appl. Sci.* **2025**, *15*, 6553. [CrossRef]
23. Gabštur, P.; Kočiško, M.; Kaščak, J.; Pollák, M. Methodology for Verification of Geometrically Complex Components Through Reverse Engineering. *Appl. Sci.* **2025**, *15*, 3963. [CrossRef]
24. Klimecka-Tatar, D.; Krynke, M. Reverse engineering tools-3D scanning-as support for precise quality control in automated special processes. *Procedia Comput. Sci.* **2025**, *253*, 1933–1942. [CrossRef]

https://doi.org/10.3390/app16062754

applied sciences

MDPI

Article

Calculation of Trusses System in MATLAB—Multibody

Štefan Ondočko [1], Jozef Svetlík [1,*], Rudolf Jánoš [1], Ján Semjon [1] and Miroslav Dovica [2]

[1] Department of Manufacturing Machinery and Robotics, Faculty of Mechanical Engineering, The Technical University of Košice, Letná 9, 04001 Košice, Slovakia; stefan.ondocko@tuke.sk (Š.O.); rudolf.janos@tuke.sk (R.J.); jan.semjon@tuke.sk (J.S.)

[2] Department of Biomedical Engineering and Measurement, Faculty of Mechanical Engineering, The Technical University of Košice, Letná 9, 04001 Košice, Slovakia; miroslav.dovica@tuke.sk

* Correspondence: jozef.svetlik@tuke.sk

Abstract: This article discusses the software tool (Simscape—Multibody program of MATLAB) primarily intended for dynamic and kinematic processes with practical applications in static calculations. Currently, there are few published scientific works utilizing this tool for tasks like basic static calculations of truss systems. We were interested in comparing the calculation using the tools we use in our work and research activities for theoretical calculation; the potential reliance on simulations in the future could help to avoid the necessity of complex theoretical calculations, which can be time-consuming and prone to errors. Despite the fact that the structure may appear simple, in practice, there may not always be time for a verification calculation in the theoretical field (proper model creation, inclusion of all conditions, etc.). The beam system is intentionally both externally and internally statically indeterminate. For this reason, it is logically necessary to also consider deformation conditions. The achieved results were interesting in terms of accuracy compared to SOLIDWORKS, which was used for computation verification. Through very simple optimization, we were able to further increase the calculation accuracy without complicating other parameters.

Keywords: FEM; static analyses; beam; bars; trusses; MATLAB; SOLIDWORKS

1. Introduction

This paper examines a statically indeterminate trusses system developed through analytical computation and the subsequent use of software tools MATLAB version 24.1.0.2628055 (R2024a) Update 4 and SOLIDWORKS Education Edition 2023 SP5.0. MATLAB—Simscape Multibody [1–4]—contains within its library Flexible Bodies Library elements named Flexible Beam [3,4] with various cross-sections. These are flexible, slender beams capable of elastic deformation. The deformation involves additions due to extension, bending, and torsion. The stiffness and inertia of the body (Flexible Beam) are calculated based on the geometry and material properties. The calculation principle employed complies with the Euler–Bernoulli beam theory [5] (pp. 174–220), [6] (pp. 8–66), [7]; hence, it is suitable for the trusses system. Numerous programs for calculating such systems exist, many in different programming languages and differing in the purpose for which they were designed. The market offers a diverse range of options. References [8,9] present an example of a statically indeterminate planar trusses system calculated using the finite element method (FEM). This is a standalone program, with the calculation verified in ANSYS [10–12]. In this paper, a model developed in the SOLIDWORKS environment using the Simulation tool is intended to verify the results. The entire design was created using the Weldments tool. The calculation also includes the visualization of results using the specified input parameters. The analytical calculations are performed in a linear domain subject to the well-known Hooke's law. The term bar refers to a body of the same cross-section fixed at both ends using a rotary link, i.e., joints. An axis of the bar passes through the center of the joints. The point of connection of at least two bars in the center of the joints (forming a rotational link) is called a gusset or node. The gussets are the nodes where the transfer of forces to the rest of the structure occurs. The bar transmits the force in its axis through tension and compression, subject to the aforementioned Bernoulli–Navier hypothesis. In the case of compression, however, this is true only to a certain extent. This hypothesis assumes the preservation of the flatness of the cross-sections before and after deformation. In other words, after deformation, the cross-sections remain perpendicular to the deformed axis of symmetry of the cross-section. The same properties apply to the material of the beam in tension and compression. The normal stress is distributed uniformly; thus, there is no shear inside the bar. It should also be mentioned that the Bernoulli–Navier hypothesis is only accurate for rather slender bars. Under compressive loading, the so-called buckling [13] (pp. 483–507) occurs when the bar loses its stability and buckles out of its axis. The buckling stability of trusses systems is difficult to handle [14–16], especially if several parameters are monitored and optimized simultaneously [17].

2. Example of Analytical Design Based on In-Plane Deformation Conditions

The system in Figure 1 is composed of five bars. It is externally and internally statically 1× indeterminate. When one degree of freedom looseness is suitably removed at node A or D, which binds the system to the world frame, the system does not turn into a mechanism. Similarly, after removing any of the bars 1 to 5, the structure remains unchanged. This means that the given system is externally and internally 1× statically indeterminate. Thus, we partially relax the system and express the unknown reactions in terms of the deformation conditions.

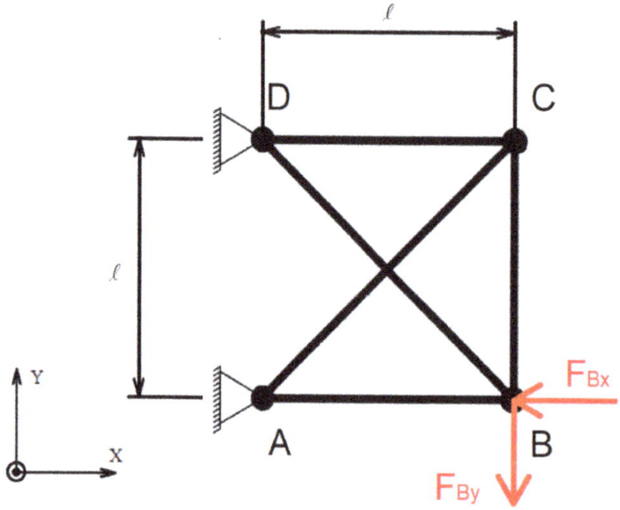

Figure 1. Structure of the trusses system loaded by the force at node B.

The example mainly involves comparing theoretical results with simulations in MATLAB—Multibody and then conducting verification in SOLIDWORKS. The structure of the trusses system loaded by the force at node B after release from the bonds at node(s) A, D, and release of the bar between nodes AB is displayed in Figure 2.

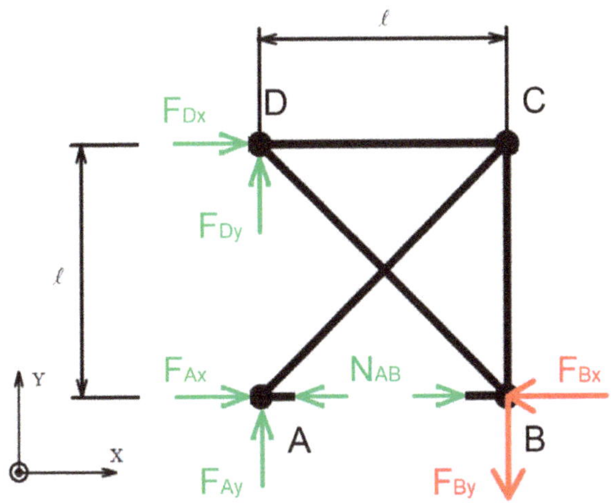

Figure 2. Structure of the trusses system loaded by the force at node B after release from the bonds at node A, D, and release of the bar between nodes AB.

For the energy change of the whole system under the N_{AB} force [13] (p. 204), [18] (p. 160) in the bar, the following applies:

$$\frac{\partial U}{\partial N_{AB}} = 0 \tag{1}$$

To determine the forces in individual nodes or gussets, we use the method of joint [19] (p. 70), [20,21]. By successively releasing the nodes, see Figures 3–6, we denote the forces acting on the bars via the index of the nodes where the bar is attached.

Node A:

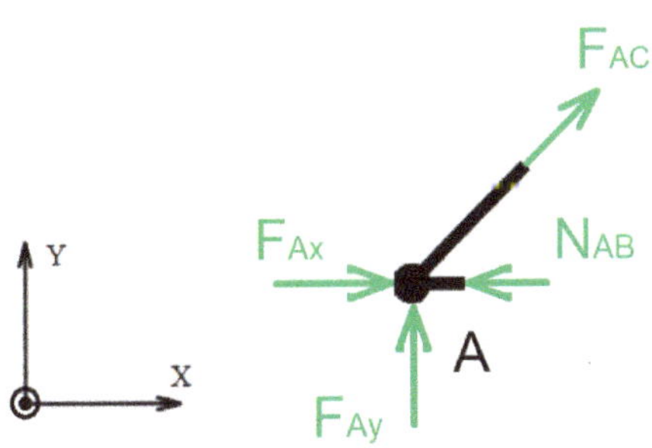

Figure 3. Reactions after release from the bonds at node A and release of bar AB, AC.

In general, the following applies to the equilibrium of forces at a node:

$$\sum \mathbf{F} = 0 \tag{2}$$

By decomposing into the individual axes of the coordinate system, the following applies:

$$\sum F_x = 0;\ F_{Ax} + \frac{F_{AC}}{\sqrt{2}} - N_{AB} = 0 \tag{3}$$

$$\sum F_y = 0;\ F_{Ay} + \frac{F_{AC}}{\sqrt{2}} = 0 \tag{4}$$

Node B:

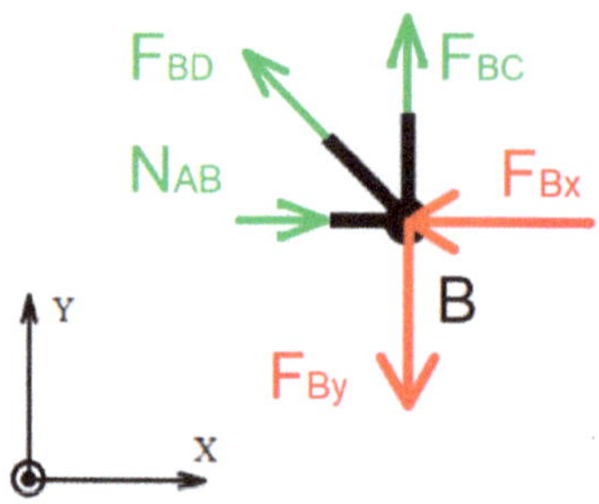

Figure 4. Reactions after release from the bonds at node B and release of bar AB, BD, BC.

$$\sum F_x = 0;\ N_{AB} - F_{Bx} - \frac{F_{BD}}{\sqrt{2}} = 0 \quad (5)$$

$$\sum F_y = 0;\ F_{BC} - F_{By} + \frac{F_{BD}}{\sqrt{2}} = 0 \quad (6)$$

Node C:

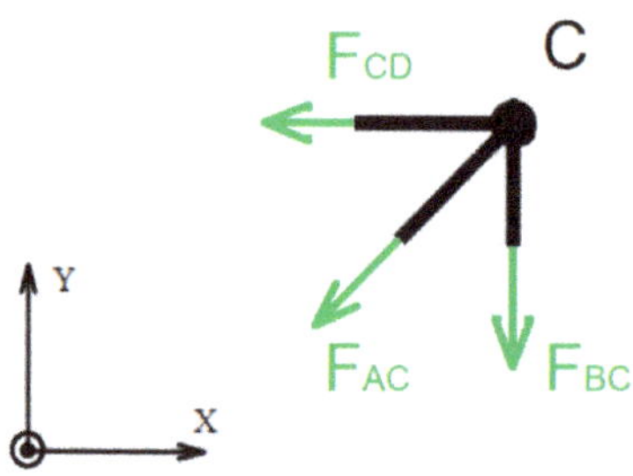

Figure 5. Reactions after release from the bonds at node C and release of bar AC, BC, CD.

$$\sum F_x = 0; -F_{CD} - \frac{F_{AC}}{\sqrt{2}} = 0 \quad (7)$$

$$\sum F_y = 0; -F_{BC} - \frac{F_{AC}}{\sqrt{2}} = 0 \quad (8)$$

Node D:

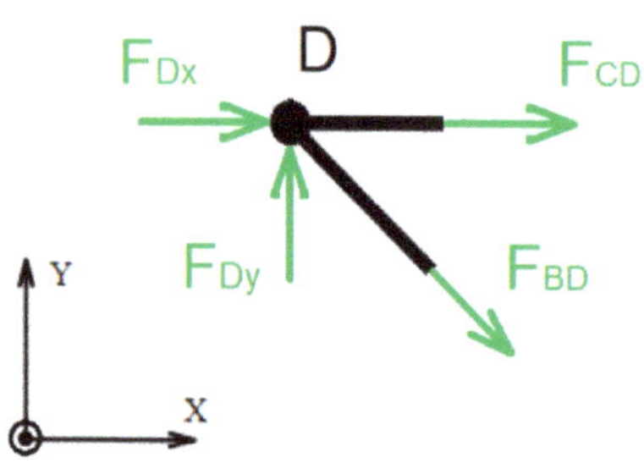

Figure 6. Reactions after release from the bonds at node D and release of bar BD, CD.

$$\sum F_x = 0;\ F_{Dx} + F_{CD} + \frac{F_{BD}}{\sqrt{2}} = 0 \quad (9)$$

$$\sum F_y = 0;\ F_{Dy} - \frac{F_{BD}}{\sqrt{2}} = 0 \tag{10}$$

Consider the system of given Equations (5)–(8), where the variables without reactions at nodes A and D are searched for:

$$N_{\mathrm{AB}} - \frac{\sqrt{2}\,F_{\mathrm{BD}}}{2} = F_{Bx} \tag{11}$$

$$F_{\mathrm{BC}} + \frac{\sqrt{2}\,F_{\mathrm{BD}}}{2} = F_{By} \tag{12}$$

$$-F_{\mathrm{CD}} - \frac{\sqrt{2}\,F_{\mathrm{AC}}}{2} = 0 \tag{13}$$

$$-F_{\mathrm{BC}} - \frac{\sqrt{2}\,F_{\mathrm{AC}}}{2} = 0 \tag{14}$$

And its transcription into matrix entry for automated computer calculation:

$$\begin{bmatrix} 1 & 0 & 0 & -\frac{\sqrt{2}}{2} & 0 \\ 0 & 0 & 1 & \frac{\sqrt{2}}{2} & 0 \\ 0 & -\frac{\sqrt{2}}{2} & 0 & 0 & -1 \\ 0 & -\frac{\sqrt{2}}{2} & -1 & 0 & 0 \end{bmatrix} \cdot \begin{bmatrix} N_{\mathrm{AB}} \\ F_{\mathrm{AC}} \\ F_{\mathrm{BC}} \\ F_{\mathrm{BD}} \\ F_{\mathrm{CD}} \end{bmatrix} = \begin{bmatrix} F_{\mathrm{Bx}} \\ F_{\mathrm{By}} \\ 0 \\ 0 \end{bmatrix} \tag{15}$$

From system (15), the unknown forces in the bars expressed in terms of the known variables F_{Bx}, F_{By}, and N_{AB} are as follows:

$$F_{\mathrm{AC}} = -\sqrt{2}\cdot(F_{\mathrm{Bx}} - N_{\mathrm{AB}} + F_{\mathrm{By}}) \tag{16}$$

$$F_{\mathrm{BC}} = F_{\mathrm{Bx}} - N_{\mathrm{AB}} + F_{\mathrm{By}} \tag{17}$$

$$F_{\mathrm{BD}} = \sqrt{2}\cdot(N_{\mathrm{AB}} - F_{\mathrm{Bx}}) \tag{18}$$

$$F_{\mathrm{CD}} = F_{\mathrm{Bx}} - N_{\mathrm{AB}} + F_{\mathrm{By}} \tag{19}$$

Based on the deformation condition (1) regarding the energy change in the entire system under the N_{AB} force in the bar, the following applies:

$$\sum \frac{F_i\cdot l_i}{E_i\cdot S_i}\cdot\frac{\partial F_i}{\partial N_{AB}} = 0 \tag{20}$$

The cross-sections S_i and Young's modulus E_i are the same in the present case. The lengths l_i can be expressed as multiples of the uniform length l. Subsequently, based on the deformation condition (1) regarding the energy change in the whole system under the force N_{AB} in the bar AB, the following applies:

$$\frac{l}{E\cdot S}\sum F_i\cdot\frac{\partial F_i}{\partial N_{AB}} = 0 \tag{21}$$

$$N_{\mathrm{AB}} - \sqrt{2}\cdot 2\cdot(F_{\mathrm{Bx}} - N_{\mathrm{AB}} + F_{\mathrm{By}}) - (F_{\mathrm{Bx}} - N_{\mathrm{AB}} + F_{\mathrm{By}}) + \sqrt{2}\cdot 2\cdot(N_{\mathrm{AB}} - F_{\mathrm{Bx}}) - (F_{\mathrm{Bx}} - N_{\mathrm{AB}} + F_{\mathrm{By}}) = 0 \tag{22}$$

$$N_{\mathrm{AB}} = 2\cdot\frac{\left(2\cdot\sqrt{2}+1\right)\cdot F_{\mathrm{Bx}} + \left(\sqrt{2}+1\right)\cdot F_{\mathrm{By}}}{\left(4\cdot\sqrt{2}+3\right)} \tag{23}$$

In general, the equations for external equilibrium are as follows:

$$\sum \mathbf{F} = 0 \tag{24}$$

$$\sum \mathbf{M} = 0 \tag{25}$$

After decomposing:

$$\sum F_x = 0;\ F_{Ax} + F_{Dx} = F_{Bx} \tag{26}$$

$$\sum F_y = 0;\ F_{Ay} + F_{Dy} = F_{By} \tag{27}$$

$$\sum M_A = 0;\ -F_{Dx} \cdot \ell = F_{By} \cdot \ell \tag{28}$$

Alternatively, its transcription into matrix entry for automated computer calculation is as follows:

$$\begin{bmatrix} 1 & 0 & 1 & 0 \\ 0 & 1 & 0 & 1 \\ 0 & 0 & -1 & 0 \end{bmatrix} \cdot \begin{bmatrix} F_{Ax} \\ F_{Ay} \\ F_{Dx} \\ F_{Dy} \end{bmatrix} = \begin{bmatrix} F_{Bx} \\ F_{By} \\ F_{By} \end{bmatrix} \tag{29}$$

From system (29), the unknown forces in the bars expressed in terms of the known variables F_{Bx}, F_{By}, and F_{Ay} are as follows:

$$F_{Ax} = F_{Bx} + F_{By} \tag{30}$$

$$F_{Dx} = -F_{By} \tag{31}$$

$$F_{Dy} = F_{By} - F_{Ay} \tag{32}$$

The force F_{Ay} can be calculated from the equilibrium of forces of node A using Equation (4) as follows:

$$F_{Ay} = -\frac{F_{AC}}{\sqrt{2}} \tag{33}$$

After substituting Equation (16) into (33):

$$F_{Ay} = -\frac{-\sqrt{2} \cdot (F_{Bx} - N_{AB} + F_{By})}{\sqrt{2}}, \tag{34}$$

and substituting (23) into (34), the equation yields:

$$F_{Ay} = F_{Dx} - 2 \cdot \frac{\left(2 \cdot \sqrt{2} + 1\right) \cdot F_{Bx} + \left(\sqrt{2} + 1\right) \cdot F_{By}}{\left(4 \cdot \sqrt{2} + 3\right)} + F_{By} \tag{35}$$

The deflections at nodes B and C of the trusses system are determined by Castigliano's first theorem [13] (pp. 202–204), [18] (36) for node B and (53) for node C. See Figures 7 and 8. This will be determined by the partial derivative of the external component of the force vector $\mathbf{F_e}$ acting in the direction of the deflection sought. It is important to note that the situation is simplified assuming the same cross-sections S, Young's modulus E, and multiples of uniform length l.

Node B:

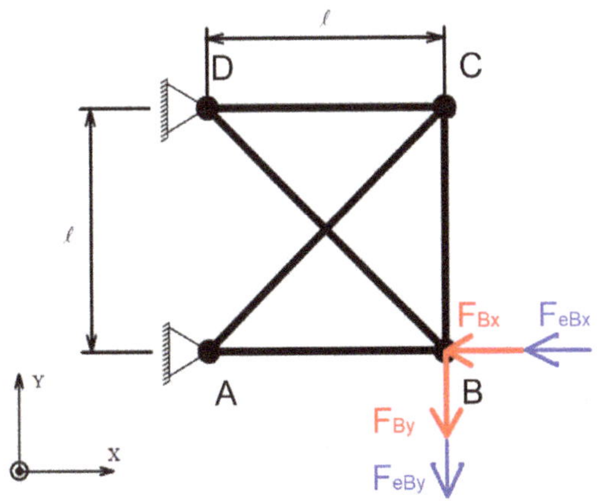

Figure 7. Structure of the trusses system loaded by the force vector $\mathbf{F}_\mathrm{B}$ at node B and the additional external force vector $\mathbf{F}_\mathrm{eB}$.

The deflection vector δ_B of node B derived from the first Castigliano theorem can be calculated using the following equation:

$$\delta_B = \frac{l}{E \cdot S}\begin{bmatrix} \sum F_i \cdot \frac{\partial F_i}{\partial F_{eBx}} \\ \sum F_i \cdot \frac{\partial F_i}{\partial F_{eBy}} \end{bmatrix} \quad (36)$$

Each component of the force F_i is expressed as a function of the component of the external force vector $\mathbf{F}_\mathrm{eB}$ as follows:

$$\delta_B = \frac{l}{E \cdot S}\begin{bmatrix} F_{eBx} + N_\mathrm{AB} \cdot \frac{\partial N_\mathrm{AB}}{\partial F_{eBx}} + \sqrt{2} \cdot F_\mathrm{AC} \cdot \frac{\partial F_\mathrm{AC}}{\partial F_{eBx}} + F_\mathrm{BC} \cdot \frac{\partial F_\mathrm{BC}}{\partial F_{eBx}} + \sqrt{2} \cdot F_\mathrm{BD} \cdot \frac{\partial F_\mathrm{BD}}{\partial F_{eBx}} + F_\mathrm{CD} \cdot \frac{\partial F_\mathrm{CD}}{\partial F_{eBx}} \\ F_{eBy} + N_\mathrm{AB} \cdot \frac{\partial N_\mathrm{AB}}{\partial F_{eBy}} + \sqrt{2} \cdot F_\mathrm{AC} \cdot \frac{\partial F_\mathrm{AC}}{\partial F_{eBy}} + F_\mathrm{BC} \cdot \frac{\partial F_\mathrm{BC}}{\partial F_{eBy}} + \sqrt{2} \cdot F_\mathrm{BD} \cdot \frac{\partial F_\mathrm{BD}}{\partial F_{eBy}} + F_\mathrm{CD} \cdot \frac{\partial F_\mathrm{CD}}{\partial F_{eBy}} \end{bmatrix} \quad (37)$$

By substituting the corresponding component of the external force vector $\mathbf{F}_\mathrm{eB}$ into Equation (23) derived from the deformation condition, the equation reads:

$$N_\mathrm{AB} = 2 \cdot \frac{\left(2 \cdot \sqrt{2} + 1\right) \cdot (F_\mathrm{Bx} + F_\mathrm{eBx}) + \left(\sqrt{2} + 1\right) \cdot (F_\mathrm{By} + F_\mathrm{eBy})}{\left(4 \cdot \sqrt{2} + 3\right)}, \quad (38)$$

Substituting N_AB into Equations (16) to (19) yields the equation as follows:

$$F_\mathrm{AC} = -\sqrt{2} \cdot \left(F_\mathrm{Bx} + F_\mathrm{eBx} - 2 \cdot \frac{\left(2 \cdot \sqrt{2} + 1\right) \cdot (F_\mathrm{Bx} + F_\mathrm{eBx}) + \left(\sqrt{2} + 1\right) \cdot (F_\mathrm{By} + F_\mathrm{eBy})}{\left(4 \cdot \sqrt{2} + 3\right)} + F_\mathrm{By} + F_\mathrm{eBy} \right) \quad (39)$$

$$F_\mathrm{BC} = F_\mathrm{Bx} + F_\mathrm{eBx} - 2 \cdot \frac{\left(2 \cdot \sqrt{2} + 1\right) \cdot (F_\mathrm{Bx} + F_\mathrm{eBx}) + \left(\sqrt{2} + 1\right) \cdot (F_\mathrm{By} + F_\mathrm{eBy})}{\left(4 \cdot \sqrt{2} + 3\right)} + F_\mathrm{By} + F_\mathrm{eBy} \quad (40)$$

$$F_\mathrm{BD} = \sqrt{2} \cdot \left(2 \cdot \frac{\left(2 \cdot \sqrt{2} + 1\right) \cdot (F_\mathrm{Bx} + F_\mathrm{eBx}) + \left(\sqrt{2} + 1\right) \cdot (F_\mathrm{By} + F_\mathrm{eBy})}{\left(4 \cdot \sqrt{2} + 3\right)} - F_\mathrm{Bx} - F_\mathrm{eBx} \right) \quad (41)$$

$$F_\mathrm{CD} = F_\mathrm{Bx} + F_\mathrm{eBx} - 2 \cdot \frac{\left(2 \cdot \sqrt{2} + 1\right) \cdot (F_\mathrm{Bx} + F_\mathrm{eBx}) + \left(\sqrt{2} + 1\right) \cdot (F_\mathrm{By} + F_\mathrm{eBy})}{\left(4 \cdot \sqrt{2} + 3\right)} + F_\mathrm{By} + F_\mathrm{eBy} \quad (42)$$

The individual elements of the matrix in Equation (37) after deriving by F_{eBx} are as follows:

$$N_{AB}\cdot\frac{\partial N_{AB}}{\partial F_{eBx}} = 4\cdot\frac{\left(2\cdot\sqrt{2}+1\right)\cdot(F_{Bx}+F_{eBx})+\left(\sqrt{2}+1\right)\cdot(F_{By}+F_{eBy})}{\left(4\cdot\sqrt{2}+3\right)}\cdot\frac{\left(2\cdot\sqrt{2}+1\right)}{\left(4\cdot\sqrt{2}+3\right)} \tag{43}$$

$$F_{AC}\cdot\frac{\partial F_{AC}}{\partial F_{eBx}} = 2\cdot\left(F_{Bx}+F_{eBx}-2\cdot\frac{\left(2\cdot\sqrt{2}+1\right)\cdot(F_{Bx}+F_{eBx})+\left(\sqrt{2}+1\right)\cdot(F_{By}+F_{eBy})}{\left(4\cdot\sqrt{2}+3\right)}+F_{By}+F_{eBy}\right)\cdot\left(1-2\cdot\frac{\left(2\cdot\sqrt{2}+1\right)}{\left(4\cdot\sqrt{2}+3\right)}\right) \tag{44}$$

$$F_{BC}\cdot\frac{\partial F_{BC}}{\partial F_{eBx}} = \left(F_{Bx}+F_{eBx}-2\cdot\frac{\left(2\cdot\sqrt{2}+1\right)\cdot(F_{Bx}+F_{eBx})+\left(\sqrt{2}+1\right)\cdot(F_{By}+F_{eBy})}{\left(4\cdot\sqrt{2}+3\right)}+F_{By}+F_{eBy}\right)\cdot\left(1-2\cdot\frac{\left(2\cdot\sqrt{2}+1\right)}{\left(4\cdot\sqrt{2}+3\right)}\right) \tag{45}$$

$$F_{BD}\cdot\frac{\partial F_{BD}}{\partial F_{eBx}} = 2\cdot\left(2\cdot\frac{\left(2\cdot\sqrt{2}+1\right)\cdot(F_{Bx}+F_{eBx})+\left(\sqrt{2}+1\right)\cdot(F_{By}+F_{eBy})}{\left(4\cdot\sqrt{2}+3\right)}-F_{Bx}-F_{eBx}\right)\cdot\left(2\cdot\frac{\left(2\cdot\sqrt{2}+1\right)}{\left(4\cdot\sqrt{2}+3\right)}-1\right) \tag{46}$$

$$F_{CD}\cdot\frac{\partial F_{CD}}{\partial F_{eBx}} = \left(F_{Bx}+F_{eBx}-2\cdot\frac{\left(2\cdot\sqrt{2}+1\right)\cdot(F_{Bx}+F_{eBx})+\left(\sqrt{2}+1\right)\cdot(F_{By}+F_{eBy})}{\left(4\cdot\sqrt{2}+3\right)}+F_{By}+F_{eBy}\right)\cdot\left(1-2\cdot\frac{\left(2\cdot\sqrt{2}+1\right)}{\left(4\cdot\sqrt{2}+3\right)}\right) \tag{47}$$

The individual elements of the matrix in Equation (37) after deriving by F_{eBy} are as follows:

$$N_{AB}\cdot\frac{\partial N_{AB}}{\partial F_{eBy}} = 4\cdot\frac{\left(2\cdot\sqrt{2}+1\right)\cdot(F_{Bx}+F_{eBx})+\left(\sqrt{2}+1\right)\cdot(F_{By}+F_{eBy})}{\left(4\cdot\sqrt{2}+3\right)}\cdot\frac{\left(\sqrt{2}+1\right)}{\left(4\cdot\sqrt{2}+3\right)} \tag{48}$$

$$F_{AC}\cdot\frac{\partial F_{AC}}{\partial F_{eBy}} = 2\cdot\left(F_{Bx}+F_{eBx}-2\cdot\frac{\left(2\cdot\sqrt{2}+1\right)\cdot(F_{Bx}+F_{eBx})+\left(\sqrt{2}+1\right)\cdot(F_{By}+F_{eBy})}{\left(4\cdot\sqrt{2}+3\right)}+F_{By}+F_{eBy}\right)\cdot\left(1-2\cdot\frac{\left(\sqrt{2}+1\right)}{\left(4\cdot\sqrt{2}+3\right)}\right) \tag{49}$$

$$F_{BC}\cdot\frac{\partial F_{BC}}{\partial F_{eBy}} = \left(F_{Bx}+F_{eBx}-2\cdot\frac{\left(2\cdot\sqrt{2}+1\right)\cdot(F_{Bx}+F_{eBx})+\left(\sqrt{2}+1\right)\cdot(F_{By}+F_{eBy})}{\left(4\cdot\sqrt{2}+3\right)}+F_{By}+F_{eBy}\right)\cdot\left(1-2\cdot\frac{\left(\sqrt{2}+1\right)}{\left(4\cdot\sqrt{2}+3\right)}\right) \tag{50}$$

$$F_{BD}\cdot\frac{\partial F_{BD}}{\partial F_{eBy}} = 2\cdot\left(2\cdot\frac{\left(2\cdot\sqrt{2}+1\right)\cdot(F_{Bx}+F_{eBx})+\left(\sqrt{2}+1\right)\cdot(F_{By}+F_{eBy})}{\left(4\cdot\sqrt{2}+3\right)}-F_{Bx}-F_{eBx}\right)\cdot\left(2\cdot\frac{\left(\sqrt{2}+1\right)}{\left(4\cdot\sqrt{2}+3\right)}\right) \tag{51}$$

$$F_{CD}\cdot\frac{\partial F_{CD}}{\partial F_{eBy}} = \left(F_{Bx}+F_{eBx}-2\cdot\frac{\left(2\cdot\sqrt{2}+1\right)\cdot(F_{Bx}+F_{eBx})+\left(\sqrt{2}+1\right)\cdot(F_{By}+F_{eBy})}{\left(4\cdot\sqrt{2}+3\right)}+F_{By}+F_{eBy}\right)\cdot\left(1-2\cdot\frac{\left(\sqrt{2}+1\right)}{\left(4\cdot\sqrt{2}+3\right)}\right) \tag{52}$$

In Equations (37) and (43)–(52), under the theorem, zero value is substituted for the vector $\mathbf{F}_{eB}$ ($\mathbf{F}_{eB} = 0$) to obtain the resulting deflection δ_B of node B.

Node C:

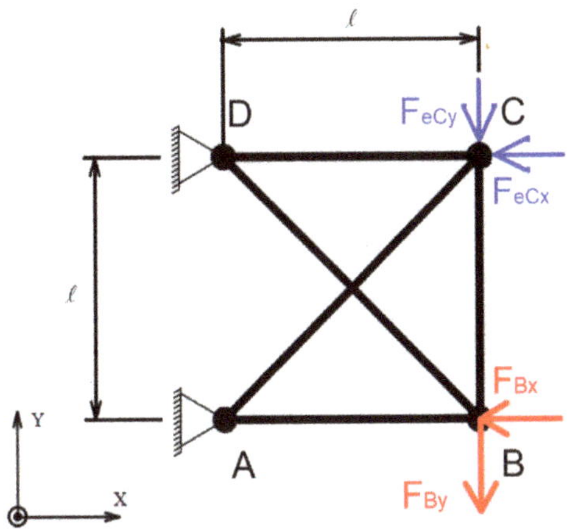

Figure 8. Structure of the trusses system loaded by the external force vector $\mathbf{F}_{eC}$ at node C.

The deflection vector δ_C of node C derived from the first Castigliano theorem can be calculated using the following equation:

$$\delta_C = \frac{l}{E\cdot S}\begin{bmatrix}\sum F_i\cdot\frac{\partial F_i}{\partial F_{eCx}}\\ \sum F_i\cdot\frac{\partial F_i}{\partial F_{eCy}}\end{bmatrix} \tag{53}$$

Each component of the force F_i is expressed as a function of the component of the external force vector $\mathbf{F}_{eC}$ as follows:

$$\delta_C = \frac{l}{E\cdot S}\begin{bmatrix}F_{eCx}+N_{AB}\cdot\frac{\partial N_{AB}}{\partial F_{eCx}}+\sqrt{2}\cdot F_{AC}\cdot\frac{\partial F_{AC}}{\partial F_{eCx}}+F_{BC}\cdot\frac{\partial F_{BC}}{\partial F_{eCx}}+\sqrt{2}\cdot F_{BD}\cdot\frac{\partial F_{BD}}{\partial F_{eCx}}+F_{CD}\cdot\frac{\partial F_{CD}}{\partial F_{eCx}}\\ F_{eCy}+N_{AB}\cdot\frac{\partial N_{AB}}{\partial F_{eCy}}+\sqrt{2}\cdot F_{AC}\cdot\frac{\partial F_{AC}}{\partial F_{eCy}}+F_{BC}\cdot\frac{\partial F_{BC}}{\partial F_{eCy}}+\sqrt{2}\cdot F_{BD}\cdot\frac{\partial F_{BD}}{\partial F_{eCy}}+F_{CD}\cdot\frac{\partial F_{CD}}{\partial F_{eCy}}\end{bmatrix} \tag{54}$$

Consider again the system of the given Equations (5)–(8) with the additional vector of the external force $\mathbf{F}_{eC}$:

$$N_{AB} - \frac{\sqrt{2}\,F_{BD}}{2} = F_{Bx} \tag{55}$$

$$F_{BC} + \frac{\sqrt{2}\,F_{BD}}{2} = F_{By} \tag{56}$$

$$-F_{CD} - F_{eCx} - \frac{\sqrt{2}\,F_{AC}}{2} = 0 \tag{57}$$

$$-F_{BC} - F_{eCy} - \frac{\sqrt{2}\,F_{AC}}{2} = 0 \tag{58}$$

And its transcription into matrix entry for automated computer calculation is as follows:

$$\begin{bmatrix}1 & 0 & 0 & -\frac{\sqrt{2}}{2} & 0 & 0 & 0\\ 0 & 0 & 1 & \frac{\sqrt{2}}{2} & 0 & 0 & 0\\ 0 & -\frac{\sqrt{2}}{2} & 0 & 0 & -1 & -1 & 0\\ 0 & -\frac{\sqrt{2}}{2} & -1 & 0 & 0 & 0 & -1\end{bmatrix}\cdot\begin{bmatrix}N_{AB}\\ F_{AC}\\ F_{BC}\\ F_{BD}\\ F_{CD}\\ F_{eCx}\\ F_{eCy}\end{bmatrix} = \begin{bmatrix}F_{Bx}\\ F_{By}\\ 0\\ 0\end{bmatrix} \tag{59}$$

From system (59), the unknown forces in the bars expressed in terms of the known variables F_{Bx}, F_{By}, N_{AB}, F_{eCx}, and F_{eCy} are as follows:

$$F_{AC} = -\sqrt{2}\cdot(F_{Bx} - N_{AB} + F_{By} + F_{eCy}) \tag{60}$$

$$F_{BC} = F_{Bx} - N_{AB} + F_{By} \tag{61}$$

$$F_{BD} = \sqrt{2}\cdot(N_{AB} - F_{Bx}) \tag{62}$$

$$F_{CD} = F_{Bx} - N_{AB} + F_{By} - F_{eCx} + F_{eCy} \tag{63}$$

Based on the deformation condition (1), where the cross-sections S_i = S, Young's modules E_i = E, and the lengths l_i are expressed as multiples of a uniform length l, the following is true for the energy change in the entire system under the N_{AB} force:

$$\frac{l}{E\cdot S}\sum F_i\cdot\frac{\partial F_i}{\partial N_{AB}} = 0 \tag{64}$$

$$N_{AB} - \sqrt{2}\cdot 2\cdot(F_{Bx} - N_{AB} + F_{By} + F_{eCy}) - (F_{Bx} - N_{AB} + F_{By}) + \sqrt{2}\cdot 2\cdot(N_{AB} - F_{Bx}) - (F_{Bx} - N_{AB} + F_{By} - F_{eCx} + F_{eCy}) = 0 \tag{65}$$

$$N_{AB} = \frac{2\cdot\left(\left(2\cdot\sqrt{2}+1\right)\cdot F_{Bx} + \left(\sqrt{2}+1\right)\cdot F_{By}\right) - F_{eCx} + \left(2\cdot\sqrt{2}+1\right)\cdot F_{eCy}}{\left(4\cdot\sqrt{2}+3\right)} \tag{66}$$

By substituting N_{AB} from Equation (66) into the system of Equations (60)–(63), the following is derived:

$$F_{AC} = -\sqrt{2}\cdot\left(F_{Bx} - \frac{2\cdot\left(\left(2\cdot\sqrt{2}+1\right)\cdot F_{Bx} + \left(\sqrt{2}+1\right)\cdot F_{By}\right) - F_{eCx} - \left(2\cdot\sqrt{2}+2\right)\cdot F_{eCy}}{\left(4\cdot\sqrt{2}+3\right)} + F_{By}\right) \tag{67}$$

$$F_{BC} = F_{Bx} - \frac{2\cdot\left(\left(2\cdot\sqrt{2}+1\right)\cdot F_{Bx} + \left(\sqrt{2}+1\right)\cdot F_{By}\right) - F_{eCx} + \left(2\cdot\sqrt{2}+1\right)\cdot F_{eCy}}{\left(4\cdot\sqrt{2}+3\right)} + F_{By} \tag{68}$$

$$F_{BD} = \sqrt{2}\cdot\left(\frac{2\cdot\left(\left(2\cdot\sqrt{2}+1\right)\cdot F_{Bx} + \left(\sqrt{2}+1\right)\cdot F_{By}\right) - F_{eCx} + \left(2\cdot\sqrt{2}+1\right)\cdot F_{eCy}}{\left(4\cdot\sqrt{2}+3\right)} - F_{Bx}\right) \tag{69}$$

$$F_{CD} = F_{Bx} - \frac{2\cdot\left(\left(2\cdot\sqrt{2}+1\right)\cdot F_{Bx} + \left(\sqrt{2}+1\right)\cdot F_{By}\right) + \left(4\cdot\sqrt{2}+2\right)\cdot F_{eCx} - 2\cdot\left(\sqrt{2}+1\right)\cdot F_{eCy}}{\left(4\cdot\sqrt{2}+3\right)} + F_{By} \tag{70}$$

The individual elements of the matrix in Equation (54) after deriving by F_{eCx} are as follows:

$$N_{AB}\cdot\frac{\partial N_{AB}}{\partial F_{eCx}} = -\frac{2\cdot\left(\left(2\cdot\sqrt{2}+1\right)\cdot F_{Bx} + \left(\sqrt{2}+1\right)\cdot F_{By}\right) - F_{eCx} + \left(2\cdot\sqrt{2}+1\right)\cdot F_{eCy}}{\left(4\cdot\sqrt{2}+3\right)^2} \tag{71}$$

$$F_{AC}\cdot\frac{\partial F_{AC}}{\partial F_{eCx}} = \left(F_{Bx} - \frac{2\cdot\left(\left(2\cdot\sqrt{2}+1\right)\cdot F_{Bx} + \left(\sqrt{2}+1\right)\cdot F_{By}\right) - F_{eCx} + \left(2\cdot\sqrt{2}+1\right)\cdot F_{eCy}}{\left(4\cdot\sqrt{2}+3\right)} + F_{By} + F_{eCy}\right)\cdot\frac{2}{\left(4\cdot\sqrt{2}+3\right)} \tag{72}$$

$$F_{BC}\cdot\frac{\partial F_{BC}}{\partial F_{eCx}} = \left(F_{Bx} - \frac{2\cdot\left(\left(2\cdot\sqrt{2}+1\right)\cdot F_{Bx} + \left(\sqrt{2}+1\right)\cdot F_{By}\right) - F_{eCx} + \left(2\cdot\sqrt{2}+1\right)\cdot F_{eCy}}{\left(4\cdot\sqrt{2}+3\right)} + F_{By}\right)\cdot\frac{1}{\left(4\cdot\sqrt{2}+3\right)} \tag{73}$$

$$F_{BD}\cdot\frac{\partial F_{BD}}{\partial F_{eCx}} = -\left(\frac{2\cdot\left(\left(2\cdot\sqrt{2}+1\right)\cdot F_{Bx} + \left(\sqrt{2}+1\right)\cdot F_{By}\right) - F_{eCx} + \left(2\cdot\sqrt{2}+1\right)\cdot F_{eCy}}{\left(4\cdot\sqrt{2}+3\right)} - F_{Bx}\right)\cdot\frac{2}{\left(4\cdot\sqrt{2}+3\right)} \tag{74}$$

$$F_{CD}\cdot\frac{\partial F_{CD}}{\partial F_{eCx}}=-\left(F_{Bx}-\frac{2\cdot\left(\left(2\cdot\sqrt{2}+1\right)\cdot F_{Bx}+\left(\sqrt{2}+1\right)\cdot F_{By}\right)+\left(4\cdot\sqrt{2}+2\right)\cdot F_{eCx}-2\cdot\left(\sqrt{2}+1\right)\cdot F_{eCy}}{\left(4\cdot\sqrt{2}+3\right)}+F_{By}\right)\cdot\frac{\left(4\cdot\sqrt{2}+2\right)}{\left(4\cdot\sqrt{2}+3\right)} \quad (75)$$

The individual elements of the matrix in Equation (54) after deriving by F_{eCy} read as follows:

$$N_{AB}\cdot\frac{\partial N_{AB}}{\partial F_{eCy}}=\left(\frac{2\cdot\left(\left(2\cdot\sqrt{2}+1\right)\cdot F_{Bx}+\left(\sqrt{2}+1\right)\cdot F_{By}\right)-F_{eCx}+\left(2\cdot\sqrt{2}+1\right)\cdot F_{eCy}}{\left(4\cdot\sqrt{2}+3\right)}\right)\cdot\frac{\left(2\cdot\sqrt{2}+1\right)}{\left(4\cdot\sqrt{2}+3\right)} \quad (76)$$

$$F_{AC}\cdot\frac{\partial F_{AC}}{\partial F_{eCy}}=\left(F_{Bx}-\frac{2\cdot\left(\left(2\cdot\sqrt{2}+1\right)\cdot F_{Bx}+\left(\sqrt{2}+1\right)\cdot F_{By}\right)-F_{eCx}-\left(2\cdot\sqrt{2}+2\right)\cdot F_{eCy}}{\left(4\cdot\sqrt{2}+3\right)}+F_{By}\right)\cdot\frac{2\cdot\left(2\cdot\sqrt{2}+2\right)}{\left(4\cdot\sqrt{2}+3\right)} \quad (77)$$

$$F_{BC}\cdot\frac{\partial F_{BC}}{\partial F_{eCy}}=-\left(F_{Bx}-\frac{2\cdot\left(\left(2\cdot\sqrt{2}+1\right)\cdot F_{Bx}+\left(\sqrt{2}+1\right)\cdot F_{By}\right)-F_{eCx}+\left(2\cdot\sqrt{2}+1\right)\cdot F_{eCy}}{\left(4\cdot\sqrt{2}+3\right)}+F_{By}\right)\cdot\frac{\left(2\cdot\sqrt{2}+1\right)}{\left(4\cdot\sqrt{2}+3\right)} \quad (78)$$

$$F_{BD}\cdot\frac{\partial F_{BD}}{\partial F_{eCy}}=\left(\frac{2\cdot\left(\left(2\cdot\sqrt{2}+1\right)\cdot F_{Bx}+\left(\sqrt{2}+1\right)\cdot F_{By}\right)-F_{eCx}+\left(2\cdot\sqrt{2}+1\right)\cdot F_{eCy}}{\left(4\cdot\sqrt{2}+3\right)}-F_{Bx}\right)\cdot\frac{2\cdot\left(2\cdot\sqrt{2}+1\right)}{\left(4\cdot\sqrt{2}+3\right)} \quad (79)$$

$$F_{CD}\cdot\frac{\partial F_{CD}}{\partial F_{eCy}}=\left(F_{Bx}-\frac{2\cdot\left(\left(2\cdot\sqrt{2}+1\right)\cdot F_{Bx}+\left(\sqrt{2}+1\right)\cdot F_{By}\right)+\left(4\cdot\sqrt{2}+2\right)\cdot F_{eCx}-2\cdot\left(\sqrt{2}+1\right)\cdot F_{eCy}}{\left(4\cdot\sqrt{2}+3\right)}+F_{By}\right)\cdot\frac{2\cdot\left(\sqrt{2}+1\right)}{\left(4\cdot\sqrt{2}+3\right)} \quad (80)$$

In Equations (54) and (71)–(80), following the theorem, a zero value is substituted into for the vector $\mathbf{F}_{eC}$ ($\mathbf{F}_{eC}$ = 0) to obtain the resulting deflection δ_C of the node C.

3. Building a Trusses Structure in MATLAB—Simscape Multibody

The Simscape Multibody environment [1–4] enables simulations of mechanical systems in space. It formulates and solves the equations of motion for the entire mechanical system and also provides visualization of the system dynamics. This allows for the examination of various dynamic and static processes on the model. Examining kinematic and dynamic quantities on advanced models can be quite challenging. Dealing with spatial assignments can be tough without an understanding of vector calculus and rotation matrix representations. The probability of error in derivation and computation is considerably high. Utilizing software tools with predefined blocks representing physical solids can assist in resolving such assignments. There are different programs available for calculating trusses systems and various tools for simulating the behavior of structures made from various materials. In such simulation program environments, there are different blocks such as "Flexible Beams". These blocks simulate a slender bar with constant cross-section with small and linear deformations, including extension, bending, and torsion. Thus, a flexible body can be represented as a system composed of multiple bodies (multibody) [22]. In Figure 9, the model of the trusses system from Figure 1 is loaded with the external force vector $\mathbf{F}_B$ at node B with the calculation results. The following values have been defined:

$\mathbf{F}_B$ = $[-2, -10]^T$ kN, bar lenghts l = 1 m, $l_{AB} = l$, $l_{BC} = l$, $l_{CD} = l$, $l_{AC} = \sqrt{2}\cdot l$, $l_{BD} = \sqrt{2}\cdot l$, bar diameters are the same with d = 0.01 m. Young's modulus E is 210 GPa, shear modulus G is 79 GPa, and density is 7800 kg/m^3. Poisson's ratio is 0.28. The values of proportional or Rayleigh damping are left "default" with respect to computational practicality b_m = 0.01 s^{-1}, b_k = 0.01 s. The Rayleigh model [23–25] defines the damping matrix $\mathbf{C}$ as a linear combination of the mass matrix $\mathbf{M}$ and the stiffness matrix $\mathbf{K}$ as follows:

$$\mathbf{C} = b_m\cdot\mathbf{M} + b_k\cdot\mathbf{K}, \quad (81)$$

with b_m being a mass coefficient and b_k being a stiffness coefficient. These are the scalar coefficients. The simulation time is also "default" in the model settings and equals t = 10 s.

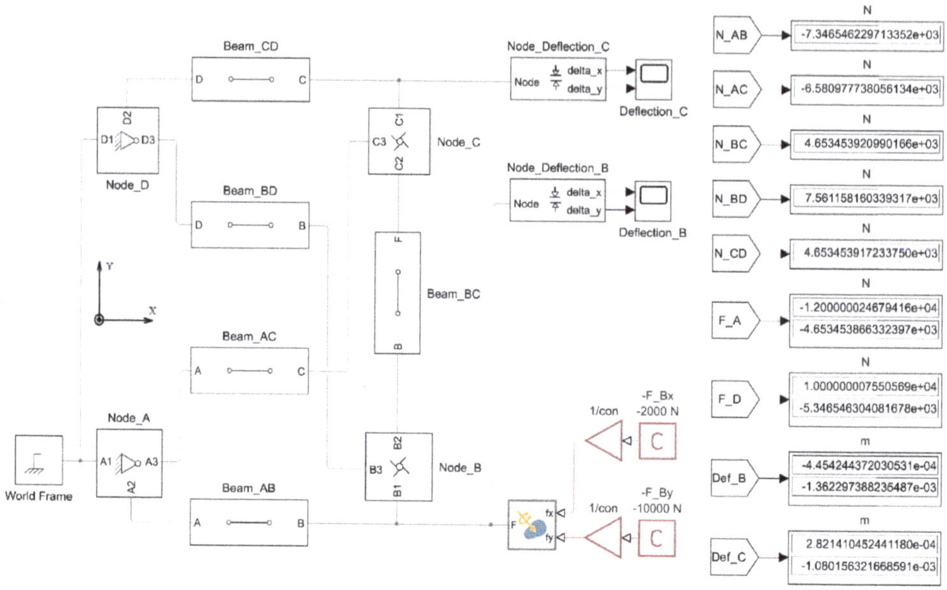

Figure 9. Structure of the trusses system loaded by the external force vector $\mathbf{F}_\mathrm{B}$ at node B in MATLAB—Simscape Multibody, based on Figure 1.

As already mentioned, for the sake of simplicity, so that we do not have to adjust and adapt the values of t, b_m, and b_k to the level of external load in the form of force $\mathbf{F}_\mathrm{B}$ and cross-section S_i, the so-called multiplication constant–*con*, is introduced. Its role is to proportionally shift the simulation to the order of units depending on the value of the external force vector $\mathbf{F}_\mathrm{B}$ as well as the uniform cross-section of the bar S_i. Practically, we can also start from the normal stress in the most stressed bar if the cross-sections of the S_i bars of a given structure would be different. This avoids unnecessary changes in the defined coefficients b_m for the mass coefficient **M**, and b_k for the stiffness coefficient **K** in Equation (81) in the Rayleigh damping model. For the multiplication constant, the following applies:

$$con = \frac{\|\mathbf{F}_\mathrm{B}\|}{1200 \cdot S_i} \tag{82}$$

This simple modification ensured an increase of two to four orders of magnitude in the accuracy of the calculation in accordance with theory within an unlimited range of external $\mathbf{F}_\mathrm{B}$ force loading. This multiplication constant can be optimized for the highest accuracy based on Equations (5) and (6), provided that the following holds with the smallest possible deviation:

$$F_{Bx} = N_\mathrm{AB} - \frac{F_{BD}}{\sqrt{2}} \tag{83}$$

$$F_{By} = F_{BC} + \frac{F_{BD}}{\sqrt{2}} \tag{84}$$

The magnitudes of the components in the force vector $\mathbf{F}_\mathrm{B}$ are known to us, and the other forces are computational outputs.

Figure 10 shows a visualization of the trusses system model in the Mechanics Explorers window of the MATLAB—Simscape Multibody environment. It is a running visualization in the case of a properly constructed model.

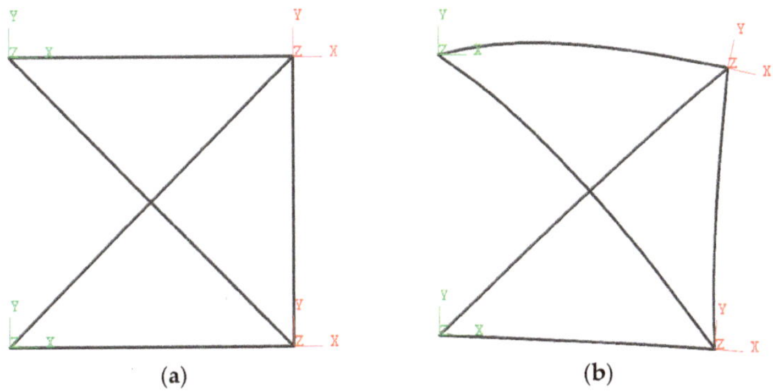

Figure 10. Visualization of the MATLAB—Simscape Multibody model of the trusses system, in the Mechanics Explorers window: (**a**) Unloaded; (**b**) Loaded with visible deformation. The movable parts (unknown calculated shift) are marked in red, and the fixed parts of the construction are marked in green.

The individual systems forming a particular trusses system should be explained. The World Frame subsystem is formed by combining the *World Frame* and *Solver Configuration* library blocks, these blocks being in the "default" configuration, see Figure 11.

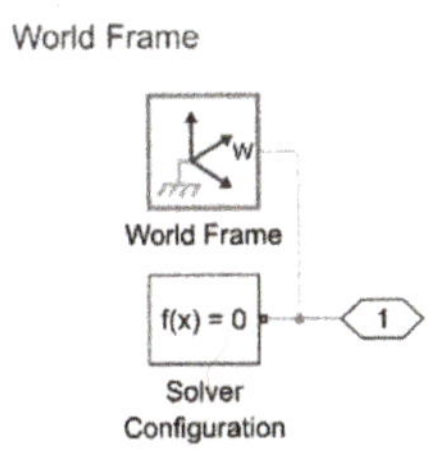

Figure 11. The World Frame subsystem is formed by combining the *World Frame* and *Solver Configuration* library blocks.

Subsystems of nodes labelled Node_A and Node_D are formed as follows in Figure 12:

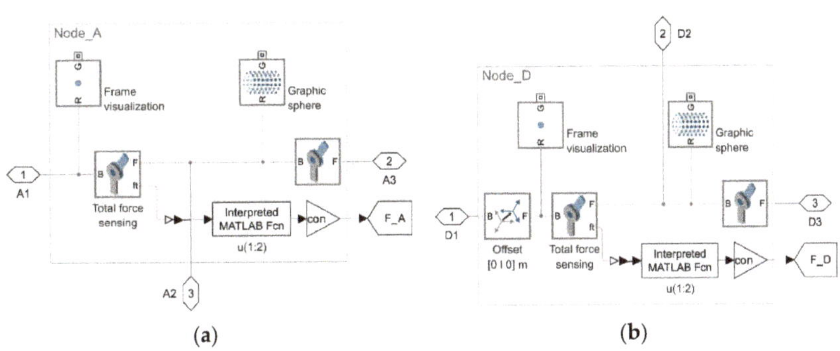

Figure 12. Detailed view of nodes A, D: (**a**) Node_A; (**b**) Node_D.

The blocks "Point"—*Graphic sphere* and "Point Cloud"—*Frame visualization* are used only for visualizations. Physically, they are not involved in the given model. The $\mathbf{F}_A$ force vector value is measured directly from the "Revolute Joint"—*Total force sensing* block. The bars are connected using "Revolute Joint" joints in the "default" setting. The block "Interpreted MATLAB Function" $-u(1{:}2)$ defines a two-dimensional force vector in the

$\mathbf{F}_A$, $\mathbf{F}_B$ plane, whose z-component is of zero value. The block *"Gain"* –*con* provides the multiplication by the multiplication constant–*con*. An explanation of constant–*con* has been provided in Equations (82)–(84).

Subsystems of nodes labelled Node_B and Node_C are formed as follows in Figure 13:

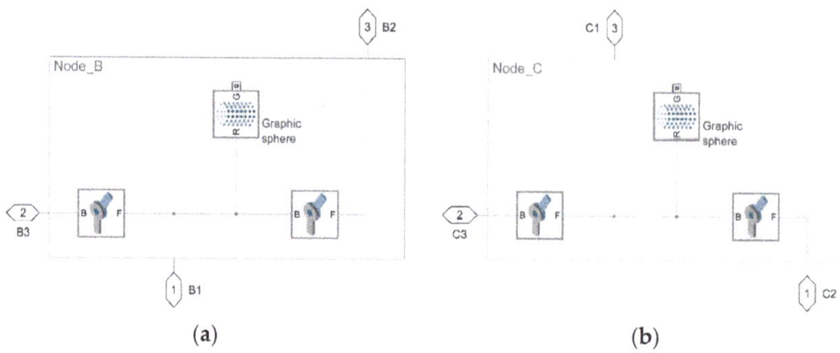

Figure 13. Detailed view of nodes B, C: (**a**) Node_B; (**b**) Node_C.

Similarly, the bars are connected using "Revolute Joint" joints in the "default" setting. The member subsystems labelled Beam_AB, Beam_AC, Beam_BC, Beam_BD, and Beam_CD are identical in structure and are formed as follows in Figure 14:

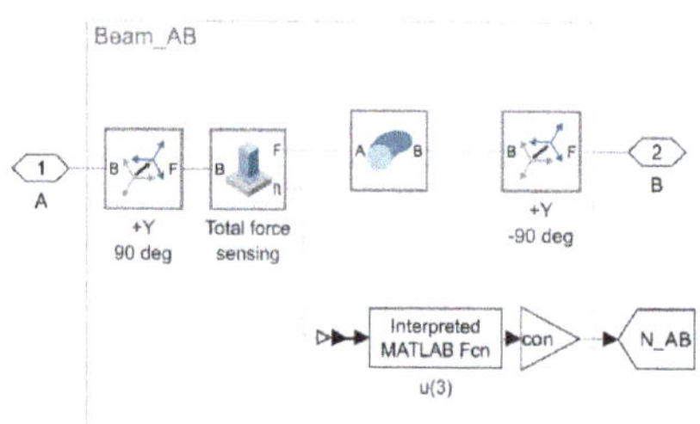

Figure 14. An example of the bar subsystem labelled Beam_AB.

The "Rigid Transform" blocks ensure that the bar is rotated into the desired plane around the y-axis of the world frame. The $\mathbf{N}_{AB}$ force vector value is measured directly from the "Weld Joint" block—*Total force sensing*. The block "Interpreted MATLAB Function" $-u(3)$ selects the z-component of the force vector in the $\mathbf{N}_{AB}$ bar. The block physically representing the bar is a "Flexible Cylindrical Beam" block type. It defines the individual parameters such as bar length l_{AB}, bar diameter d, Young's modulus E, shear modulus G, density, and proportional Rayleigh damping values b_m, b_k.

Subsystems labelled Node_Deflection_B, Node_Deflection_C are designed to measure the deflection of nodes B, C, and are formed as follows in Figure 15:

The "Transform Sensor"—*Translation* $[x, y, z]$ blocks provide deflection measurement relative to the world frame. The remaining blocks select only those components of the vector that define the deflection in the x, y plane relative to the original position.

Figure 16 illustrates the control measurement of the reaction forces to optimize the multiplication constant–*con* based on Equations (83) and (84).

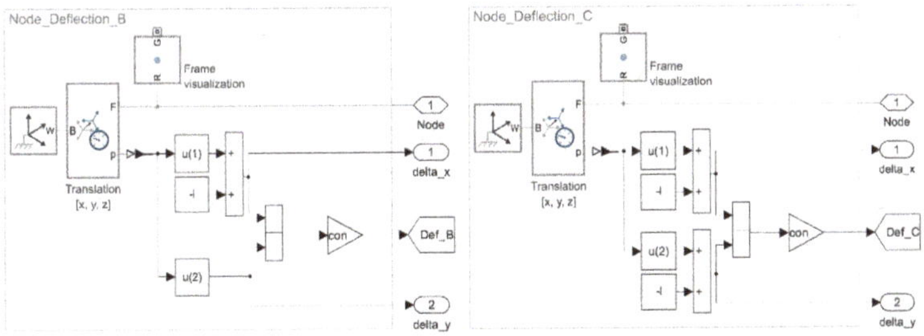

(a) (b)

Figure 15. Detailed view of the node deflection measurement subsystems labelled Node_Deflection_B and Node_Deflection_C: (**a**) Node_B node deflection; (**b**) Node_C node deflection.

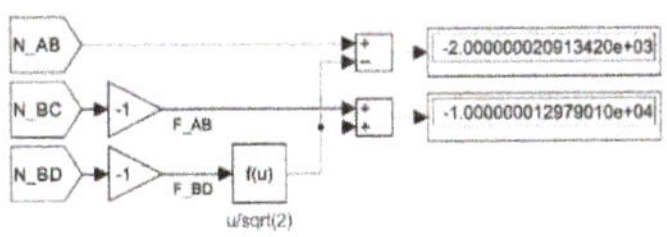

Figure 16. Subsystem of control measurement of reaction forces for optimizing the multiplication constant–*con* based on Equations (83) and (84).

4. Building a Trusses Structure in the SOLIDWORKS—Weldments Tool

To verify the calculation of the given trusses system, the SOLIDWORKS environment was used, and a similar model was developed using the "Weldments" tool, which allows the welded structure to be designed as a single multibody part. This tool yielded the most accurate results when simulated on bars under bending, torsion, tension–compression, and continuous loads against the theory.

Load $\mathbf{F}_B = [-2, -10]^T$ kN, bar lengths $l = 1$ m, $l_{AB} = l$, $l_{BC} = l$, $l_{CD} = l$, $l_{AC} = \sqrt{2}{\cdot}l$, $l_{BD} = \sqrt{2}{\cdot}l$, bar diameters are the same with $d = 0.01$m full profile. The material used is "Plain Carbon Steel" with Young's modulus E is 210 GPa, shear modulus G of 79 GPa and density of 7800 kg/m^3. Poisson's ratio is 0.28.

A detailed procedure for developing such a model is not provided in this section as it is available in [26–29]. It should be noted that the node where the diagonal bars intersect is not a fixed joint. This is also evident when loaded with the visible deformation in Figure 17b. A comparison of the individual results from MATLAB—Multibody, SOLIDWORKS—Weldments tool, and the theoretical calculations are summarized in the table in Section 5.

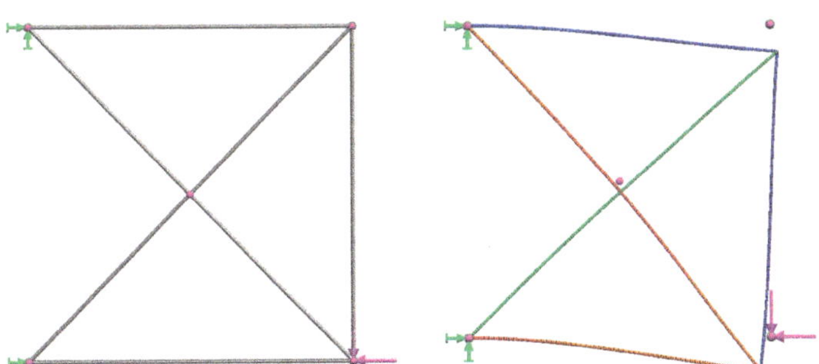

(a) (b)

Figure 17. The figure displays a similar trusses structure developed with the Weldments tool in SOLIDWORKS: (**a**) Unloaded with indicated reactions at nodes A, D; (**b**) Loaded with visible deformation of the bars.

5. Results

A comparison of the individual results from MATLAB—Multibody, SOLIDWORKS—Weldments tool, and the theoretical calculation is presented in the Table 1 overview.

Firstly, the respective values are substituted into the individual equations and the results are quantified:

5.1. Deriving from the Theory

By substituting the respective values into the individual equations for N_{AB}, F_{AC}, F_{BC}, F_{BD}, and F_{CD}, the following results can be derived:

$$N_{AB} = 2\cdot\frac{\left(2\cdot\sqrt{2}+1\right)\cdot(-2000)+\left(\sqrt{2}+1\right)\cdot(-10,000)}{\left(4\cdot\sqrt{2}+3\right)} = -7.346546206455528\cdot 10^{+03} \tag{85}$$

From the equilibrium in the nodes, the reaction forces F_i in the bars are derived:

$$F_{AC} = -\sqrt{2}\cdot(-2000 - F_{AB} - 10,000) = 6.580977466707121\cdot 10^{+03} \tag{86}$$

$$F_{BC} = -2000 - F_{AB} - 10,000 = -4.653453793544472\cdot 10^{+03} \tag{87}$$

$$F_{BD} = \sqrt{2}\cdot(F_{AB} + 2000) = -7.561158157023830\cdot 10^{+03} \tag{88}$$

$$F_{CD} = -2000 - F_{AB} + 10,000 = -4.653453793544472\cdot 10^{+03} \tag{89}$$

Consequently, the N_i forces in the bars are as follows:

$$N_{AC} = -F_{AC} = -6.580977466707121\cdot 10^{+03} \tag{90}$$

$$N_{BC} = -F_{BC} = 4.653453793544472\cdot 10^{+03} \tag{91}$$

$$N_{BD} = -F_{BD} = 7.561158157023830\cdot 10^{+03} \tag{92}$$

$$N_{CD} = -F_{CD} = 4.653453793544472\cdot 10^{+03} \tag{93}$$

5.2. The MATAB and SOLIDWORKS Results

For ease of overview, the individual results are arranged in Table 1.

Table 1. A comparison of theoretical results of forces N_{AB}, N_{AC}, N_{BC}, N_{BD}, N_{CD} acting in bars, nodes A, B, F_{Ax}, F_{Ay}, F_{Bx}, F_{By} (not reactions) and deflections δ_{Bx}, δ_{By}, δ_{Cx}, δ_{Cy} with the MATLAB—Multibody, SOLIDWORKS—Weldments tool results.

Parameter [Unit]	Theoretical Calculation	MATLAB Not Optimized	MATLAB Optimized	SOLIDWORKS
N_{AB} [N]	-7.34654620×10^3	-7.34654003×10^3	-7.34654622×10^3	-7.34590527×10^3
N_{AC} [N]	-6.58097746×10^3	-6.58097926×10^3	-6.58097773×10^3	-6.58031299×10^3
N_{BC} [N]	4.65345379×10^3	4.65345692×10^3	4.65345392×10^3	4.65308789×10^3
N_{BD} [N]	7.56115815×10^3	7.56116105×10^3	7.56115816×10^3	7.55990186×10^3
N_{CD} [N]	4.65345379×10^3	4.65345484×10^3	4.65345391×10^3	4.65298145×10^3
F_{Ax} [N]	-1.20000000×10^4	-1.19999982×10^4	-1.20000002×10^4	-1.19988418×10^4
F_{Ay} [N]	-4.65345379×10^3	-4.65344188×10^3	-4.65345386×10^3	-4.65380371×10^3
F_{Dx} [N]	1.00000000×10^4	9.99999827×10^3	1.00000000×10^4	9.99884180×10^3
F_{Dy} [N]	-5.34654620×10^3	-5.34655811×10^3	-5.34654630×10^3	-5.34619678×10^3
F_{Bx} [1] [N]	-2.00000000×10^3	-1.99999177×10^3	-2.00000002×10^3	-
F_{By} [1] [N]	-1.00000000×10^4	-1.00000051×10^4	-1.00000001×10^4	-
δ_{Bx} [m]	$-4.45424435 \times 10^{-4}$	$-4.45424989 \times 10^{-4}$	$-4.45424437 \times 10^{-4}$	$-4.45385580 \times 10^{-4}$
δ_{By} [m]	$-1.36229734 \times 10^{-3}$	$-1.36229743 \times 10^{-3}$	$-1.36229738 \times 10^{-3}$	$-1.36211328 \times 10^{-3}$
δ_{Cx} [m]	$2.82141018 \times 10^{-4}$	$2.82140499 \times 10^{-4}$	$2.82141045 \times 10^{-4}$	$2.82112334 \times 10^{-4}$
δ_{Cy} [m]	$-1.08015632 \times 10^{-3}$	$-1.08015649 \times 10^{-3}$	$-1.08015632 \times 10^{-3}$	$-1.07999449 \times 10^{-3}$

[1] Verification of the calculation of the magnitude of the external loading force $\mathbf{F}_B$ based on the equilibrium of node B, Equations (83) and (84). It serves only for the optimization of the multiplication–*con*.

It can be noticed that the accuracy in MATLAB—Multibody is guaranteed to six decimal places, while in SOLIDWORKS—Weldments tool, it is guaranteed to two decimal places. It is important to note that SOLIDWORKS also offers many other deliverables and extensions that would have to be completed in MATLAB. Additionally, the calculation in MATLAB has been optimized using *–con* to ensure the calculation is not tedious and delivers interesting results. If the simulation had not been proportionally shifted to lower values using the *–con* constant, the computation would have been tedious, and the accuracy would have been approximately four decimal places. Alternatively, it would be necessary to increase the values of damping b_k and adjust the simulation time accordingly.

In MATLAB, for a more complex structure, individual blocks representing a repeating part can be created, and the entire structure can be further developed for static or dynamic assessment—see Figure 18.

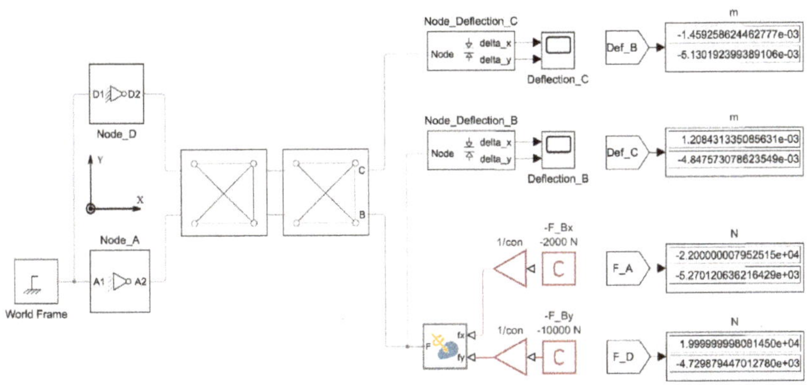

Figure 18. An example of a system composed of repeating subsystems to develop a more complex structure.

Figure 19 presents a visualization of the trusses system model from Figure 18 in the Mechanics Explorers window of the MATLAB—Simscape Multibody environment.

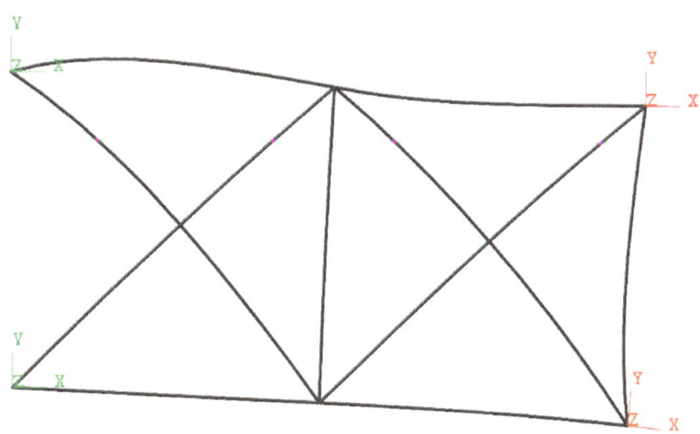

Figure 19. Visualization of the model from Figure 18 of the loaded trusses system from MATLAB—Simscape Multibody, in the Mechanics Explorers window, with the visible deformation.

6. Discussion

The Simscape Multibody toolbox from MATLAB is primarily designed to simulate mechanical systems in space, detecting their kinematic and dynamic quantities. The market offers numerous applications and programs for solving trusses systems. When individual tests were conducted on the flexible beam models available in the library, very good results were obtained, with less errors than SOLIDWORKS. Deformation includes additions due to extension, bending, and torsion. The stiffness and inertia properties of the body are calculated based on its geometry and material properties. Developing the model involves correctly assembling a few blocks to calculate the forces and deflections of the mechanical system. Regarding deflections, Multibody is an interesting choice compared to the classical mathematical solution of the equations. While testing the trusses system, the possibility of achieving better results in the simplest possible way caught our attention. The proportional damping type achieved the best results with the Rayleigh damping model. For more accuracy in the calculation than two to four decimal places, the damping needs to be adjusted to the load rate, bar length, cross-section, and simulation time. In order to simplify the entire issue, we have relied on the basic idea that all processes in the body are reversible and linear, following principles such as Hooke's law. From a practical standpoint, the model should preferably be founded on the default settings of the components and the model. We intended to avoid adjusting and adapting the values of the simulation time t, the coefficients of proportionality b_m, b_k to the level of the external load in the form of the force $\mathbf{F}_B$ and the cross-section S_i. Instead, we considered proportionally shifting the simulation based on the external force vector $\mathbf{F}_B$ and the uniform cross-section S_i, which introduced a multiplication constant–*con*. We adjusted its value to the equilibrium of forces at node B, where the load is known. It would also be beneficial to consider further enhancements, such as "deep learning", to identify a better optimum for different systems. Additionally, it is theoretically feasible to begin with normal stress in the most heavily stressed bar if the cross-sections of the S_i bars in a particular structure were different. In practical terms, the current level of accuracy in the calculation may not be significant for the designer. However, it indicates the possibility of conducting a more thorough model assessment. It is not always efficient to derive the equations or to have other suitable software available. Therefore, it is beneficial to be aware of this possibility. There are only a few published works using Simscape Multibody for trusses systems.

The calculations were also verified in SOLIDWORKS using the Simulation tool. The entire structure was developed using the Weldments tool. Developing the model was not time-consuming, and it is possible to analyze various parameters and outputs, such as forces at different points and the second moments of area. The agreement against the theoretical calculations was accurate to two decimal places, which is sufficient in many cases.

7. Conclusions

Our contribution focuses on further possibilities of utilizing built-in tools that are part of the software package and that we use in our work and research activities. We were interested in the results in terms of accuracy and to what extent the computation differs from the theoretical calculation using the tools we employ. The potential reliance on simulations in the future could help avoid the need for complex theoretical calculations. As we can see in Section 2, the computation is time-consuming and risky in terms of error occurrence, despite the fact that the structure is not complex. In practice, there may not always be time for a validation calculation in the theoretical domain (proper model creation, inclusion of all deficiencies, incorrect assumptions, etc.).

As mentioned in Section 4 of the discussion, the Simscape Multibody Toolbox from MATLAB is primarily intended for simulating mechanical systems in space and determining their kinematic and dynamic quantities. We have noticed that it could also be used for the static assessment of truss systems, for which there are countless programs available, most of which need to be purchased. Currently, there are hardly any published scientific studies regarding this tool for rod systems and their static tasks. Therefore, with this article, we want to highlight the existence of a potential solution such as this. Many aspects, including the assessment of buckling stability, can be further developed if necessary.

We also consider the introduction of simple proportionality in the form of the constant "con" interesting, as it essentially shifts the entire calculation into an area without overload, and thus earlier into a stable state, where the process can practically be considered static.

Author Contributions: Draft concept, theoretical analysis, Š.O.; model design and data processing in MATLAB, Š.O.; overview, supervision, approval, J.S. (Jozef Svetlík); project, administration methodology, J.S. (Jozef Svetlík), J.S. (Ján Semjon), R.J. and M.D. All authors have read and agreed to the published version of the manuscript.

Funding: This paper was prepared with the support of the grant projects KEGA 043TUKE—4/2024 Creation of prospective educational tools for the field of additive manufacturing with the implementation of progressive elements of virtual reality, KEGA 005TUKE—4/2025 Synthesis of machine vision with elements of artificial intelligence in the context of industrial and service robotics for the study program Robotics and robot technology, VEGA 1/0191/24 Development, optimization, and application of coordinate measurement strategies of geometric parameters and structure of additive products.

Institutional Review Board Statement: Not applicable.

Informed Consent Statement: Not applicable.

Data Availability Statement: The raw data supporting the conclusions of this article will be made available by the authors on request.

Conflicts of Interest: The authors declare no conflicts of interest. The funders had no role in the design of the study; in the collection, analyses, or interpretation of data; in the writing of the manuscript, or in the decision to publish the results.

Nomenclature

Abbreviations	
FEM	Finite Element Method
Parametres, variables	
A, B, C, D	Node, Gusset
U	Potential strain energy of the trusses system
F, N	Normal force in a bar
F	Force vector
M	Moment
l	Bar length
d	Bar diameter
S	cross-section
E	The Young's modulus of a bar
G	Shear modulus
Density	Density
μ	Poisson's ratio
δ_B, δ_C	Deflection vector at node B or C in the plane
F_{eB}, F_{eC}	Additional external force vector at node B or C
C	Damping matrix of the Rayleigh damping model
K	Stiffness matrix of the Rayleigh damping model
M	Mass matrix of the Rayleigh damping model
b_m	Mass coefficient proportional to the mass matrix M of the Rayleigh damping model
b_k	Stiffness coefficient proportional to the stiffness matrix K of the Rayleigh damping model
t	Simulation time
con	Multiplication constant
x, y, z	Coordinate system axes, or position vector elements
u(i)	*i*—the element of the vector u of the input variable for the function in SIMULINK

References

1. Mathworks; Simscape. *MultibodyTM Getting Started Guide*; Mathworks: Natick, MA, USA, 2024.
2. Mathworks. MATLAB for Artificial Intelligence. Available online: https://www.mathworks.com/help/pdf_doc/sm/sm_gs.pdf (accessed on 14 August 2024).

3. Mathworks; Simscape. *MultibodyTM User's Guide;* Mathworks: Natick, MA, USA, 2024.
4. Mathworks. MATLAB for Artificial Intelligence. Available online: https://www.mathworks.com/help/pdf_doc/sm/sm_ug.pdf (accessed on 14 August 2024).
5. Bauchau, O.A.; Craig, J.I. Euler—Bernoulli beam theory. In *Structural Analysis;* Springer: Dordrecht, The Netherlands, 2009.
6. Öchsner, A. Euler—Bernoulli beam theory. In *Classical Beam Theories of Structural Mechanics;* Springer: Cham, Switzerland, 2021; pp. 7–66.
7. Euler-Bernoulli Beams: Bending, Buckling, and Vibration. Available online: https://ocw.mit.edu/courses/2-002-mechanics-and-materials-ii-spring-2004/bc25a56b5a91ad29ca5c7419616686f7_lec2.pdf (accessed on 1 August 2024).
8. Šarišský, P. Zpracování Algoritmu pro Řešení Deformace Prutových Soustav v Reálném Čase [Development of an Algorithm for Real—Time Deformation Analysis of Trusses Systems]. Bachelor's Thesis, Brno University of Technology, Brno-Královo Pole, Czech Republic, 2017.
9. Bakošová, A.; Krmela, J.; Handrik, M. Computing of truss structure using MATLAB. *Manuf. Technol.* **2020**, *20*, 279–285. [CrossRef]
10. ANSYS. *ANSYS Fluent User Guide;* ANSYS: Canonsburg, PA, USA, 2017; Available online: http://www.ansys.com (accessed on 14 August 2024).
11. Madenci, E.; Guven, I. Fundamentals of ANSYS. In *The Finite Element Method and Applications in Engineering Using ANSYS®;* ANSYS: Canonsburg, PA, USA, 2006; pp. 15–35.
12. Thompson, M.K.; Thompson, J.M. *ANSYS Mechanical APDL for Finite Element Analysis;* Butterworth—Heinemann: Oxford, UK, 2017.
13. Trebuňa, F.; Šimčák, F. Odolnosť Prvkov Mechanických Sústav [Resistance of Mechanical System Components]. Technical University of Košice: Košice, Slovakia, 2004.
14. Jones, R.M. *Buckling of Bars, Plates, and Shells;* Bull Ridge Corporation: Blacksburg, VA, USA, 2006.
15. Cai, Q.; Feng, R.; Zhang, Z. Topology optimization of truss structure considering nodal stability and local buckling stability. *Structures* **2022**, *40*, 64–73. [CrossRef]
16. Gambhir, M.L. *Stability Analysis and Design of Structures;* Springer Science & Business Media: Dordrecht, The Netherlands, 2004.
17. Manguri, A.; Saeed, N.; Szczepanski, M.; Jankowski, R. Buckling and shape control of prestressable trusses using optimum number of actuators. *Sci. Rep.* **2023**, *13*, 3838. [CrossRef] [PubMed]
18. Janíček, P.; Ondráček, E.; Vrbka, J. *Mechanika těles. Pružnost a Pevnost I [Mechanics of Bodies. Elasticity and Rigidity I];* Vysoké učení Technické: Brno, Czech Republic, 1992.
19. Karnovsky, I.A.; Lebed, O. *Advanced Methods of Structural Analysis;* Springer Nature: Berlin/Heidelberg, Germany, 2021.
20. Hughes, A.J.; Merrill, C. Method of joints: Theory and practice of designing, building, and testing trusses. *Technol. Eng. Teach.* **2020**, *80*, 28.
21. Florian, Z.; Pellant, K.; Suchánek, M. *Technická Mechanika I—Statika [Technical Mechanics I—Statics];* Vysoké učení Technické v Brně: Brno, Czech Republic, 2004.
22. Miller, S.; Soares, T.; Weddingen, Y.V.; Wendlandt, J. *Modeling Flexible Bodies with Simscape Multibody Software. An Overview of Two Methods for Capturing the Effects of Small Elastic Deformations;* MathWorks: Natick, MA, USA, 2017.
23. Králik, J. *Modelovanie Konštrukcií v Metóde Konečných Prvkov [Finite Element Modelling of Structures];* ANSYS: Canonsburg, PA, USA, 2009.
24. Adhikari, S.; Phani, A.S. Rayleigh's classical damping revisited. In Proceedings of the International Conference on Civil Engineering in the New Millennium: Opportunities and Challenges, Howrah, India, 11–14 January 2007.
25. Rahman, M.H.; Gupta, C. Computation of Rayleigh damping coefficient of a rectangular submerged floating tunnel (SFT). *SN Appl. Sci.* **2020**, *2*, 936. [CrossRef]
26. Planchard, D. *Solidworks 2014 Reference Guide;* SDC Publications: Mission, KS, USA, 2014.
27. Kurowski, P. *Engineering Analysis with SolidWorks Simulation 2015;* SDC Publications: Mission, KS, USA, 2015.
28. Solidworks. Available online: https://www.solidworks.com/ (accessed on 1 August 2024).
29. Mechanical Engineer. Available online: https://shantgananian.github.io (accessed on 1 August 2024).

applied sciences MDPI

Article

Technical Diagnostics of Industrial Robots Using Vibration Signals: Case Study on Detecting Base Unfastening

Daria Fedorova [1,*], Vladimír Tlach [1], Ivan Kuric [1], Tomáš Dodok [1], Ivan Zajačko [1] and Karol Tucki [2]

[1] Department of Automation and Production Systems, Faculty of Mechanical Engineering, University of Zilina, Univerzitna 8215/1, 010 26 Zilina, Slovakia; vladimir.tlach@fstroj.uniza.sk (V.T.); ivan.kuric@fstroj.uniza.sk (I.K.); tomas.dodok@fstroj.uniza.sk (T.D.); ivan.zajacko@fstroj.uniza.sk (I.Z.)

[2] Institute of Mechanical Engineering, Department of Production Engineering, Warsaw University of Life Sciences, Nowoursynowska 164, 02-787 Warszawa, Poland; karol_tucki@sggw.edu.pl

* Correspondence: daria.fedorova@fstroj.uniza.sk

Abstract: In the domain of modern manufacturing digitalization, artificial intelligence tools are increasingly employed for condition monitoring and technical diagnostics. However, the majority of existing methodologies primarily concentrate on the technical diagnosis of rotating machines, with a noticeable lack of research addressing these issues in sequential machines. In this paper, we deal with the selection of suitable vibration signal characteristics for the detection of an industrial robot's release from its base during a handling operation. Statistical methods, including one-way ANOVA and *t*-tests, were used to identify the most significant features, which allowed us to isolate vibration metrics with significant predictive potential. These selected features were then used as inputs to various machine learning models to evaluate the hypothesis that these parameters can reliably indicate fastening releasing events. The results show that the optimized parameters significantly improve the detection accuracy, thus providing a reliable basis for future applications in predictive maintenance and monitoring. The findings represent an advance in robotic condition monitoring, providing a structured approach to feature selection that improves the reliability of disconnection detection in automated systems with potential applicability in various industrial environments.

Keywords: vibrodiagnostics; detection; feature extraction; feature selection; ensemble model; neural network; technical diagnostic; industrial robot

1. Introduction

Condition monitoring is an effective and widely used tool in modern manufacturing. It contributes to improving production productivity and reliability. Condition monitoring of a variety of industrial equipment is used today to ensure that it is possible to detect any problems in time to prevent unplanned downtime and extensive damage.

Vibrodiagnostics plays a significant role in monitoring the condition of machinery and equipment. Vibration signatures are an attractive parameter for condition monitoring as they are easily measurable and allow up-to-date information about the machine's condition without direct intervention in the production process. By analysing vibration patterns, anomalies such as mechanical damage, misalignment, or component failures can be detected in the machine long before they develop into critical problems [1,2].

In addition to the traditional statistical methods of evaluation and processing of such signals, methods or algorithms based on artificial intelligence are also gaining popularity, as they can monitor and evaluate non-linear dependencies of the analysed parameters [3–5].

Lorenz et. all [5] in the study defines several key techniques that are currently used in vibration signal classification, which include Fuzzy Logic [6,7], Artificial Neural Networks [8–13], and traditional signal processing techniques [14–16]. Recent research in this field has also explored the application of various machine learning and AI techniques for vibration-based condition monitoring and fault diagnosis of rotating machinery [17–22].

The key advantages of these techniques include improved diagnostic accuracy, reduced analysis effort, and the ability to handle complex, nonlinear relationships in vibration data [18–20]. A growing trend can be seen with the implementation of models based on various machine learning principles, even in industrial cases [3,23–26].

The diversity of model architectures used is also noticeable. LSTM and recurrent neural networks have been successfully used for RUL prediction [10,27] and maintenance task scheduling [28]. Deep convolutional neural networks, in addition to the traditional visual data processing domain, have been successfully used for bearing fault detection [29–31], gear faults [2,32], and shaft misalignments [19].

Researchers have also explored the integration of AI-based methods with traditional vibration analysis techniques, such as time-frequency analysis to enhance the diagnosis of rotating machinery under variable speed conditions [33,34]. These hybrid approaches have improved performance in extracting relevant features and identifying fault signatures [33,34].

However, there are still some gaps and challenges in this area of research. Data-driven models typically rely on signal processing, which demands increased computational capacity and can pose significant challenges in real manufacturing environments. Moreover, it is essential to investigate the integration of machine learning with various data sources and domain-specific knowledge to improve the performance and reliability of these techniques [5,26]. Furthermore, the interpretability and explainability of AI-based vibration diagnosis methods is an area that requires further research [18,20]. In addition, a more comprehensive investigation is necessary to evaluate the robustness of these techniques in response to environmental and operational changes, as well as to ascertain their applicability to a broader range of rotating machine types [20,22,26].

2. Motivation

As shown above, vibrodiagnostics and condition monitoring are closely related and represent an important and relevant domain of interest for science and practical application. However, the primary research is dedicated to the field of rotating machinery and components, for which there are validated methods of vibration signal processing in the time domain, frequency domain, and time-frequency domain [2].

At the same time, it is evident that studies focusing on the vibrodiagnostics of sequential machines, including industrial robots, are significantly fewer than those addressing the vibrodiagnostics of rotational components. In the case of industrial robotics-related research, it is possible to see examples of successful applications of combined approaches based on deep neural networks for anomaly detection [35] and models used for audio-based fault detection [36]. There are also methods based on optical flow processing for detecting robot arm motion errors [37] and classification models based on the motion parameters of the joints of the industrial robot arm [38]. In previous research, we have also validated the ability of the LSTM neural network to classify the robot motion speed based on the vibration signal obtained from the MEMS accelerometer [39]. Within the broader field of vibrodiagnostics, industrial robot vibrodiagnostics also face similar challenges. These include difficulties in interpreting the designed models, the need for signal pre-processing, and complicated integration into real industrial environments. This is often compounded by the computational complexity and the inherently complex nature of operations performed under unstable conditions, such as load variations and variations in operating speed.

In modern manufacturing, the field of application of industrial robots is much more extensive than just handling heavy products. It also includes applications requiring high precision, such as machining, measuring, precision assembly, etc. In precision applications, it is important to maintain stable values of the performance criteria of an industrial robot [40]. Therefore, early identification of changes in these parameters can prevent the occurrence of collision situations or the production of non-conforming products.

The objective of our research is to contribute to the resolution of problems associated with vibrodiagnostics, with a particular focus on industrial robots. We have focused our attention on the selection of appropriate vibration signal characteristics, intending to facilitate the application of AI and ML-based models into real production environments. The selected features should be straightforward to compute, while simultaneously providing sufficient attributes to enable the detection of anomalies in the operation of an industrial robot using machine learning and artificial intelligence models.

Based on the above, in the present paper, we consider the selection of the most effective metrics and properties of vibration signals that could be used in a follow-up implementation in a production environment for the detection of potential faults and predictive maintenance of industrial robots. The case study deals with the detection of the release of a FANUC industrial robot from its base during a handling operation for the application of the obtained knowledge in the predictive maintenance of sequential machines.

Based on the aforementioned, the hypothesis of the given research is that, based on the selected parameters of the vibration signal, it is possible to use the selected machine learning algorithms to detect the unfastening of the robot at its base with an accuracy of at least 95%.

3. Preparation of the Experiment

Tracking robot fastening release is crucial for ensuring production quality, preventing downtime, and improving maintenance processes. This ultimately leads to more reliable and cost-effective operations. Traditional methods of detecting robot fastening loosening often rely on manual checks or basic feedback from sensors.

However, recent advances in machine learning and signal processing represent a transformative approach to this challenge. By analysing the specific parameters of vibration signals generated during robot operation, machine learning algorithms can detect even subtle changes in the robot's fastening. In our case, the main goal of the experiment is to find out what vibration signal metrics can provide relevant and objective data about the robot's fastening to its base.

3.1. Robotic Workstation and Work Cycle

The experiments took place at the assembly workstation (Figure 1) in the laboratory at the Department of Automation and Production Systems, University of Žilina. The industrial robot Fanuc LR Mate 200iC was responsible for carrying out manipulation tasks within this workplace. The assembly and subsequent manipulation tasks of the industrial robot were designed to closely simulate real production process conditions.

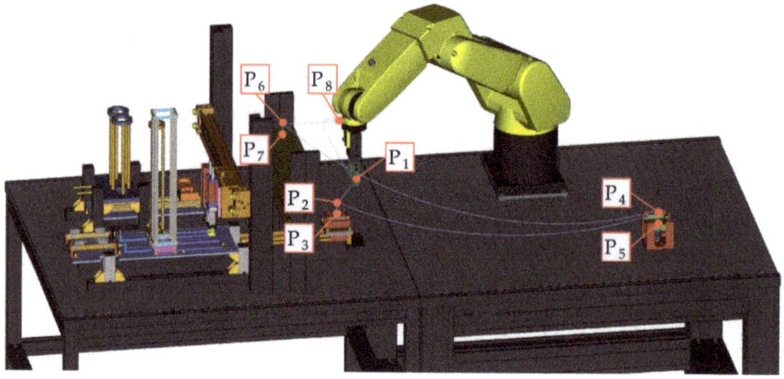

Figure 1. Robotic workstation.

During the working cycle, the robot takes the pre-assembled products from the assembly station (Figure 1, points P_2 and P_3), moves them to the fixture (Figure 1, points P_4, P_5), and finally places them in the storage rack (Figure 1, points P_6, P_7, P_8). In real production, cycle time plays a significant role because reducing cycle time is crucial for increasing overall productivity in the production process. For this reason, the program for the industrial robot is designed to ensure that all movements are carried out at the maximum speed of the TCP (Tool Center Point) of the robot. During the process of removing a part and inserting it into the fixture and storage rack, the robot's end member moves at low speeds to ensure accuracy and safety. The specific values of the robot's TCP speeds when moving between points are detailed in Table 1. The total working cycle time of the robot is 17.93 s.

Table 1. Movement speeds of the robot's TCP.

Path of Movement	Speed of the Robot's TCP	Operation Description
P_1–P_2	100%	movement above the assembly station
P_2–P_3	5%	product gripping process
P_3–P_7	5%	product gripping process
P_3–P_4	100%	movement above the fixture
P_4–P_5	5%	screwing process
P_5–P_4	2.5%	screwing process
P_4–P_5	2.5%	screwing process
P_5–P_4	5%	screwing process
P_4–P_6	100%	moving to the storage rack
P_6–P_7	5%	product storage process
P_7–P_6	5%	product storage process
P_6–P_8	100%	movement to home position
P_8–P_1	100%	movement to home position

The Fanuc LR Mate 200iC (FANUC CORPORATION, Oshino-mura, Yamanashi Prefecture, Japan) industrial robot is fastened to the work surface using a mounting plate attached to a table made of modular profiles. The robot is fastened to the mounting plate with four M10×25 screws (torqued to 64 Nm) [41], and the plate is then clamped to the workbench using four M8×20 screws (torqued to 30 Nm) [41] and T-nuts. The clamping method is illustrated in Figure 2.

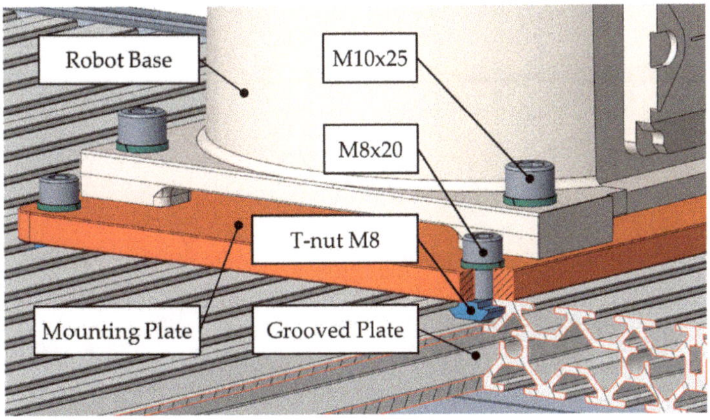

Figure 2. Clamping method used for industrial robot.

3.2. Vibration Monitoring Sensors Description and Positioning

During the experiment, we simulated the release of the robot fastening by loosening the M8×20 screws, which secure the robot and the clamping plate to the workbench. We used Balluff BCM0002 (Balluff GmbH, Neuhausen, Germany) condition monitoring sensors for the experiment. These sensors can detect not only vibrations, but also other operational data related to the equipment or its environment, such as surface temperature, humidity, and atmospheric pressure. The essential technical specifications of the sensor used can be found in Table 2.

Table 2. Technical characteristics for vibrations of condition monitoring sensor.

Specs	Balluff BCM0002
Evaluation values time domain	RMS, Peak to Peak, Max, Crest Factor, Skewness, Kurtosis
Frequency range	2...1800 Hz (±10%) 2...2500 Hz (±3 dB)
Measuring principle	MEMS
Measuring range	−16...16 g
Number of measuring axes	3
Interface	IO-Link 1.1

For the measurements, three sensors, which were placed on the mounting plate via a Balluff BAM03FA magnetic bracket (see Figure 3a) were used. This type of magnetic bracket is used specifically for this sensor type and the installation has been carried out according to the sensor manufacturer's instructions. The positioning of the sensors is shown in Figure 3b. The position of the sensors was determined with a view to install them as close as possible to the object to be traced while ensuring that the accelerometer axes were aligned with the robot's World Coordinate System (WCS).

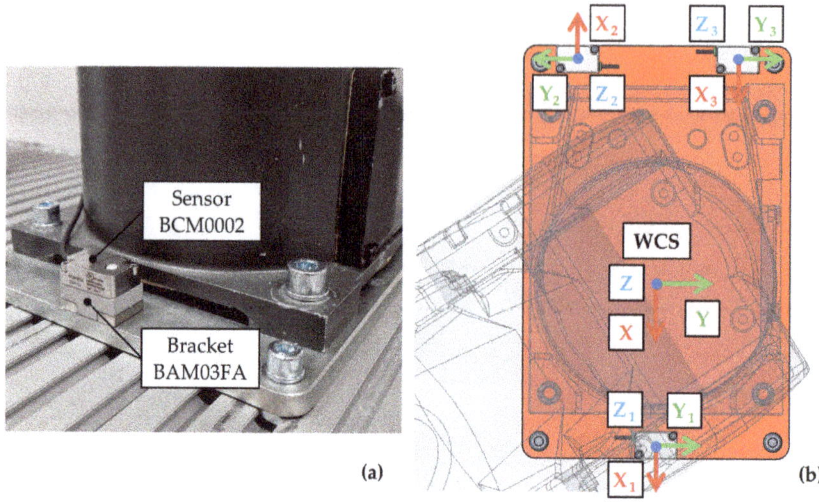

Figure 3. Sensor clamping method (**a**), and placement of sensors on the clamping plate, including orientation of accelerometer axes (**b**).

4. Experimental Methodology

In the following section, we focus on the methodology for the experimental investigation of release detection for industrial robots during manipulation operations. This research methodology comprises several interdependent stages, each described in detail in the individual subsections. These range from the procedures for data collection and pre-processing to the selection of the crucial characteristics of the vibration signal for subsequent use in the implementation of the chosen machine learning techniques.

The primary objective of this chapter is to present a comprehensive framework for industrial robot unfastening detection that combines advanced signal processing technologies and artificial intelligence methods.

4.1. Data Gathering

Selected industrial condition monitoring sensors offer the possibility of sensing a number of various parameters that need to be configured via a web-based interface. In the presented research, we are dealing with an industrial robot, which belongs to the class of sequential machines, along with presses, pneumatic linear actuators, and linear drives. For sequential machines, three vibration monitoring profiles are available. These profiles vary according to the speed of the machine. In the case of our research, in this context, the profile was set to sense at very fast process speeds. The sensor calculates the RMS and peak-to-peak values of the vibration velocity for the three axes within the configured time window of the vibration module.

Both methods have their advantages and disadvantages and specific areas of application. The RMS method has proven to be more effective in monitoring the long-term effects of vibration on various equipment and structures. This method provides a comprehensive overview of the total vibration intensity, which is crucial for failure prediction. For this research, we decided to use the peak-to-peak (P-P) method. This represents the measurement from the minimum peak to the maximum peak of the signal (Figure 4). An increase in this value may indicate the occurrence of spikes or faults. Measuring this metric is useful in applications where it is important to know the maximum vibration amplitude, which aids in real-time fault detection [42].

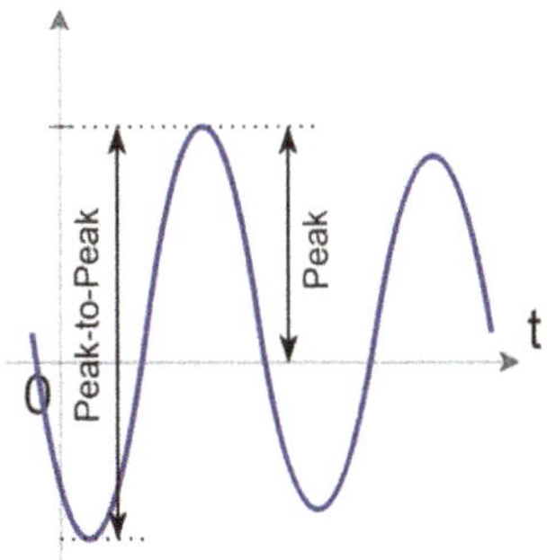

Figure 4. Peak-to-peak measurement.

The vibration velocity data is collected by the IO-Link hub, which subsequently communicates with the PC using Matlab script. One BCM sensor cyclically transmits four parameters at the same time (20 bytes of real-time data) through the IO-Link interface. These are divided into five slots, each with 4 bytes. The first four slots contain float32 numbers, which contain information about the P-P values of the vibration velocity for a certain set time window (20 ms). The next process is to decode the values and store them in a .csv file for post-processing. The experimental data were collected from three BCM sensors. Each sensor captures peak-to-peak velocity values in three axes (X, Y, and Z axes).

Abnormal robot operation data includes vibration patterns in cases where the robot attachment to the base is loosened due to multiple forces in the operating process. This condition was simulated by manually loosening all screws after 6 h of operation. Therefore, in the case of our research, all data collected before the loosening moment is considered normal data. Since the fact that the exact moment when the base started loosening is clearly defined, we can consider that all data after this event correspond to anomalous conditions.

4.2. Data Preparation

Data preprocessing is essential for improving the quality and performance of machine learning models because it ensures that the input data is in an optimized format for the algorithms, thereby minimizing errors and maximizing prediction accuracy. In general, data preprocessing includes such steps as normalization, removal of missing values and duplicate records, correction or removal of defective data that does not indicate an error in the operation of the device but errors in the data collection process, etc. The proposed methods of preparing the data obtained in the experiments will be described in more detail below.

4.2.1. Data Preprocessing

The data collected for the experiment represent the P-P values of the acceleration in the X, Y, and Z axes for each of the three sensors that are located on the robot's base. The dataset includes 10 input parameters: a timestamp and nine records of the recorded vibrations (in the X, Y, and Z axes for the 3 sensors). In addition, the moment, when the robot's fastening on the pedestal was released is also known (after 6 h of continuous operation).

The data set was divided into normal operation data and data corresponding to abnormal conditions. The following data were processed into sequences. Each sequence corresponds to one robot operating cycle. The length of one sequence is 17.93 s, which corresponds to 52 callbacks from the BCM0002 sensor. For solving the following classification task, each sequence was also labelled "Normal" if the sequence corresponds to a fault-free operation of the machine and "Anomaly" if the sequence corresponds to the operation of the robot on a loose base. Examples of a normal and abnormal sequence are shown in Figure 5.

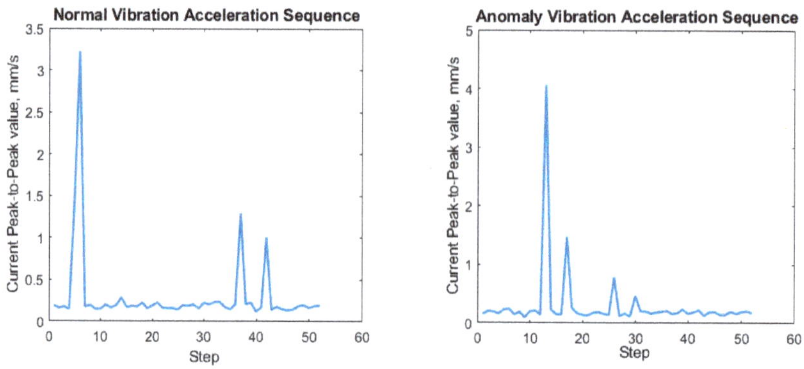

Figure 5. Examples of normal and anomaly sequences.

Since the sensor transmits the pre-processed signal within the selected time window by the set profile, it is not necessary to use additional data processing, such as signal filtering, interpolation, or removal of invalid values.

4.2.2. Feature Extraction

Feature extraction is a crucial step in the machine learning process because it transforms the raw data into a set of relevant and meaningful attributes that better represent the underlying patterns necessary for analysis. By isolating these key features, feature extraction helps to reduce the dimensionality of the data, making it more manageable and computationally efficient for machine learning algorithms to process. In addition, effective feature extraction helps reduce the risk of overfitting when the model might otherwise tend to merely over-replicate erroneous patterns.

There are many techniques to extract features. In particular, the frequently used ones are [43]:

- Time Domain Features—signal amplitude, zero-crossing rate, sum of squared amplitudes (signal energy), signal entropy.
- Frequency-Domain Features—Fourier Transform, Power Spectral Density (PSD), Spectral Centroid, Spectral Bandwidth.
- Time-Frequency Features—Wavelet Transform, Short-Time Fourier Transform (STFT).
- Statistical Features—Statistical measures that describe the distribution and shape of the signal data (mean, variance, skewness, kurtosis, etc.), autocorrelation.
- Other the application-specific techniques.

To extract signal features, we used the Diagnostic Feature Designer tool from Matlab software. It allows rapid feature extraction of signals using selected techniques and the subsequent selection of the most relevant features using multiple evaluation algorithms that use specific criteria to determine which features are the most effective.

As mentioned previously, a large number of criteria are defined for rotating machines and components such as gearboxes, bearings, and rotors of electric motors [5]. However, for sequential mechanisms, there is a lack of research in the relevant field. For this reason, it is difficult to predict what signal characteristics will best indicate the state of the device. At the same time, the dataset contains nine parameters, so calculating all possible properties that the software offers for all signals and their subsequent analysis would be time-consuming and computationally intensive. Therefore, we decided to use only statistical features, and for the first attempt, feature extraction was performed for one sensor (using data from all three axes).

For the sensor located at the front of the robot's base (see Figure 3), the basic statistical features of the signal were computed. These properties can be applied to any kind of signal, including the time-synchronized average (TSA) vibration signal. Any changes in these features can indicate changes in the state of the system. The calculated features are:

- clearance factor,
- crest factor,
- impulse factor,
- kurtosis,
- mean,
- RMS,
- skewness,
- std.

Next, selecting the features that most accurately represent the predicted variable is necessary. From an initial point of view, a feature can be effective if it separates data groups with different state labels (normal and anomaly) with sufficient clarity. For an initial evaluation of the effectiveness of the feature, histograms were plotted (Figure 6) to show the separation of data between the group of normal and anomalous sequences.

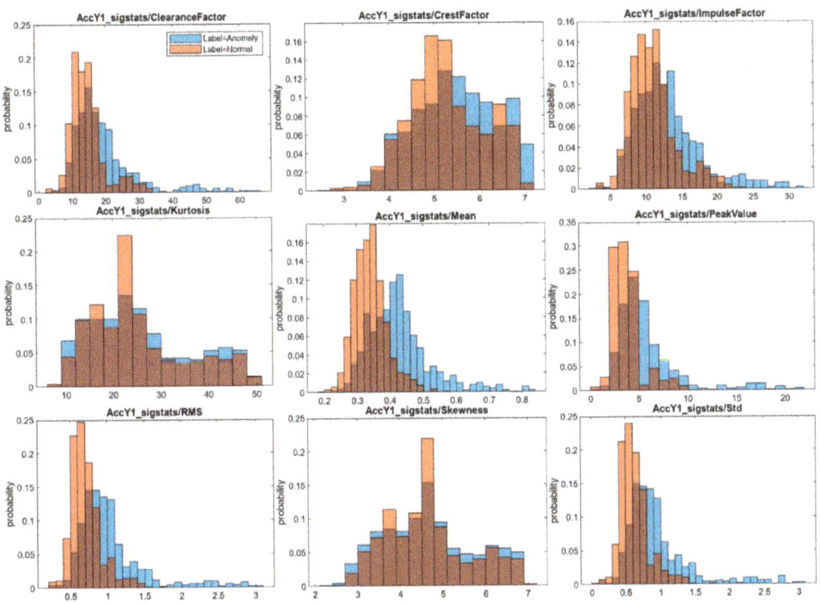

Figure 6. Feature histograms for normal and anomaly data.

Figure 6 shows histograms representing the distribution of various statistical properties of the dataset. The width of one column in each case is chosen individually depending on the range of change in the value of a given property for a better graphical representation. From the analysis of these bar charts, it is possible to observe that even though none of the metrics show a perfect distribution of normal and abnormal sequences, it is possible to observe the following dependencies:

- ClearanceFactor, PeakFactor, Mean, Std, RMS, and PeakValue properties have the lowest overlap (shown in brown in Figure 6) between them, and a slight rightward shift in the distribution of anomalous values can be seen for each of these properties. At first glance, these properties seem to be able to describe the investigated signal more explicitly.

- In contrast, the feature factors Skewness and Kurtosis show a large rate of overlap and an almost identical distribution of data along the horizontal axis. The CrestFactor feature, while characterized by less overlap, also shows the same distribution along the horizontal axis.

It can be concluded that the most significant metrics are with high probability among the first group of features. Since the graphical representation only allows a first insight into the aspect in which each feature represents the target categorical variable, to select the right features for training machine learning models and to increase the efficiency of the training process, analysis and selection of appropriate features will be performed in the next step.

4.3. Feature Ranking Methods

Selecting relevant features for machine learning is another crucial step in developing an effective machine learning model. This means selecting only those attributes that have a significant impact on the output of the model.

After feature extraction, the experiment dataset contains nine new features for each axis of each sensor; thus, the total number of potential features for designing and training machine learning algorithms is 81 (or 27 for a single sensor). However, some of these features may be redundant, some may be correlated, and some may be irrelevant to the model. In this case, using the whole dataset for training the model may lead to an overfitting. With overfitting, the machine learning model loses the ability to generalize the patterns and characteristics that were detected in the training data, leading to low accuracy of the trained model on the new data. To avoid this, we performed simple statistical feature selection algorithms: *t*-test and one-way ANOVA.

4.3.1. *T*-Test

T-test can be applied between two classes to compare their means and conclude whether they are from the same distribution. In doing so, it is assumed that the distribution of the parameter under examination is normal. A high *t*-test value for a given parameter indicates that the investigated parameter (feature) varies widely for each class and, hence, represents a suitable choice for training a machine learning classification algorithm. The computation of the T-value for each pair of parameters is according to the Formula (1), where $\overline{x}_1$ and $\overline{x}_2$ are the means of the two groups being compared, s_1^2 and s_2^2 are the pooled standard errors of the two groups, and n_1 and n_2 are the number of observations in each of the groups.

$$T_{score} = \frac{\overline{x}_1 - \overline{x}_2}{\sqrt{\left(\frac{s_1^2}{n_1} + \frac{s_2^2}{n_2}\right)}} \tag{1}$$

When selecting appropriate features using a *t*-test, the null hypothesis and alternative hypothesis are critical in determining the significance of each feature with the target variable.

The null hypothesis usually assumes that there is no significant difference in the mean values of an element between different groups of the target variable. That means that the attribute provides no discriminating power between groups and any observed differences in means are caused by random chance. In the present research, the null hypothesis is established as follows:

H0: *There is no significant difference between anomalous and normal data for the values of the investigated feature.*

The alternative hypothesis suggests that there is a statistically significant difference in the mean values of the feature between the groups. This indicates that this feature could potentially be useful for distinguishing between classes and should be considered for inclusion in the model. The alternative hypothesis is stated as follows:

H1: *The difference between the anomalous and normal data in the investigated feature is substantial, making it meaningful for classification tasks.*

After calculating T_{score} values for each parameter, the features were sorted by importance. The histograms in Figure 7 show all calculated features for each axis, sorted by importance.

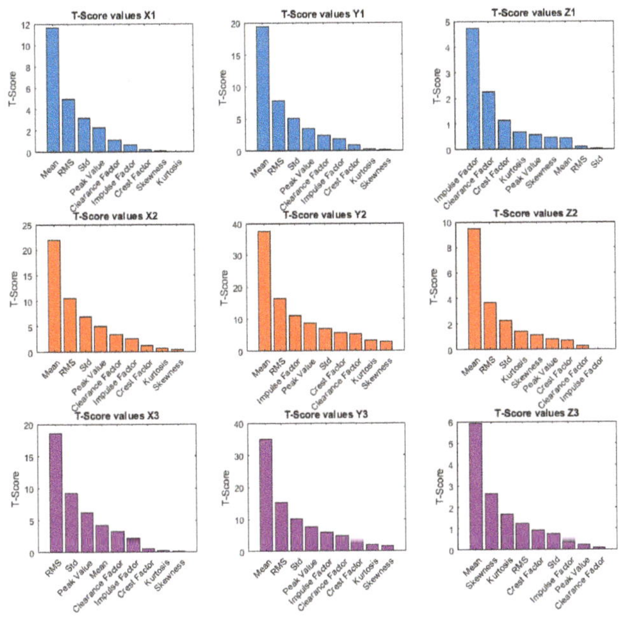

Figure 7. T-score values for each sensor.

After the performed calculation, it is important to compare T_{score} values with the defined critical value of the T_{score} to decide if the null hypothesis will be rejected for the selected feature. A lower calculated T_{score} than the T_{score} critical value leads to the rejection of the null hypothesis, therefore supporting the alternative hypothesis. This process allows the selection of features that have a statistically significant relationship with the target variable, enhancing the predictive power of the model. The table below (Table 3) shows the parameters for which the alternative hypothesis was accepted. In the present research, the number of degrees of freedom is assumed to be $df = 1498$, the significance level α is $\alpha = 0.05$, and the critical value of the T_{score} is $T_{score} = 1.9515$.

Based on the *t*-test, for sensor 1 for the X_1 axis (Figure 7), the Mean feature has the highest T_{score}, which means that it is the most significant in discriminating between classes. RMS, STD, and Peak Value also show a relatively high T_{score}. For the Y_1 axis, the results are similar to the case of the X_1 axis: Mean has the highest T_{score}. This parameter is followed by

RMS and Std metrics. The Z_1 axis shows a different pattern—in this case, the Impulse Factor, Clearance Factor, and Skewness parameters were shown to be the most significant factors.

Table 3. Accepting an alternative or null hypothesis for the specified signal features based on a *t*-test.

	X_1	Y_1	Z_1	X_2	Y_2	Z_2	X_3	Y_3	Z_3
Mean	H1	H1	H0	H1	H1	H1	H1	H1	H1
RMS	H1	H1	H0	H1	H1	H1	H1	H1	H0
Std	H1	H1	H0	H1	H1	H1	H1	H1	H0
Peak Value	H1	H1	H0	H1	H1	H0	H1	H1	H0
Clearance Factor	H0	H1	H1	H1	H1	H0	H1	H1	H0
Impulse Factor	H0	H0	H1	H1	H1	H0	H1	H1	H0
Crest Factor	H0	H0	H0	H0	H1	H0	H0	H1	H0
Skewness	H0	H0	H0	H0	H1	H0	H0	H0	H1
Kurtosis	H0	H0	H0	H0	H1	H0	H0	H1	H0

Parameters for which the alternative hypothesis was accepted are marked in green. Parameters for which the null hypothesis is accepted are marked in red.

For sensor 2, for the X_2 and Z_2 axes (Figure 7), the Mean parameter again leads with the highest T_{score}, followed by the RMS and Std parameters. For the Y_2 axis, the Mean parameter also leads, followed by the STD and Impulse Factor as significant features.

For sensor 3, the most significant parameter for the X_3 axis (Figure 7) is RMS, followed by Std and Peak Value. For the Y_3 axis, Mean, RMS, and Std lead as the most significant parameters. Mean and Skewness are the most significant parameters for the Z_3 axis.

4.3.2. One-Way ANOVA

Analysis of variance is used to compare sample means for two or more classes and determine if observed differences may be due to chance or reflect true differences between classes. Like the *t*-test, the null hypothesis (that there is no difference between the means of the groups) and the alternative hypothesis (that at least one group is significantly different from the overall mean of the dependent variable) are established. A significant difference in the means indicates that the function has good discriminative capacity and is likely to be useful in discriminating between classes in a classification task.

One-way ANOVA was performed in the context of the present paper with the following parameters: $df = 1498$, the significance level α is $\alpha = 0.05$, and the critical value of F_{score} is $F_{score} = 3.8477$. Figure 8 shows bar graphs of the F_{score} for each of the characteristics under investigation.

A higher F_{score} indicates that the group means are more spread out, suggesting that a significant difference may exist between at least one pair of groups. Similar to the *t*-test, if the calculated F_{score} for a given variable is less than a specified critical value, the null hypothesis is confirmed for that trait and the parameter is considered to have low discriminatory power. Table 4 shows for which parameters the null hypothesis was rejected, and the alternative hypothesis was accepted.

The results of the one-way ANOVA analysis for sensor 1 (Figure 8) are identical to the *t*-test results and again demonstrate that within the X_1 axis and Y_1 axis Mean, Std, and RMS are the features with the highest resolution for the anomalous and normal data. In contrast to X_1 and Y_1, for the Z_1 axis, Impulse Factor has the highest F_{score}, indicating that it is crucial for class distribution in this dataset, but RMS and Std have a very low F_{score}, indicating low relevance of the given features.

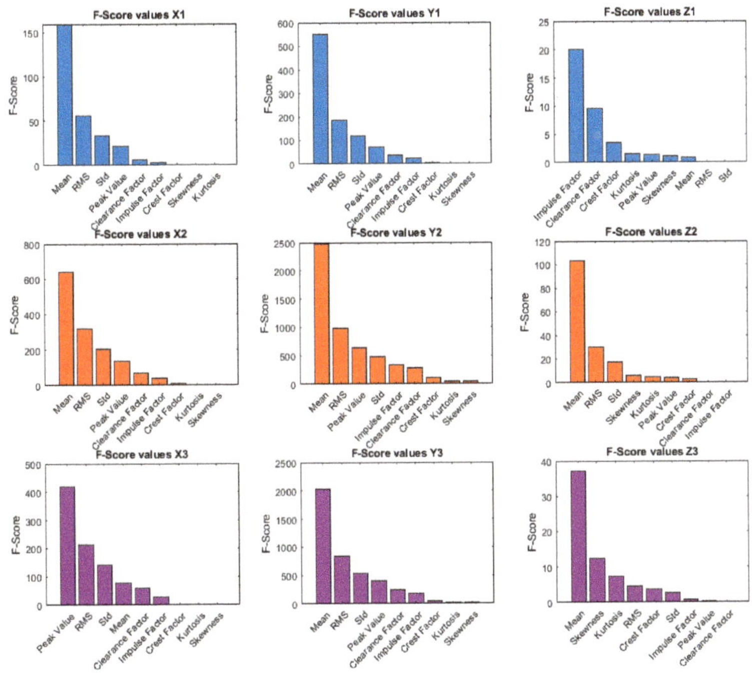

Figure 8. F-score values for each sensor.

Table 4. Accepting an alternative or null hypothesis for the specified signal features based on one-way ANOVA.

	X_1	Y_1	Z_1	X_2	Y_2	Z_3	X_3	Y_3	Z_3
Mean	H1	H1	H0	H1	H1	H1	H1	H1	H1
RMS	H1	H1	H0	H1	H1	H1	H1	H1	H1
Std	H1	H1	H0	H1	H1	H1	H1	H1	H0
Peak Value	H1	H1	H0	H1	H1	H1	H1	H1	H0
Clearance Factor	H1	H1	H1	H1	H1	H0	H1	H1	H0
Impulse Factor	H0	H1	H1	H1	H1	H0	H1	H1	H0
Crest Factor	H0	H1	H0	H1	H1	H0	H0	H1	H0
Skewness	H0	H0	H0	H0	H1	H1	H0	H1	H1
Kurtosis	H0	H0	H0	H0	H1	H1	H0	H1	H1

Parameters for which the alternative hypothesis was accepted are marked in green. Parameters for which the null hypothesis is accepted are marked in red.

For sensor 2 in all three axes (Figure 8), the Mean parameter has the highest F_{score}, followed by the RMS and Std parameters for the X_2 and Z_2 axes, and the RMS and Peak Value parameters for the Y_2 axis.

Sensor 3 differs from the standard displays, especially within the X_3 axis (Figure 8). Peak Value has the highest F_{score}, indicating that it is the most important feature of this data set. This is followed by RMS and Std, while Skewness and Kurtosis have minimal significance. For the Y_3 axis, the dominant feature is again Mean, while for the Z_3 axis dominant features are also Skewness and Kurtosis, as in the case of the *t*-test.

4.4. Feature Selection

After the one-way ANOVA and *t*-test analyses were performed, the analysis of statistical results along with the analysis of the obtained histograms were followed to select relevant features for the training of the following machine learning algorithms. The selection of the features also considered the fact that, for the potential application of the classification model in practice in terms of optimizing data processing times, it is necessary to select the same parameters for each axis.

It can be concluded that among the listed signal properties, none can be used to explicitly identify anomalies in the signal waveform from all three sensors in each axis, suggesting the need to use a set of properties to identify the errors. However, based on the above, some significant dependencies in data can be observed.

Mean is consistently one of the most important characteristics in most data sets and often shows the highest T_{score} in the case of *t*-test and F_{score} in the case of one-way ANOVA. It can also be seen in the bar graphs in Figures 7 and 8 that the values of the statistical score for a given trait are much higher in most cases than the values of any other feature. RMS and Std are also often significant in distinguishing between groups of anomalous and normal data. These features are likely to be crucial for classification tasks.

Peak Value in addition to Mean, RMS, and Std, has a relatively high F_{score}, suggesting that it is also a potential key feature for classification. For the above parameters, the alternative hypothesis was accepted in most cases (6 features out of 9 for *t*-test and 7 out of 9 for one-way ANOVA). The acceptance of the null hypothesis for Mean, Std, and RMS in the case of Z_1 is likely due to the sensor location in the middle of the front face of the base of the robot and, hence, significantly less variation in the anomalous data compared to sensors that are located near the fixing screws.

After generalizing the above, we decided to use Mean, Std, RMS, and Peak Value parameters as key metrics for designing and training machine learning algorithms for the detection classification of normal and abnormal vibration sequences. The selection of the key features allowed us to reduce the dimensionality from the original 27 parameters for a single sensor (81 parameters for the whole dataset) to 12 parameters for a single sensor (36 parameters for the whole dataset). In addition, the Balluff BCM0002 sensors, when properly configured with the sensing profiles, allow direct sensing of three of the selected metrics, namely, Mean, RMS, and Std of the signal, which is also a benefit for feature implementation of the trained models.

4.5. Machine Learning Models for Unfastening Detection

For the next step, we designed several artificial neural network models and popular ensemble learning models because their classification abilities have been demonstrated in many previous studies. A brief description of the structures of the models used is given in Tables 5 and 6.

The design and training of the machine learning models were performed using Classification Learner in MATLAB software. The Classification Learner application simplifies the training of models for data classification. This tool allows rapid testing of multiple classifiers simultaneously using a comprehensive environment for exploring data, selecting features, specifying validation schemes, training models, and evaluating results.

All the proposed algorithms belong to the class of supervised learning algorithms, which require the use of a pre-labelled dataset for training and testing purposes. The data used in the present research contains two classes namely "Normal" and "Anomaly" and are divided into training data (70% of the original dataset) and test data (30% of the original dataset).

Table 5. Ensemble models.

Model Name	Learner Type	Ensemble Method	Number of Learners	Specification
Boosted Trees	Decision Tree	AdaBoost	30	Number of splits: 20
Bagged Trees	Decision Tree	Bag	30	Number of splits: 20
Subspace Discriminant	Discriminant	Subspace	30	Subspace dimension: 6
Subspace KNN	Nearest Neighbors	Subspace	30	Subspace dimension: 6
RUSBoosted Trees	Decision Tree	RUSBoost	30	Number of splits: 20

Table 6. Neural network models.

Model Name	Number of Fully Connected Layers	First Fully Connected Layer Size	Next Fully Connected Layers Size	Activation
Narrow Neural Network	1	10	-	ReLU
Medium Neural Network	1	25	-	ReLU
Wide Neural Network	1	100	-	ReLU
Bilayered Neural Network	2	10	20	ReLU
Trilayered Neural Network	3	10	20, 30	ReLU

5. Detection of the Robot Mounting Release from the Base Using Machine Learning Algorithms

Detecting loose robot fasteners is key to ensuring the reliability and safety of the entire automated process. Traditional methods for detecting loose fasteners tend to rely on manual inspection or fixed rule-based systems that are error-prone and limited in adaptability. In contrast, machine learning algorithms can use large data sets to learn complex patterns related to mounting anomalies, enabling more robust and scalable detection solutions.

As mentioned, choosing key signal features reduced the total number of training parameters from 81 to 36 for the whole dataset. However, the fact that the selected metrics are the same for each sensor must be considered. Therefore, it may be irrational to design and train machine learning models using data from all three sensors.

Correct placement of the minimum required number of sensors also plays an essential role in condition monitoring. The manufacturer of the Balluff BCM0002 sensors suggests methods for sensor placement, shown in Figure 9. These methods are common for monitoring rotating components.

However, we investigate the release detection of a sequential machine. For this reason, we decided to place one sensor in accordance with the manufacturer's recommendation (sensor 1 in Figure 3). The two remaining sensors were placed at the locations where the most variation in attachment release is expected to occur. For the above reasons, we decided to divide experiment into two steps:

- Firstly, the selected models are trained and tested separately on data from each sensor.
- Next, for the final analysis, data from sensors that enable the models to demonstrate the highest accuracy will be utilized.

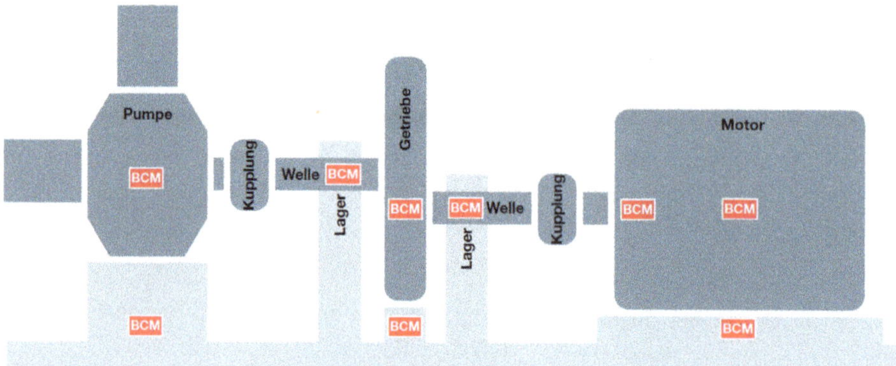

Figure 9. Examples of placement of BCM-type sensors for correct vibration measurement [44].

In the next section, we test designed machine learning techniques for detecting loose assembly elements in robotic assembly processes. We will use the signal features selected in the previous section for training and verification of the selected models. Our goal is to prove the main hypothesis of the current research and to establish basic concepts for future research in the field of technical diagnosis of sequential mechanisms using machine learning algorithms.

5.1. Testing Chosen Machine Learning Models for Unfastening Detection

Models that were introduced in the Section 4.5 were trained on data from each of the sensors. To avoid overfitting and to examine the accuracy of each model, a cross-validation method was chosen.

The classification results after training are shown in Tables 7 and 8. The accuracy of all models based on sensor 1 data is significantly lower than that of sensor 2 and sensor 3, for which the accuracy is comparable and above 90%. This fact can be explained by the configuration of the workstation and the robot's behaviour: all the tasks that the robot performs during the working cycle take place in front of the robot's base, and thus vibrations caused by the releasing of the robot's fastening are more visible on sensors 2 and 3, which are located at the rear part of the mounting plate (see Figure 3).

Table 7. Ensemble model accuracy.

	Boosted Trees	Bagged Trees	Subspace Discriminant	Subspace KNN	RUSBoosted Trees
Sensor 1	87.58%	87.91%	87.47%	85.82%	88.92%
Sensor 2	92.64%	96.92%	93.30%	96.26%	96.81%
Sensor 3	94.25%	96.59%	92.97%	95.82%	95.15%

Table 8. Neural network model accuracy.

	Narrow Neural Network	Medium Neural Network	Wide Neural Network	Bilayered Neural Network	Trilayered Neural Network
Sensor 1	90.09%	91.55%	90.88%	89.89%	89.01%
Sensor 2	97.36%	96.81%	97.47%	96.48%	96.92%
Sensor 3	97.25%	97.69%	97.25%	98.13%	96.92%

For all models used, except Boosted Trees and Subspace Discriminant at a given stage, the main research hypothesis is assumed to be met using data from sensors 2 and 3 (classification accuracy is above 95%). Therefore, we excluded Boosted Trees and Subspace Discriminant models from subsequent analysis. Next, since sensors 2 and 3 differ only in their axis orientation, and the prediction results after validation on the training data are largely the same, we decided to focus only on sensor 2 for the following analysis and testing using new data.

5.2. Testing Results and Discussion

To prove that the selected machine learning algorithms can detect the robot's releasing at the base with an accuracy of 95% or higher by using specific parameters of the vibration signal, we tested the chosen machine learning models using new data gathered from sensor 2. The test data considered a total of 590 observations, including 297 normal sequences and 293 anomaly sequences. The resulting testing accuracy is shown in Table 9, sorted by highest accuracy.

Table 9. Tested classification accuracy.

Model Type	Preset	Accuracy % (Test)
Neural Network	Wide Neural Network	98.47
Neural Network	Narrow Neural Network	98.35
Neural Network	Trilayered Neural Network	97.97
Neural Network	Bilayered Neural Network	97.97
Neural Network	Medium Neural Network	97.29
Ensemble	RUSBoosted Trees	96.95
Ensemble	Bagged Trees	96.61
Ensemble	Subspace KNN	93.22

As can be seen, all models except Subspace KNN achieved a resulting accuracy on the test data above 95%. However, the percentage accuracy shows only the number of correct predictions (both positive and negative) out of all predictions (Formula (2)). Even though the test dataset can be considered consistent since the numbers of normal and abnormal samples are almost the same, we took a closer look at the performance of the models since, in real operations, misclassification of an anomalous sequence as normal can have serious consequences.

$$Accuracy = \frac{TP + TN}{TP + TN + FP + FN} \tag{2}$$

To examine the performance of the models using the new data, we used confusion matrices compiled for the Wide Neural Network model, which was the model with the highest accuracy, and the Subspace KNN model, which proved to have the lowest percentage accuracy.

Using the confusion matrix helps to summarize the prediction results and gives an overview of how well the models perform in different classes. The confusion matrix allows a comprehensive view of the models using three more metrics, namely, recall, precision, and specificity, in addition to accuracy. Confusion matrices for the selected models are shown in Figure 10.

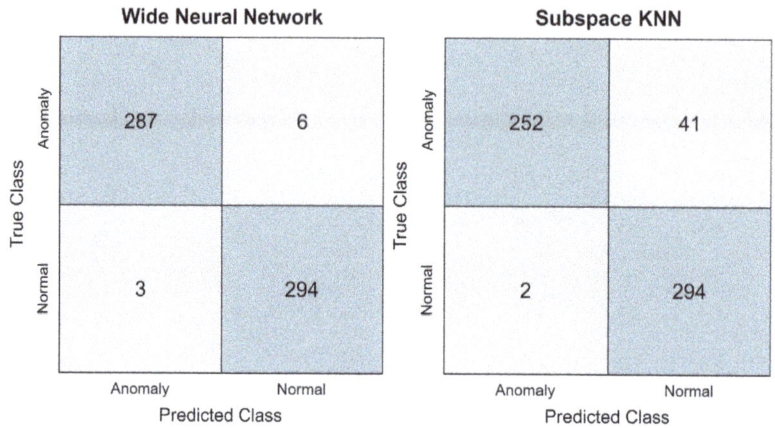

Figure 10. Confusion matrixes for selected models.

The resulting accuracy of the model (Precision) is given by Equation (3), where *TP*, or true positive, refers to a sample that belongs to the class "Normal" and is correctly classified. *FP*, or false positive, represents the number of samples that have been classified as normal even though they represent anomalous data. This metric shows how many of the predicted positive cases were truly positive. High precision means a low rate of false positive cases [45].

$$Precision = \frac{TP}{TP + FP} \quad (3)$$

Recall for the positive class shows the number of samples correctly predicted as belonging to the positive class out of all samples that truly belong to the positive class. This metric is represented by Equation (4), where *FN*, or false negative, refers to a sample that belongs to the normal class but is misclassified as belonging to the anomalous class [45].

$$Recall = \frac{TP}{TP + FN} \quad (4)$$

Another particularly useful metric for the defined problem is specificity, which is the number of samples correctly assigned to the anomalous class out of all samples in the dataset that truly belong to the anomalous class. Specificity is calculated using Equation (5), where *TN*, or true negative, refers to the sample that belongs to the anomalous class and is correctly classified [45].

$$Specificity = \frac{TN}{TN + FP} \quad (5)$$

These metrics were calculated for all tested models. The results are presented in Table 10.

Table 10. Calculated evaluation metrics.

Evaluation Metric	Wide Neural Network	Narrow Neural Network	Trilayered Neural Network	Bilayered Neural Network	Medium Neural Network	RUSBoosted Trees	Bagged Trees	Subspace KNN
Accuracy	98.47%	98.47%	97.97%	97.97%	97.29%	96.95%	96.61%	92.71%
Precision	98.00%	97.68%	97.03%	97.03%	96.37%	96.66%	95.41%	87.80%
Recall	98.99%	99.33%	98.99%	98.99%	98.32%	97.31%	97.98%	99.33%

As we mentioned at the beginning of the section, the Wide Neural Network (WNN) model shows significantly better accuracy (98.47%) compared to the Subspace KNN model (92.70%), indicating that the WNN is more reliable in correctly classifying both anomalies and normal sequences.

WNN has also a higher precision (98.00%), which means that it better minimizes false positives compared to Subspace KNN, which has a lower precision (87.76%). This suggests that Subspace KNN may more often mislabel normal sequences as an anomaly sequence.

Both models perform well in the recall, with Subspace KNN slightly outperforming WNN (99.32% vs. 98.99%). This indicates that both models are highly effective in detecting anomalies, with KNN being slightly better at avoiding false negatives.

In case of specificity WNN excels here with a specificity of 97.95%, which is much higher than the 86.01% for Subspace KNN. This means that the Subspace KNN model is more prone to false positives and more likely to mislabel anomalous sequences as normal.

Based on the above, the WNN model trained using the vibration signal metrics chosen in 5.3.3 performs well in accuracy, precision, and specificity, making it more reliable in detecting true anomalies and minimizing false positives. While Subspace KNN performs slightly better in recall, indicating its strength in detecting almost all true normal sequences, the model suffers from lower accuracy and specificity, indicating a higher rate of classifying anomalous samples as normal. This factor is crucial in practical implementation, as much more expense may be incurred just by incorrectly fitting anomalous data into the normal category. For this specific anomaly detection task, the WNN model appears to be a more balanced and accurate choice overall.

6. Conclusions and Feature Work

The proposed methodology was demonstrated through a case study involving the detection of a release fault in a FANUC industrial robot during handling operations. This scenario highlights the potential of our approach for improving predictive maintenance strategies, particularly for industrial robots.

In addressing a particular problem—detecting a robot's releasing at its base—Mean, Std, RMS, and Peak Value were shown to be the best-performing signal metrics, even under the realistic conditions of manipulation tasks concerning change in robot movement speed within the technological process. Moreover, we confirmed the hypothesis of the experiment: the chosen metrics allow the detection of anomalies in the vibration signal with an accuracy of higher than 95%, and this accuracy has been achieved for several neural network models and ensemble-learning techniques.

Furthermore, to validate the hypothesis and reveal any weaknesses in the proposed models, a classification problem was solved to detect the anomaly—detection of the robot's release during the reloading process. Based on the Mean, Std, Peak Value, and RMS metrics of the vibration signal sequences, the performance of two machine learning models, Wide Neural Network (WNN) and Subspace K-Nearest Neighbors (KNN), were compared. The evaluation was based on four key metrics: accuracy, precision, recall, and specificity. In the analysis of the results, it was shown that although accuracies above 90% can still be considered high, the amount of incorrectly captured anomalous samples plays a key role in the choice of the right AI model. In fact, in addition to reducing the overall accuracy of the model, low values of the recall and specificity metrics indicate a low discriminative ability of the model, and therefore implementation of a similar model in practice can lead to increased costs and risks in production.

By leveraging carefully chosen vibration signal properties, our approach facilitates early fault detection, minimizes unplanned downtime, and enhances the reliability of industrial robots. These metrics will play a key role in the early detection of anomalies, enabling timely maintenance interventions and minimizing machine downtime.

In our future work, we will focus on further validating the proposed models and metrics across various robotic systems and operating conditions. Additionally, the application of these techniques, aimed at developing intelligent real-time fault detection and prediction systems for industrial equipment and improving the system's ability to generalize across different production environments, contribute to the advancement of modern manufacturing.

Author Contributions: Conceptualization, D.F.; funding acquisition, I.K. and I.Z.; investigation, D.F. and V.T.; methodology, D.F. and V.T.; project administration, I.Z. and K.T.; supervision, I.K.; visualization, T.D.; writing—original draft preparation, D.F.; writing—review and editing, V.T. and T.D. All authors have read and agreed to the published version of the manuscript.

Funding: This research was funded by Research of Implementation Methods and Means of Artificial Intelligence in Automated Quality Control Systems of Products with Volatile Quality Parameters, 1/0470/23 and Transformation and Innovation Consortium of Robotics and Automation of the Slovak Republic (Acronym Robotic), 09I02-03-V01-00010.

Institutional Review Board Statement: Not applicable.

Informed Consent Statement: Not applicable.

Data Availability Statement: The data presented in this study are available on request from the corresponding author.

Conflicts of Interest: The authors declare no conflicts of interest.

References

1. Peťková, V. *Teória a Aplikácia Vybraných Metód Technickej Diagnostiky*; Knapíková, A., Ed.; Technická univerzita v Košiciach: Nitra, Slovakia, 2010; ISBN 978-80-553-0483-0.
2. Lei, Y.; Lin, J.; Zuo, M.J.; He, Z. Condition Monitoring and Fault Diagnosis of Planetary Gearboxes: A Review. *Measurement* **2014**, *48*, 292–305. [CrossRef]
3. Fink, O. Data-Driven Intelligent Predictive Maintenance of Industrial Assets. In *Women in Industrial and Systems Engineering*; Springer: Cham, Switzerland, 2020; pp. 589–605.
4. Carvalho, T.P.; Soares, F.A.A.M.N.; Vita, R.; Francisco, R.d.P.; Basto, J.P.; Alcalá, S.G.S. A Systematic Literature Review of Machine Learning Methods Applied to Predictive Maintenance. *Comput. Ind. Eng.* **2019**, *137*, 106024. [CrossRef]
5. Lorenz, A.; Siewertsen, B.; Kyhe Clemmensen, V.; Blaamann Petersen, J.; Friederich, J.; Lazarova-Molnar, S. Vibration Data Analysis for Fault Detection in Manufacturing Systems—A Systematic Literature Review. In Proceedings of the ICIEA 2022-17th IEEE Conference on Industrial Electronics and Applications, Chengdu, China, 16–19 December 2022; pp. 851–857. [CrossRef]
6. Li, C.; Cerrada, M.; Cabrera, D.; Sanchez, R.V.; Pacheco, F.; Ulutagay, G.; Valente De Oliveira, J. A Comparison of Fuzzy Clustering Algorithms for Bearing Fault Diagnosis. *J. Intell. Fuzzy Syst.* **2018**, *34*, 3565–3580. [CrossRef]
7. Helmi, H.; Forouzantabar, A. Rolling Bearing Fault Detection of Electric Motor Using Time Domain and Frequency Domain Features Extraction and ANFIS. *IET Electr. Power Appl.* **2019**, *13*, 662–669. [CrossRef]
8. Kumar, H.S.; Srinivasa Pai, P.; Sriram, N.S.; Vijay, G.S. ANN Based Evaluation of Performance of Wavelet Transform for Condition Monitoring of Rolling Element Bearing. *Procedia Eng.* **2013**, *64*, 805–814. [CrossRef]
9. Guo, X.; Chen, L.; Shen, C. Hierarchical Adaptive Deep Convolution Neural Network and Its Application to Bearing Fault Diagnosis. *Measurement* **2016**, *93*, 490–502. [CrossRef]
10. Sampaio, G.S.; Vallim Filho, A.R.d.A.; da Silva, L.S.; da Silva, L.A. Prediction of Motor Failure Time Using An Artificial Neural Network. *Sensors* **2019**, *19*, 4342. [CrossRef]
11. Magar, R.; Ghule, L.; Li, J.; Zhao, Y.; Farimani, A.B. FaultNet: A Deep Convolutional Neural Network for Bearing Fault Classification. *IEEE Access* **2021**, *9*, 25189–25199. [CrossRef]
12. Sepulveda, N.E.; Sinha, J. Parameter Optimisation in the Vibration-Based Machine Learning Model for Accurate and Reliable Faults Diagnosis in Rotating Machines. *Machines* **2020**, *8*, 66. [CrossRef]
13. Almutairi, K.M.; Sinha, J.K. Experimental Vibration Data in Fault Diagnosis: A Machine Learning Approach to Robust Classification of Rotor and Bearing Defects in Rotating Machines. *Machines* **2023**, *11*, 943. [CrossRef]

14. Gilles, J. Empirical Wavelet Transform. *IEEE Trans. Signal Process.* **2013**, *61*, 3999–4010. [CrossRef]
15. Tan, L.; Jiang, J. Digital Signal Processing: Fundamentals and Applications. In *Digital Signal Processing: Fundamentals and Applications*, 3rd ed.; Academic Press: Cambridge, MA, USA, 2019; pp. 1–903. [CrossRef]
16. Liu, Z.; Zhang, L.; Carrasco, J. Vibration Analysis for Large-Scale Wind Turbine Blade Bearing Fault Detection with an Empirical Wavelet Thresholding Method. *Renew. Energy* **2020**, *146*, 99–110. [CrossRef]
17. Yuan, X.; He, Y.; Wan, S.; Qiu, M.; Jiang, H. Remote Vibration Monitoring and Fault Diagnosis System of Synchronous Motor Based on Internet of Things Technology. *Mob. Inf. Syst.* **2021**, *2021*, 3456624. [CrossRef]
18. Mey, O.; Neufeld, D. Explainable AI Algorithms for Vibration Data-Based Fault Detection: Use Case-Adadpted Methods and Critical Evaluation. *Sensors* **2022**, *22*, 9037. [CrossRef]
19. Mey, O.; Neudeck, W.; Schneider, A.; Enge-Rosenblatt, O. Machine Learning-Based Unbalance Detection of a Rotating Shaft Using Vibration Data. In Proceedings of the IEEE International Conference on Emerging Technologies and Factory Automation, ETFA 2020, Vienna, Austria, 8–11 September 2020; pp. 1610–1617. [CrossRef]
20. Toh, G.; Park, J. Review of Vibration-Based Structural Health Monitoring Using Deep Learning. *Appl. Sci.* **2020**, *10*, 1680. [CrossRef]
21. Rahman, A.; Hoque, M.E.; Rashid, F.; Alam, F.; Ahmed, M.M. Health Condition Monitoring and Control of Vibrations of a Rotating System through Vibration Analysis. *J. Sens.* **2022**, *2022*, 4281596. [CrossRef]
22. Tiboni, M.; Remino, C.; Bussola, R.; Amici, C. A Review on Vibration-Based Condition Monitoring of Rotating Machinery. *Appl. Sci.* **2022**, *12*, 972. [CrossRef]
23. Amihai, I.; Gitzel, R.; Kotriwala, A.M.; Pareschi, D.; Subbiah, S.; Sosale, G. An Industrial Case Study Using Vibration Data and Machine Learning to Predict Asset Health. In Proceedings of the 2018 20th IEEE International Conference on Business Informatics, CBI 2018, Vienna, Austria, 11–14 July 2018; Volume 1, pp. 178–185. [CrossRef]
24. Zeba, G.; Dabić, M.; Čičak, M.; Daim, T.; Yalcin, H. Technology Mining: Artificial Intelligence in Manufacturing. *Technol. Forecast Soc. Chang.* **2021**, *171*, 120971. [CrossRef]
25. Baron, P.; Kočiško, M.; Hlavatá, S.; Franas, E. Vibrodiagnostics as a Predictive Maintenance Tool in the Operation of Turbo Generators of a Small Hydropower Plant. *Adv. Mech. Eng.* **2022**, *14*, 16878132221101023. [CrossRef]
26. Zuth, D.; Blecha, P.; Marada, T.; Huzlik, R.; Tuma, J.; Maradova, K.; Frkal, V. Vibrodiagnostics Faults Classification for the Safety Enhancement of Industrial Machinery. *Machines* **2021**, *9*, 222. [CrossRef]
27. Rivas, A.; Fraile, J.M.; Chamoso, P.; González-Briones, A.; Sittón, I.; Corchado, J.M. A Predictive Maintenance Model Using Recurrent Neural Networks. In *Proceedings of the Advances in Intelligent Systems and Computing*; Springer: Berlin/Heidelberg, Germany, 2020; Volume 950, pp. 261–270.
28. Wu, S.J.; Gebraeel, N.; Lawley, M.A.; Yih, Y. A Neural Network Integrated Decision Support System for Condition-Based Optimal Predictive Maintenance Policy. *IEEE Trans. Syst. Man Cybern. Part A Syst. Hum.* **2007**, *37*, 226–236. [CrossRef]
29. Zan, T.; Wang, H.; Wang, M.; Liu, Z.; Gao, X. Application of Multi-Dimension Input Convolutional Neural Network in Fault Diagnosis of Rolling Bearings. *Appl. Sci.* **2019**, *9*, 2690. [CrossRef]
30. Kumar, P.; Hati, A.S. Deep Convolutional Neural Network Based on Adaptive Gradient Optimizer for Fault Detection in SCIM. *ISA Trans.* **2021**, *111*, 350–359. [CrossRef] [PubMed]
31. Wu, C.; Jiang, P.; Ding, C.; Feng, F.; Chen, T. Intelligent Fault Diagnosis of Rotating Machinery Based on One-Dimensional Convolutional Neural Network. *Comput. Ind.* **2019**, *108*, 53–61. [CrossRef]
32. Jing, L.; Zhao, M.; Li, P.; Xu, X. A Convolutional Neural Network Based Feature Learning and Fault Diagnosis Method for the Condition Monitoring of Gearbox. *Measurement* **2017**, *111*, 1–10. [CrossRef]
33. Ding, C.; Huang, W.; Shen, C.; Jiang, X.; Wang, J.; Zhu, Z. Synchroextracting Frequency Synchronous Chirplet Transform for Fault Diagnosis of Rotating Machinery under Varying Speed Conditions. *Struct. Health Monit.* **2023**, *23*, 1403–1422. [CrossRef]
34. Xu, Y.; Fu, C.; Hu, N.; Huang, B.; Gu, F.; Ball, A.D. A Phase Linearisation–Based Modulation Signal Bispectrum for Analysing Cyclostationary Bearing Signals. *Struct. Health Monit.* **2020**, *20*, 1231–1246. [CrossRef]
35. Fathi, K.; van de Venn, H.W.; Honegger, M. Predictive Maintenance: An Autoencoder Anomaly-Based Approach for a 3 DoF Delta Robot. *Sensors* **2021**, *21*, 6979. [CrossRef]
36. Olsson, E.; Funk, P.; Bengtsson, M. Fault Diagnosis of Industrial Robots Using Acoustic Signals and Case-Based Reasoning. In *Lecture Notes in Computer Science (Including Subseries Lecture Notes in Artificial Intelligence and Lecture Notes in Bioinformatics)*; Springer: Berlin/Heidelberg, Germany, 2004; Volume 3155, pp. 686–701. [CrossRef]
37. Fan, S.; Zhang, Y.; Zhang, Y.; Fang, Z. Motion Process Monitoring Using Optical Flow–Based Principal Component Analysis-Independent Component Analysis Method. *Adv. Mech. Eng.* **2017**, *9*, 1687814017733231. [CrossRef]
38. Costa, M.A.; Wullt, B.; Norrlöf, M.; Gunnarsson, S. Failure Detection in Robotic Arms Using Statistical Modeling, Machine Learning and Hybrid Gradient Boosting. *Measurement* **2019**, *146*, 425–436. [CrossRef]

39. Fedorova, D.; Tlach, V.; Kuric, I. Classification of Industrial Robot Movements Based on Vibration Monitoring Data Using Lstm Neural Network. In *Proceedings of the Projektowanie, Badania i Eksploatacja*; Rysiński, J., Ed.; Wydawnictwo Naukowe Uniwersytetu Bielsko-Bialskiego: Bielsko-Biała, Poland, 2023; pp. 169–178.
40. Kuric, I.; Tlach, V.; Císar, M.; Sagová, Z.; Zajačko, I. Examination of Industrial Robot Performance Parameters Utilizing Machine Tool Diagnostic Methods. *Int. J. Adv. Robot. Syst.* **2020**, *17*, 1729881420905723. [CrossRef]
41. FANUC Corporation. *FANUC Robot LR Mate 200iC/ARC Mate 50iC MECHANICAL UNIT, OPERATOR'S MANUAL*; B-82584EN/08; FANUC Corporation: Yamanashi, Japan, 2011.
42. Igba, J.; Alemzadeh, K.; Durugbo, C.; Eiriksson, E.T. Analysing RMS and Peak Values of Vibration Signals for Condition Monitoring of Wind Turbine Gearboxes. *Renew. Energy* **2016**, *91*, 90–106. [CrossRef]
43. Zhang, C.; Mousavi, A.A.; Masri, S.F.; Gholipour, G.; Yan, K.; Li, X. Vibration Feature Extraction Using Signal Processing Techniques for Structural Health Monitoring: A Review. *Mech. Syst. Signal Process.* **2022**, *177*, 109175. [CrossRef]
44. Balluff GmbH. *BCM R15E-002-DI00-01,5-S4 Condition Monitoring Sensor-User's Guide*; Balluff GmbH: Neuhausen, Germany, 2023.
45. Hossin, M.; Sulaiman, M.N. A Review on Evaluation Metrics for Data Classification Evaluations. *Int. J. Data Min. Knowl. Manag. Process* **2015**, *5*, 1–11. [CrossRef]

applied sciences

MDPI

Article

Comprehensive Analysis of Deposition Parameters and Energy-Dispersive X-Ray Spectroscopy Characterization in Cataphoretic Coating Processes

Patrik Fejko [1], Damián Peti [1], Jozef Dobránsky [1,*], Miroslav Gombár [2] and Peter Michalík [1]

[1] Faculty of Manufacturing Technologies with a Seat in Prešov, Technical University of Kosice, Štúrova St. 31, 080 01 Prešov, Slovakia; patrik.fejko@tuke.sk (P.F.); damian.peti@tuke.sk (D.P.); peter.michalik@tuke.sk (P.M.)

[2] Faculty of Mechanical Engineering, University of West Bohemia Pilsen, Univerzitní 8, 306 14 Plzeň, Czech Republic; gombar.mirek@gmail.com

* Correspondence: jozef.dobransky@tuke.sk; Tel.: +421-55-602-6350

Abstract: This research examines the inter-relationship between the deposition time, degreasing temperature, and applied voltage in the cataphoretic painting process, focusing on their cumulative effects on the thickness of the formed layers. A series of experiments was conducted, systematically varying deposition time effects through voltage levels (200 V to 300 V) and degreasing temperatures (40 °C to 80 °C). The results demonstrate that the maximum layer thickness is achieved at longer cataphoretic times, with significant thickness increments observed at optimal voltage levels. Conversely, the study reveals that lower degreasing temperatures lead to increased layer thickness, while elevated temperatures tend to diminish it. Notably, the thickness variations are consistent across different voltage applications, with a discernible threshold at which the layer thickness stabilizes. Additionally, energy-dispersive X-ray spectroscopy (EDX) was utilized to characterize the elemental composition of the cataphoretic layer, providing deeper insights into the coating structure and its relationship to process parameters. This work provides valuable insights into the optimization of cataphoretic processes, offering a framework for enhancing the quality and uniformity of coatings in industrial applications. The findings underscore the importance of the precise control over process parameters to achieve the desired material characteristics, thereby advancing the field of surface engineering and coating technologies.

Keywords: cataphoretic painting; layer thickness; surface coating; design of experiments; energy-dispersive X-ray spectroscopy

1. Introduction

Surface treatment is one of the critical aspects of the manufacturing process that fundamentally affects the quality, functionality, and durability of the final product. For this reason, it is essential that this phase is carried out with a high degree of precision and expertise. A thorough and high-quality surface treatment ensures the long-term durability of the material and its reliability in use. Surface treatments provide protection against mechanical damage, weathering, and increased adhesion to other materials. The detrimental impact of metal corrosion on the economy, the environment, and safety requires urgent preventive measures [1–3]. The use of organic coatings on metal surfaces has been shown to effectively retard the corrosion process, preventing the metal from encountering corrosive media. However, the effectiveness of these coatings is influenced by various parameters, including the intrinsic nature of the coatings themselves, the continuity of the coating phase, and the thickness of the coatings applied [4]. These parameters are not easily controlled by traditional methods, such as spraying the solution, dipping, or brushing. Consequently, the parameters in question could be controlled by the electrophoretic deposition (EPD) method, which has been used to prepare coatings of controllable thickness at the electrode by applying a DC electric field between the cathode and the anode. The EPD process can be divided into two distinct types, cathodic and anodic deposition, depending on the nature of the materials to be deposited [5]. For corrosion resistance purposes, cathodic deposition has advantages over anodic deposition. In anodic electrodeposition, the inevitable electrolysis of water in the anode, which produces hydrogen ions and oxygen, weakens the effectiveness of corrosion protection [6]. Recently, the use of nanoparticles as inorganic nanofillers in polymeric protective coatings has attracted special attention because of their unique barrier properties. The homogeneous dispersion of nanoparticles is a key factor in the preparation of polymer nanocomposites.

A study by Zivkovic [7] investigated the effect of cerium and zirconium nanoparticles on the corrosion resistance of cataphoretic epoxy coatings on AA 6060 alloy. The properties of the coatings were analyzed by various spectroscopic and microscopic techniques, while the corrosion stability was evaluated by electrochemical impedance spectroscopy in NaCl solution. The results confirmed the importance of cataphoretic electro spraying as an effective corrosion protection method [8]. Cataphoretic electrodeposition is a widespread technique used in many fields, especially in the automotive industry, to protect the body of automobiles [9]. Manufacturers of organic coatings for the automotive industry test new products and materials according to production parameters to verify their properties before mass production. The key is selecting the appropriate test to differentiate between similar coatings and provide results in a short time. Accelerated ageing tests are used to predict the durability of coatings to speed up development. To improve the adhesion and corrosion protection of cataphoretic coatings, metal substrates are commonly subjected to surface pretreatment.

Research by Zanello [10] has focused on the relationship between the electrochemical properties of metal substrates and the properties of electrophoretic clear coatings. The study investigated three types of substrates: active (bare steel), passive (nickel), and noble (gold). The aim was to determine whether the electrochemical behavior of the substrate influences coating properties, such as barrier properties, adhesion, and water absorption. The results showed that the properties of the coating are influenced by the type of substrate and its electrochemical behavior. Another study investigated the effect of graphene filler and thermochromic pigments on cataphoretic acrylic coatings. The combination was found to improve abrasion resistance and aesthetic stability after UV-B exposure but negatively affect corrosion protection. Research has revealed both positive and negative effects of this synergy [11]. The cataphoretic deposition of coatings allows even complex surfaces to be evenly covered due to electrostatic attraction. The process is highly efficient, with minimal material loss (below 10%), low environmental impact, and low emissions.

A study by Darowicki [12] presents a method for the on-line monitoring of the cataphoretic coating process using dynamic electrochemical impedance spectroscopy (DEIS). This technique overcomes the limitations of classical impedance spectroscopy for non-steady-state processes, allowing them to monitor impedance changes during deposition and to estimate the time to reach steady state after polarization switch-off. DEIS offers the potential to improve real-time process control and optimization. Sternadel's [13] work describes those electrophoretic paints, also known as cataphoretic coatings, organic coatings that dissolve in water. These coatings have an electrical charge, which allows them to be applied to metal surfaces [14]. This process considers the opposite electrical charge of the metal sample, which affects the coating formulation [15,16]. Cataphoretic deposition represents a modern technology that produces durable and resistant surfaces. These corrosion-resistant surfaces not only meet customer expectations in terms of aesthetic appearance but also increase efficiency and meet environmental standards. These results are the result of a century of experience and research and the refinement of techniques that have also led to their theoretical evaluation. Cataphoretic deposition uses electrophoretic coatings or paints that are applied to surfaces. These coatings are of organic origin, dispersed in water, and carry an electrical charge. The process allows them to be deposited on metal surfaces with an opposite charge, thus providing effective and uniform coverage [17–20]. The preparation of the material surface prior to cataphoretic deposition is considered to be both a crucial and challenging phase of anticorrosion technology, involving processes such as degreasing and zinc phosphating [21]. Zinc, known for its cost-effectiveness, is an important element used in the production of anticorrosion coatings (Zn, Cu, Ni, Cr) [22], while the deposition of zinc coatings requires only minimal financial investment [23].

The optimization of the cataphoretic painting process is imperative for several fundamental reasons, which considerably enhance manufacturing efficiency and product quality. The uniformity thus achieved is of critical importance in meeting the stringent industrial engineering standards and customer requirements. Furthermore, the optimization of the process contributes to environmental sustainability by minimizing waste and energy use. A well-optimized process fosters innovation, strengthens competitive advantage, and enhances customer satisfaction through the reliable and timely delivery of high-quality products. The authors of this study acknowledge its significance in advancing the methodology for cataphoretic painting, leading to the formation of an anticorrosion layer.

2. Materials and Methods

2.1. Material Selection—VDA 239-100 CR4

The choice of material plays a pivotal role in investigations that necessitate exceptional surface integrity and meticulous specimen preparation, especially within the framework of complex experimental methodologies. For materials adhering to standards such as "VDA 239-100 CR4", which prescribes the performance criteria for cold-rolled sheet steel commonly utilized in automotive applications, the uniformity of material properties is imperative to guarantee the precision and reproducibility of experimental findings.

The material selected for the designed experiment (DoE) is CR4 steel (Table 1), classified as a cold-rolled low-carbon steel in accordance with the VDA 239-100 standard [24]. This steel grade, conforming to the specifications of DIN EN 10130 [25], particularly grades DC05 and DC06, is characterized by high ductility, low strength, and an optimal strength-to-use ratio. These properties render it highly suitable for applications requiring intricate forming processes, such as those in the automotive and general industrial sectors. Furthermore, CR4 steel is ideal for producing body panels and components demanding high surface quality and precise dimensional accuracy.

Table 1. Mechanical properties of VDA 239-100 CR4.

Property	Value
Yield strength ($Rp_{0,2}$)	140–180 MPa
Tensile strength (Rm)	270–330 MPa
Hardness by Brinell HBW	267
Elongation (A_{80})	≥39%
r-value ($r_{90/20}$)	≥1.9
r-value m/20	≥1.6
n-value (n_{10-20}/Ag)	≥0.20

Its surface properties and uniform composition make it particularly advantageous for the cataphoretic painting process, as it ensures consistent adhesion of the coating layer and uniform thickness distribution. The material's attributes also make it an optimal candidate for use in the design of experiments (DoEs) [26], as its predictable behavior under controlled conditions allows for the reliable investigation of the effects of process parameters on coating quality. The specimen dimensions, as specified in the Chemcore compendium, were determined to be 105 mm × 190 mm with a thickness of 0.80 mm.

2.2. Experimental Conditions

2.2.1. Sample Preparation

The experiment includes 88 samples, with samples 83 and 84 having different preparation (no degreasing and activation/phosphating steps). Sample preparation was carried out in KTL s.r.o., Prešov, Slovakia with the participation of the managing director. The material was cleaned of impurities by degreasing with Pragolod 57 N prior to coating application, with temperature, concentration, and time influencing the efficiency of the process (Table 2).

Table 2. Chemical composition of Pragolod 57 N.

Chemical Element	Concentration [%]
Sodium carbonate (Na_2CO_3)	20–30
Sodium metasilicate ($Na_2SiO_3 \times 5H_2O$)	20–30
Sodium hydroxide (NaOH)	20–30
Fatty tallowamine (POE), ethoxylate (5EO)	3–5

After degreasing, the samples were rinsed with demineralized water. They were then subjected to an activation process on an automated line where the crystalline structure of the material was modified, improving the ability to bind phosphate layers. Phosphating was carried out in accordance with internal standards, with the samples being exposed to the phosphate solution for three and seven minutes. The phosphating was followed by further rinsing, and the samples were prepared for cataphoretic coating. Parameters such as temperature, voltage, and current were monitored throughout the process. The coating was applied in the range of 15–30 μm to achieve the desired corrosion protection. After rinsing, the specimens underwent the polymerization process in the oven, while the temperature and time were monitored.

2.2.2. The Process of Cataphoretic Coating

Cataphoretic coating is one of two methods of electrophoretic coating, the other being anaphoresis. The cataphoretic process uses cationic coatings, which are water soluble and based on epoxies or acrylic resins with very low organic solvent content (about 2%). These coatings contain paint particles in the form of polymer cations. During coating, the product is immersed in the coating solution and attached as a cathode. By means of a direct current between the cathode (the object to be painted) and the anode, an electric field is created, which causes the polycations to move towards the cathode. On its surface, these cations react with hydroxyl ions from water, causing their exclusion on the surface of the object in the form of a coating. As the thickness of the layer increases, the resistance of the coating increases, reducing the rate of further paint deposition, causing this process to take place mainly in areas of lesser layer thickness, i.e., hard-to-reach areas, such as corners, edges, or cavities. This mechanism ensures that the coating is uniform over the entire surface, even in areas that are harder to reach. Once the desired coating thickness is reached, the shedding stops. The thickness of the coating depends primarily on the tension being applied and is usually between 15 and 30 μm. Once the process is complete, the excess varnish is rinsed off, and the coating formed must be fired at temperatures between 155 and 180 °C, which leads to polymerization and guarantees the final properties [27]. The parameters influencing the coating thickness and uniformity are further analyzed in relation to the planned experiment, as referenced in Table 3, which presents the values of input variable factors.

Table 3. Values of input variable factors.

Factor Code	Variable	Unit	Level of Factors				
			−2.05464	−1	0	1	+2.05464
x_1	k_{ODM}	$g{\cdot}L^{-1}$	15	25	35	45	55
x_2	t_{ODM}	min	3	6	8	10	13
x_3	T_{ODM}	°C	40	50	60	70	80
x_4	t_{FOSF}	min	1	3	5	7	9
x_5	t_{KTL}	min	3	4	5	6	7
x_6	U_{KTL}	V	200	226	250	274	300
x_7	t_{POLY}	min	13	17	20	23	27
x_8	T_{POLY}	°C	150	176	200	224	250

k_{ODM}—degreasing concentration, t_{ODM}—degreasing deposition time, T_{ODM}—degreasing temperature, t_{FOSF}—phosphating deposition time, t_{KTL}—cataphoresis deposition time, U_{KTL}—cataphoresis voltage, t_{POLY}—polymerization time, T_{POLY}—polymerization temperature.

2.3. Measurement Methods

The Elcometer 456 is a non-destructive device that can be utilized to measure dry film thickness on ferromagnetic and non-ferromagnetic materials with a high degree of accuracy. This instrument has been shown to provide rapid measurements within a range of ±1% of the target value. The Elcometer 456 complies with the relevant industry standards, but calibration prior to use is necessary to ensure optimal performance and accuracy. The T456CF1S flat ferromagnetic probe measures non-magnetic coatings on magnetic substrates in the range of 0–1500 μm with an accuracy of ±1–3% or ±2.5 μm. The instrument is used in quality control, surface treatment, and manufacturing. Elcometer Ltd. is a leading manufacturer of high-quality instruments for non-destructive testing, inspection, and measurement. The company is headquartered in Manchester, UK.

Three-dimensional printing technology was used to fabricate a device that ensured precise and quick measurements at consistent points on each sample (Figure 1B). Data were collected in four rows and eight columns, with each point measured five times for accuracy. A control measurement with a standard film thickness of 24.1 μm was conducted, resulting in 192 measurements per sample. The measurements followed the standard "Coating substances, determination of coating thickness STN EN 2808:2019" [28].

2.4. Analysis of Prediction Model

Outlier analysis was performed using Statistica 14 software to identify and exclude anomalous values from both data columns, focusing on measurements exhibiting significant deviation from the primary dataset (Figure 2). After data filtration, the mean thickness of the deposited layer was calculated using the Statistica 14 software, and the results were tabulated. This process was replicated for measurements obtained from the reference standard to ensure consistency.

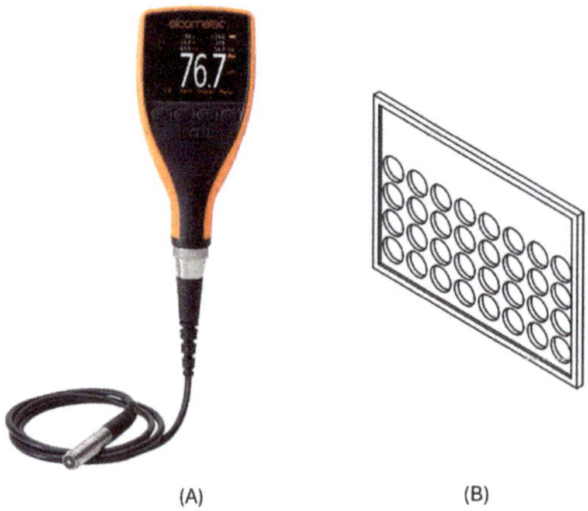

Figure 1. Elcometer 456 (**A**) and 3D-printed device for precise and expeditious measurements of samples (**B**).

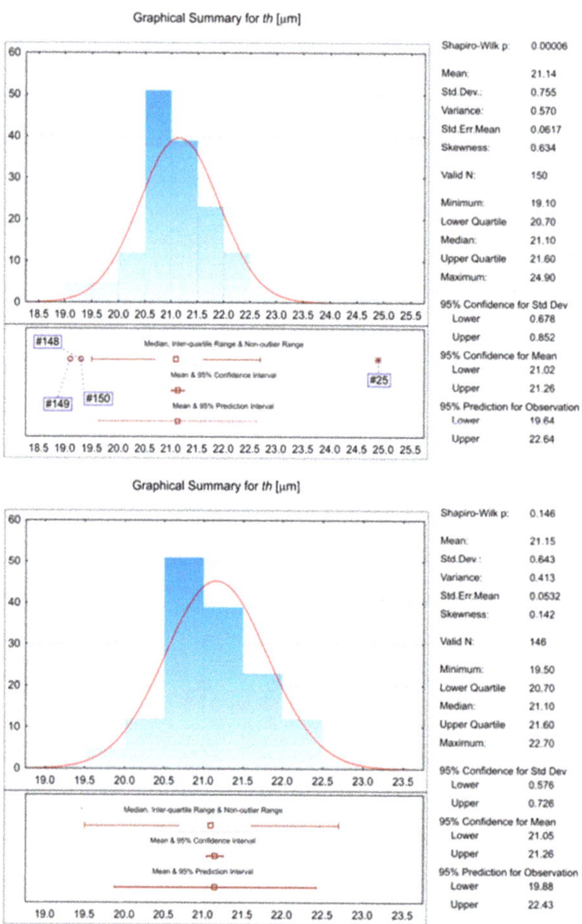

Figure 2. Filtering of gross errors.

The compiled data were further refined by calculating the layer thickness through control measurements conducted on the reference standard. A manufacturer-calibrated foil with a certified thickness of 24.1 μm served as the benchmark for these measurements. The control measurement value was subtracted from the foil's thickness to determine the deviation, which was then incorporated into the filtered thickness data. This methodology enabled a precise and reliable determination of the average thickness of the deposited layer.

Table 4 presents the model suitability analysis and evaluates the influence of various factors on the thickness of the formed cataphoretic layer. The table further reveals that the proportion of variability in the measured thicknesses, represented by the coefficient of determination (R^2), is approximately 75%. The adjusted coefficient of determination, shown in Table 3, quantifies the model's adequacy in describing the relationship between the input, controlled, chemical, and physical factors and the system's response (i.e., the thickness of the formed layer). This value indicates that 70.79% of the variability is accounted for by the model. Based on the experimentally derived thickness values of the cataphoretic layer, the average thickness (mean of response) is determined to be 17.71 μm.

Table 4. Summary of the model's suitability.

Source	Value
RSquare	0.751148
RSquare Adj	0.707869
Root Mean Square Error	1.922936
Mean of Response	17.71049
Observations (or Sum Wgts)	82

As presented in Table 5, the variability attributable to random error is markedly smaller than the variability in the measurements accounted for by the model. The significance level (Prob > F), derived from the Fisher–Snedecor test, further substantiates the appropriateness of the model. With a chosen significance threshold of $\alpha = 0.05$, the results demonstrate that the model is statistically significant, indicating its reliability in describing the observed phenomena.

Table 5. Analysis of variance (ANOVA).

Source	DF	Sum of Squares	Mean Square	F Ratio	Prob > F
Model	12	770.1278	64.1773	17.3561	<0.0001
Error	69	255.1402	3.6977	-	-
C. Total	81	1025.268	-	-	-

Based on the lack-of-fit error presented in Table 6, it can be concluded that the significance level is 0.1312, which allows for the acceptance of the null hypothesis at the chosen significance threshold of $\alpha = 0.05$. This suggests that the variance of residuals is less than or equal to the variance within groups, thereby confirming the suitability of the model. Following the assumptions outlined and validated in Tables 5 and 6, the parameter estimates for the model are provided in Table 7. The significance of individual effects and their interactions is assessed at a significance level of $\alpha = 0.05$. Table 5 illustrates the estimated effect sizes for each component of the model, which quantifies the influence of various factors on the thickness of the formed layer. Furthermore, the interaction between variables such as concentration, cathodic deposition time, degreasing temperature, and immersion time in the degreasing agent is shown to have a substantial impact on the outcome.

Table 6. Lack-of-fit error.

Source	DF	Sum of Squares	Mean Square	F Ratio	Prob > F	Max RSq
Lack Of Fit	30	135.09267	4.50309	1.4629	0.1312	0.8829
Pure Error	39	120.04749	3.07814	-	-	-
Total Error	69	255.14016	-	-	-	-

Table 7. Model parameters.

Term	Estimate	Std Error	t Ratio	Prob > |t|	Lower 95%	Upper 95%
Intercept	15.24063	0.480734	31.7	<0.0001	14.28159	16.19966
x_1	−2.08862	0.225926	−9.24	<0.0001	−2.53933	−1.63791
x_5	0.903898	0.225926	4	0.0002	0.453188	1.354608
x_6	0.805741	0.225926	3.57	0.0007	0.355031	1.256451
x_3	−0.58184	0.225926	−2.58	0.0122	−1.03255	−0.13113
$x_1 \cdot x_1$	3.453701	0.680272	5.08	<0.0001	2.096597	4.810806
$x_1 \cdot x_5$	−0.51656	0.240367	−2.15	0.0351	−0.99608	−0.03704
$x_1 \cdot x_2$	−0.68031	0.240367	−2.83	0.0061	−1.15983	−0.20079
$x_3 \cdot x_2$	0.71	0.240367	2.95	0.0043	0.230481	1.189519
$x_1 \cdot x_5 \cdot x_3$	0.51625	0.240367	2.15	0.0352	0.036731	0.995769
$x_1 \cdot x_1 \cdot x_1 \cdot x_1$	−0.47839	0.168911	−2.83	0.0061	−0.81536	−0.14142
$x_1 \cdot x_5 \cdot x_6 \cdot x_2$	−0.68906	0.240367	−2.87	0.0055	−1.16858	−0.20954
$x_1 \cdot x_5 \cdot x_3 \cdot x_2$	0.94625	0.240367	3.94	0.0002	0.466731	1.425769

x_1—degreasing concentration, x_2—degreasing deposition time, x_3—degreasing temperature, x_5—cataphoresis deposition time, x_6—cataphoresis voltage.

Based on the estimated parameters, it is possible to construct the equation in a coded scale:

$$\widehat{y} = 15.261 - 2.089 \cdot x_1 + 0.904 \cdot x_5 + 0.806 \cdot x_6 - 0.582 \cdot x_3 + 3.454 \cdot x_2 - 0.517 \cdot x_1 \cdot x_5 - 0.680 \cdot x_1 \cdot x + 0.710 \cdot x_3 \cdot x_2 + 0.516 \cdot x_1 \cdot x_5 \cdot x_3 - 0.478 \cdot x_4 - 0.689 \cdot x_1 \cdot x_5 \cdot x_6 \cdot x_2 + 0.946 \cdot x_1 \cdot x_5 \cdot x_3 \cdot x_2 \tag{1}$$

To establish the predictive relationship in the natural scale, it is important to note that the factors used during the analysis were encoded in the coded scale through design of experiments normalization:

$$x_d(i) = \frac{x_i - \frac{x_{max}+x_{min}}{2}}{\frac{x_{max}-x_{min}}{2}} \tag{2}$$

where $x_d(i)$ represents the coded variable according to the design of experiments, and $x(i)$ denotes the original base variable, with i representing the set of integers corresponding to the number of base factors. x_{max} is the maximum value of the original variable x(i), and x_{min} is the minimum value of the original variable $x(i)$. Using the coded equation and the statistical equation, it is possible to define the predictive equation in the natural scale that describes the investigated relationship representing the thickness of the formed layer th:

$$\begin{aligned} th = {} & 0.2348441563 \cdot DEGR_{con} + 0.34283975 \cdot DEGR_{tim} - 0.3041498125 \cdot DEGR_{temp} + 3.10826525 \cdot KTL_{tim} \\ & + 0.11258322 \cdot KTL_U - 0.01334178812 \cdot DEGR_{con}{}^2 + 0.00041859125 \cdot DEGR_{con} - 0.0000029899375 \cdot \\ & DEGR_{con}{}^4 - 0.02196685 \cdot DEGR_{con} \cdot DEGR_{tim} + 0.0062359375 \cdot DEGR_{con} \cdot DEGR_{temp} + 0.0484984375 \cdot \\ & DEGR_{tim} \cdot DEGR_{temp} - 0.07589475 \cdot DEGR_{con} \cdot KTL_{tim} - 0.00275624 \cdot DEGR_{con} \cdot KTL_U - 0.10614625 \cdot \\ & DEGR_{con} \cdot DEGR_{tim} \cdot KTL_U - 0.0012471875 \cdot DEGR_{con} \cdot DEGR_{temp} \cdot KTL_{tim} - 0.0082796875 \cdot DEGR_{tim} \\ & \cdot DEGR_{temp} \cdot KTL_{tim} + 0.000551248 \cdot DEGR_{con} \cdot KTL_{tim} \cdot KTL_U + 0.00241171 \cdot DEGR_{tim} \cdot KTL_{tim} \cdot \\ & KTL_U + 0.0002365625 \cdot DEGR_{con} \cdot DEGR_{tim} \cdot DEGR_{temp} \cdot KTL_{tim} - 0.000068906 \cdot DEGR_{con} \cdot DEGR_{tim} \\ & \cdot KTL_{tim} \cdot KTL_U + 8.664520102 \end{aligned} \tag{3}$$

where $DEGR_{con}$—degreasing concentration, $DEGR_{tim}$—de-greasing time, $DEGR_{temp}$—degreasing temperature, KTL_{tim}—time of cataphoresis, and KTL_U—cataphoresis voltage.

To ensure the complexity of the conducted analysis and to confirm the accuracy and suitability of the chosen model, it is necessary to examine the residuals. The difference between the observed values and the predicted values calculated using the predictive model must be assessed. The achieved significance level of the Shapiro–Wilk test [29], illustrated in Figure 3, indicates a Gaussian distribution of the residuals, meaning that the value is greater than 0.05. Therefore, it can be concluded that the predictive model is statistically and numerically valid.

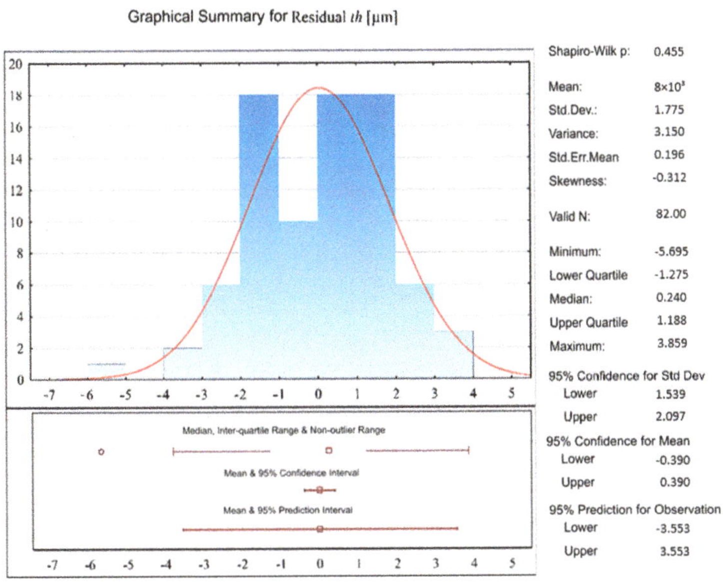

Figure 3. Analysis of residual th.

3. Results and Discussion

3.1. Influence of Deposition Time and Cataphoretic Voltage on Coating Characteristics

Figure 4 illustrates the significant impact of varying both the duration of deposition and the applied voltage on the thickness of the resultant layer. It has been identified that concentration is also a critical factor under these conditions. At a concentration of 15 $g \cdot L^{-1}$, the maximum achievable layer thickness is attained with a coating voltage of 300 V. Following a deposition period of 3 min, this maximum thickness measures 21.42 μm. As the deposition time is extended, the thickness further increases, reaching 22.02 μm after 7 min. The minimum recorded thickness of 18.43 μm occurs at a coating voltage of 200 V with a 3 min deposition time. With an increase in time, the thickness also rises, achieving 21.78 μm after 7 min. The disparity between the maximum and minimum thicknesses at 3 min of cathodic electrocoating is 2.99 μm, while this difference diminishes to 0.24 μm after 7 min. When the concentration is adjusted to 35 $g \cdot L^{-1}$, a relatively uniform thickness distribution is observed across the entire spectrum of tested voltages. In every instance, the thickness exhibits a proportional increase with time. Notably, at this concentration, the lowest thickness values are recorded at a coating voltage of 200 V, where the thickness measures 14.82 μm after 3 min and reaches 16.63 μm following an extended deposition time of 7 min.

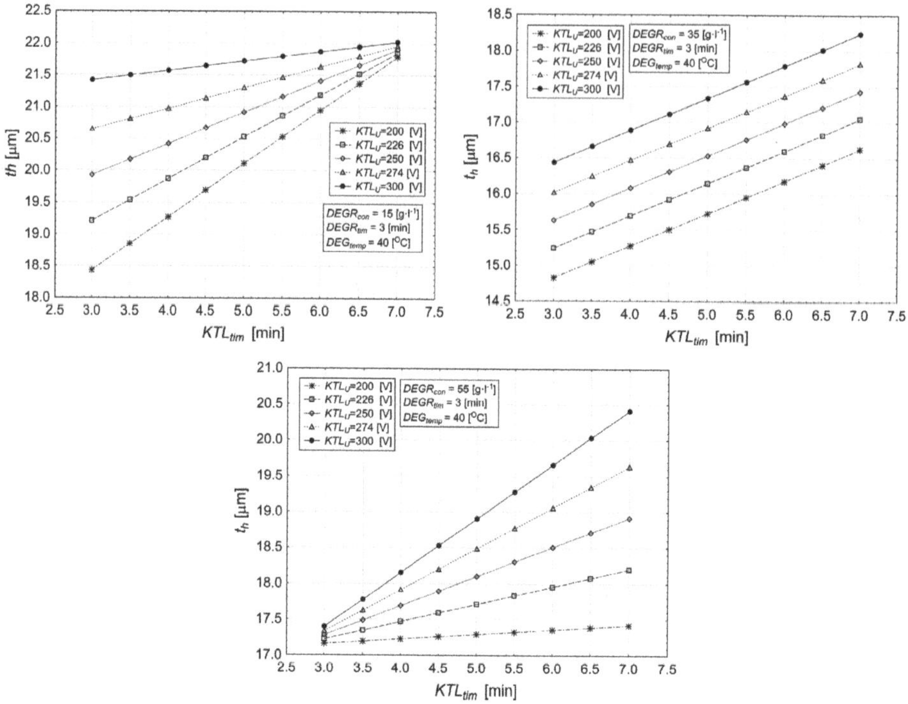

Figure 4. The influence of KTL deposition time and KLT voltage on the change of the formed layer at the minimum values of the design of experiments methodology.

The subsequent layer thickness values increase by approximately 0.42 μm as a result of changes in the specified coating voltage and duration. A maximum thickness of 16.43 μm is achieved at a coating voltage of 300 V after 3 min of deposition. As the deposition time increases to 7 min, the thickness reaches 18.24 μm. When the degreasing concentration is adjusted to 55 g·L^{-1}, and the cathodic electrocoating (KTL) process lasts for 3 min, the layer thickness becomes more uniform, with a variance of 0.23 μm. The lowest thickness of 17.16 μm is produced at a coating voltage of 200 V, and by extending the deposition time to 7 min, the thickness increases to 17.42 μm.

The greatest thickness is observed at a voltage of 300 V, where after 3 min of deposition, the thickness reaches 17.39 μm, and with 7 min of deposition, it reaches a peak of 20.41 μm. The results also confirm the conclusion of Garcia et al. [30] and Kamas et al. [31].

Figure 5 illustrates that, by varying the deposition time in cathodic electrocoating and KTLU at the maximum experimental values, while simultaneously adjusting the concentration to 15 g·L^{-1}, the minimum thickness is achieved at a coating voltage of 200 V, where after a deposition time of 3 min, the thickness reaches 21.03 μm. At this voltage, with increasing time, the thickness decreases, reaching 19.57 μm after 7 min. A similar reduction in thickness is observed at voltages of 226 V and 250 V. However, at 250 V, the decrease in thickness over the KTL deposition period is only 0.08 μm. In contrast, voltages of 275 V and 300 V result in an increase in thickness throughout the entire deposition time. The difference between the maximum and minimum thickness at a 3 min deposition is 0.24 μm. A voltage of 300 V ensures the greatest thickness, reaching 22.56 μm after 7 min of KTL deposition. When the concentration is changed to 35 g·L^{-1}, the thickness levels increase uniformly with rising voltage.

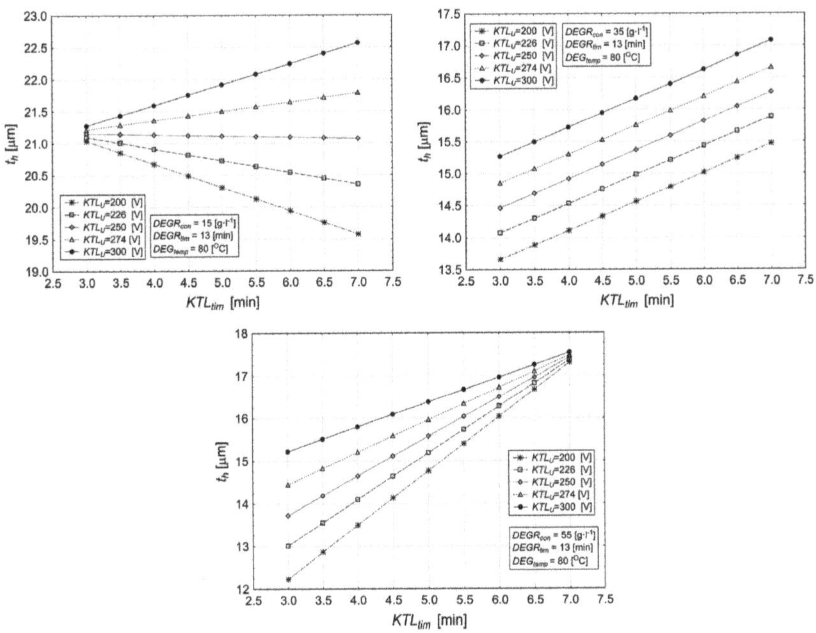

Figure 5. The influence of KTL deposition time and KLT voltage on the change of the formed layer at the maximum values of the design of experiments methodology.

The minimum thickness is observed at 200 V, where the thickness is 13.65 µm after 3 min and 15.46 µm after 7 min. The thickness increases by 0.42 µm as the voltage rises to the defined value. At a voltage of 300 V and a 3 min deposition time, the thickness reaches 15.27 µm, and with time, it continues to increase, reaching 17.07 µm after 7 min. When increasing the concentration of Pragolod 57 N to 55 $g{\cdot}L^{-1}$, the thickness grows consistently across the deposition time range. The minimum thickness of 12.23 µm is observed at 200 V after 3 min. The difference between the minimum and maximum layer thickness is 2.99 µm. The maximum thickness is reached at 300 V, where after 7 min, it measures 17.54 µm, representing the largest possible layer thickness within this interval. The difference between the maximum and minimum thickness at 7 min of deposition is 0.24 µm. The results also confirm the conclusion of Krenicky et al. [32] and Kania et al. [33].

3.2. Influence of Deposition Time and Degreasing Temperature on Coating Characteristics

Figure 6 illustrates the effect of KTL deposition time and the degreasing solution temperature, along with voltage variation, on the resulting layer thickness. These factors consistently lead to an increase in layer thickness. As the degreasing temperature decreases, the rate of layer growth also diminishes. The difference between the maximum and minimum thickness remains constant across the three voltage changes. For a KTL time of 3 min, this difference is 3.45 µm, while for a 7 min deposition, it is 1.72 µm. The thickness variation with a gradual change in temperature over a 3 min KTL period is 0.87 µm, and for a 7 min deposition, it is 0.47 µm. At an applied voltage of 200 V, the minimum thickness of 14.98 µm after 3 min is achieved with a degreasing temperature of 80 °C. The maximum thickness of 18.43 µm occurs with minimal degreasing at 40 °C. At this lower temperature, the maximum layer thickness of 21.78 µm is reached after 7 min of deposition. When the voltage is increased to 250 V, the smallest thickness of 16.48 µm is again observed at a degreasing temperature of 80 °C. At this temperature, a minimum thickness of 20.18 µm is also reached when the deposition time is extended to 7 min. The maximum thickness is achieved at a degreasing temperature of 40 °C throughout the entire KTL deposition period. At 300 V, the highest thickness values are recorded, with the difference between 250 V at 80 °C and a deposition time of 3 min being 1.49 µm. However, for a 7 min KTL deposition, this difference is reduced to only 0.11 µm. The maximum thicknesses are achieved with degreasing at 40 °C, with a thickness of 21.42 µm after 3 min and 22.02 µm after 7 min.

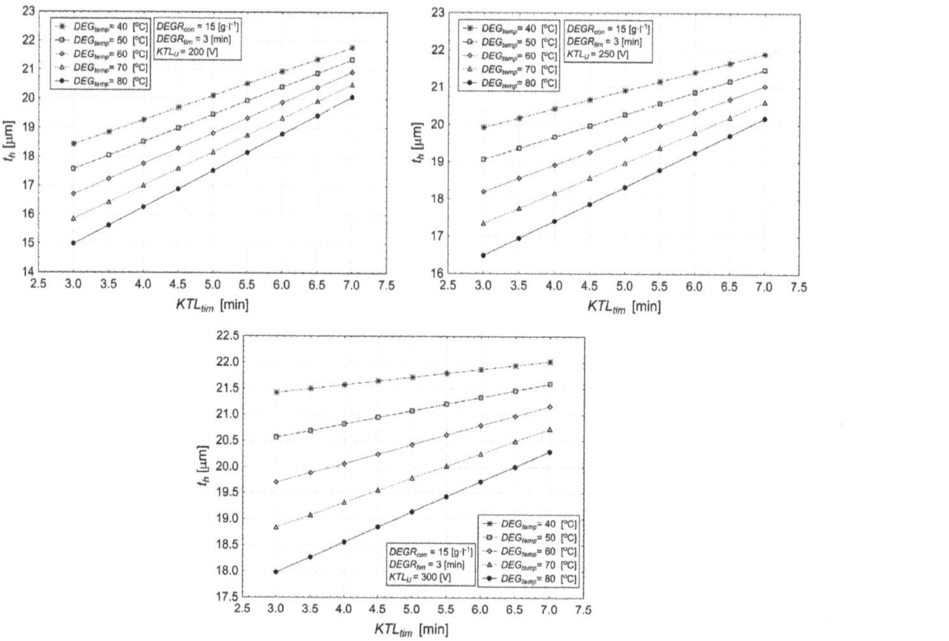

Figure 6. The effect of KTL deposition time and degreasing temperature on the change of the formed layer at the minimum values of the design of experiments methodology.

Figure 7 illustrates the relationship between the KTL deposition time, degreasing temperature, and voltage variation and their combined effect on the resulting layer thickness. At a voltage of 200 V, the minimum thickness of 12.23 µm is observed at a degreasing temperature of 80 °C. At this temperature, the thickness increases over time, and around 4.8 min of KTL deposition, the layers intersect, resulting in maximum values of 17.30 µm after 7 min of deposition. The maximum thickness of 14.89 µm, recorded after 3 min of deposition, occurs at a degreasing temperature of 40 °C. When the thickness reaches approximately 14.57 µm, the graph intersects, indicating that this degreasing temperature reaches the minimum thickness values. After 7 min of deposition, the thickness measures 14.12 µm. Only at this temperature does the thickness decrease with increasing deposition time. When the voltage is raised to 250 V, the smallest thickness, 13.72 µm, is observed at a degreasing temperature of 80 °C. This temperature, along with increasing time, contributes to an increase in thickness, and at around 4.75 min, it intersects with other degreasing temperatures, reaching the maximum layer thickness. After 7 min, the thickness reaches 17.42 µm. A rising trend in layer formation is also observed at degreasing temperatures of 60 °C and 70°C. However, at a degreasing temperature of 50 °C, the layer thickness decreases as time increases. At 40 °C, a maximum thickness of 16.39 µm is achieved at 3 min of KTL deposition. At the point where the graphs intersect, the thickness is approximately 15.40 µm. The smallest thickness of 14.24 µm is recorded after 7 min of deposition. This behavior agreed with data observed by authors Hamera et al. [34], Wang et al. [35], and Swierczynski [36].

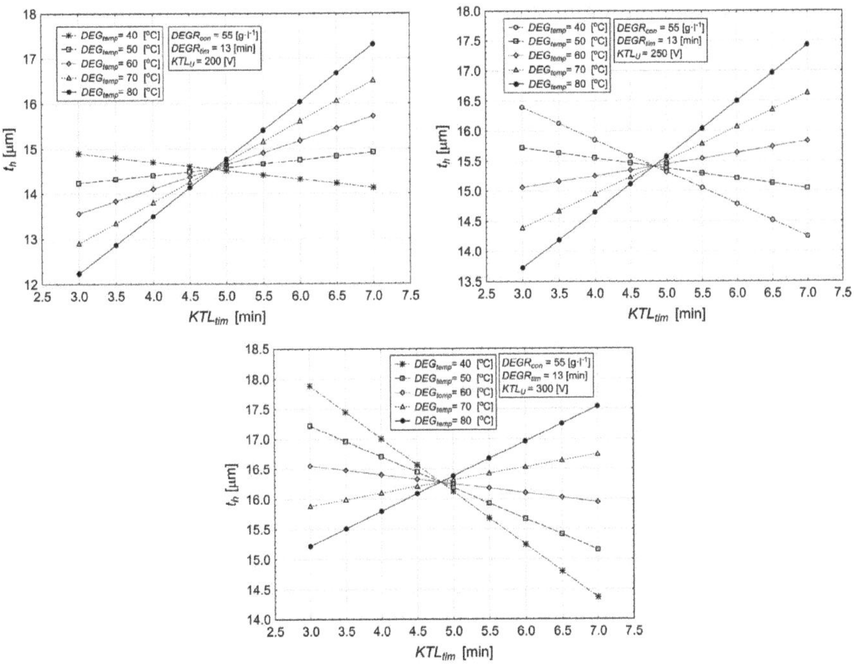

Figure 7. The effect of KTL deposition time and degreasing temperature on the change of the formed layer at maximum values of the design of experiments methodology.

When the voltage is increased to 300 V, the minimum thickness, 15.22 μm, is reached after 3 min of deposition at a degreasing temperature of 80 °C. At degreasing temperatures of 70 °C and 80 °C, the thickness continues to increase with time. At 4.8 min, the graph intersects, and from that point onward, the degreasing temperature of 80 °C produces the maximum thickness, reaching 17.54 μm after 7 min of deposition. The maximum thickness after 3 min, 17.88 μm, is achieved at a degreasing temperature of 40 °C. When the thickness falls to around 16.27 μm, this temperature produces the minimum thickness as time increases. After 7 min of deposition, the layer thickness is 14.36 μm. The created layer thickness does not decrease with increasing time at degreasing temperatures of 40 °C, 50 °C, or 60 °C.

3.3. *Influence of Cataphoretic Deposition and Degreasing Deposition on Coating Characteristics*

Figure 8 depicts the impact of time factors, specifically KTL deposition time and degreasing deposition time, along with concurrent changes in degreasing temperature, on the resulting coating thickness. At a degreasing temperature of 40 °C, the thickness consistently increases across all displayed time intervals as the deposition time lengthens. The lowest recorded thickness of 17.85 μm occurred after 13 min of degreasing deposition and 3 min of KTL deposition. From this point, the thickness increased with time, and the graph intersected around 5.3 min. The maximum thickness values were achieved with a 13 min degreasing deposition time, reaching 22.24 μm after 7 min of KTL deposition. The highest thickness of 18.43 μm was measured after 3 min of KTL deposition and 3 min of degreasing. At approximately 20.35 μm, another intersection occurs, and from this point forward, the thickness reaches its minimum values. At 7 min of KTL deposition, the thickness reaches 21.78 μm with a degreasing time of 3 min. When the degreasing temperature is increased to 60 °C, a maximum thickness of 19.44 μm is observed after 3 min of KTL deposition, corresponding to 13 min of degreasing deposition. The minimum thickness at this point is 16.70 μm, which was observed after 3 min of degreasing.

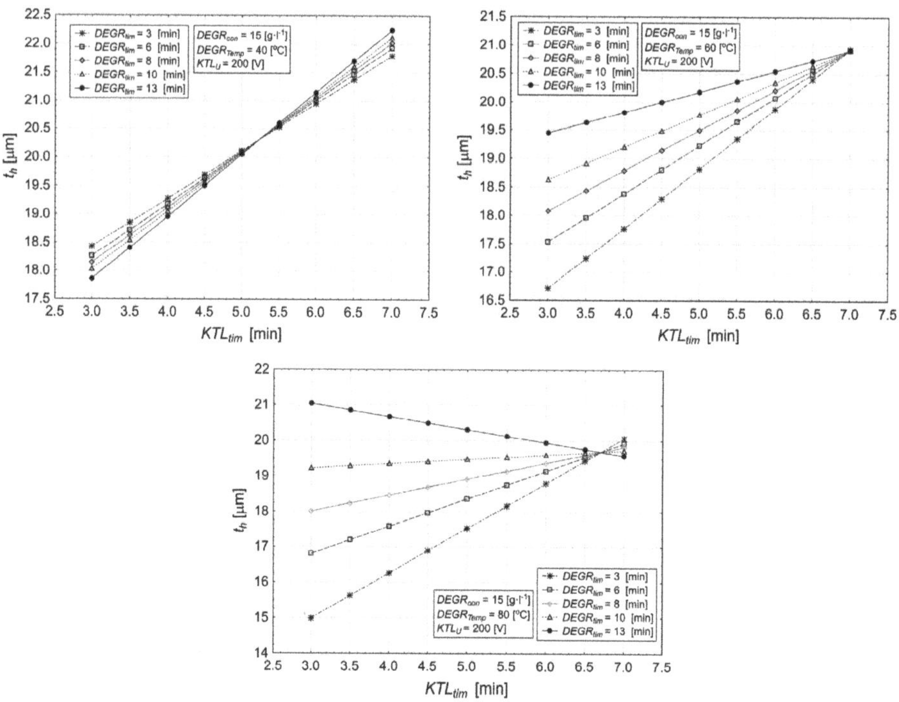

Figure 8. The influence of the cataphoretic deposition time and the degreasing deposition time on the change of the created layer at the minimum values of the design of experiments methodology.

Thickness increases across all defined degreasing times as deposition time progresses. At 7 min of KTL deposition, the thicknesses converge to a uniform value of approximately 20.91 μm, representing the maximum thickness achievable under these conditions. At a degreasing temperature of 80 °C, the minimum thickness of 14.98 μm occurs after 3 min of degreasing and 3 min of KTL deposition. Increasing thickness is observed for degreasing times of 3, 6, 8, and 10 min. The maximum thickness of 21.03 μm is reached after 13 min of degreasing and 3 min of KTL deposition. As time increases, this thickness begins to decrease, with an intersection occurring at approximately 19.63 μm. A degreasing time of 13 min with 7 min of KTL deposition results in a minimum thickness of 19.57 μm, while the maximum thickness of 20.06 μm for 7 min of KTL deposition is achieved with 3 min of degreasing.

Figure 9 illustrates the impact of deposition time during degreasing and cataphoretic coating at maximum experimental values, along with concurrent changes in degreasing temperature, on the resulting thickness. At a degreasing temperature of 40 °C, a thickness difference of 0.49 μm is observed after 3 min of coating. The maximum thickness of 17.88 μm is achieved with a 13 min deposition time. When the KTL deposition time reaches approximately 3.3 min, a cross-point occurs, where the thickness is about 17.62 μm. From this point, the maximum thickness is formed during a 3 min deposition. If the KTL process extends to 7 min, the thickness reaches 20.41 µm. The results also confirm the conclusions of Rossi et al. [37] and Calovi et al. [38].

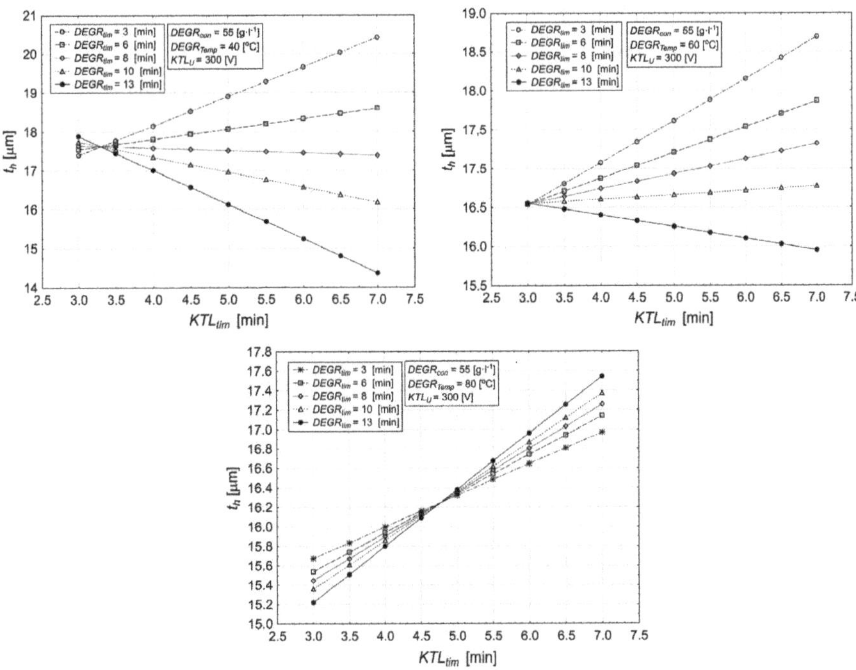

Figure 9. The influence of cataphoretic deposition time and degreasing temperature on the change of the formed layer at maximum values of the design of experiments methodology.

The minimum thickness of 14.36 µm for the aforementioned KTL process is reached after 13 min of degreasing. The smallest thickness variation of 0.26 µm throughout the KTL process is observed with an 8 min degreasing deposition time. When the degreasing temperature is raised to 60 °C, a nearly uniform thickness of 16.54 µm is reached after 3 min of KTL deposition. Under these conditions, a reduction in thickness is only observed with a 13 min degreasing deposition time, where the minimum thickness of 15.95 µm is reached after 7 min of KTL deposition. The maximum thickness is produced during a 3 min degreasing deposition. At this degreasing time, the largest thickness increase of 2.15 µm is observed as the KTL deposition time increases, with the thickness reaching a maximum of 18.68 µm at 7 min of KTL deposition. When the degreasing temperature is increased to 80 °C, a graph is produced with a range of values of 0.45 µm at a minimum coating time of 3 min, with thickness values increasing as the deposition time increases. The smallest thickness of 15.22 µm is achieved after 13 min of deposition. The highest thickness values are observed with a 3 min KTL deposition time. After approximately 4.8 min of KTL deposition, a cross-point occurs on the graph, with a thickness of around 16.28 µm at this point. As the deposition time increases, the maximum thickness is achieved with a 13 min deposition duration. After reaching 7 min of KTL, the resulting layer exhibits a thickness of 17.54 µm. During this specified deposition time, the minimum thickness of 16.96 µm was obtained with a 3 min degreasing process. The results also confirm the conclusions of Su et al. [39], Li et al. [40], and Kania et al. [41].

3.4. Energy-Dispersive X-Ray Spectroscopy (EDX) of Cataphoretic Layer

Energy-dispersive X-ray spectroscopy (EDX) is a very effective technique for analyzing elemental composition and assessing the quality of the interface between cataphoretic coatings and the substrate. When evaluating coating adhesion, EDX provides detailed information on the chemical composition of the coating and its interactions with the substrate. Since adhesion between coatings depends on their chemical and structural properties, EDX analysis allows the identification of whether there have been undesirable changes in composition that could affect the strength and stability of the bond. EDX also allows the detection of potential defects or inhomogeneities that may indicate adhesion problems, as well as assessing the uniformity of elemental distribution throughout the coating.

Figure 10 illustrates the results of energy-dispersive X-ray spectroscopy (EDX) conducted on a sample from a cataphoretic painting process, combining a scanning electron microscopy (SEM) image with the corresponding EDX spectrum. The SEM image presents a cross-sectional view of the sample, clearly showing two distinct regions: the upper layer, representing the cataphoretic paint coating, and the lower layer, corresponding to the metallic substrate. A vertical line, labeled "Line Data 1", marks the analyzed region, crossing the interface between the paint layer and the substrate, enabling a detailed elemental composition analysis.

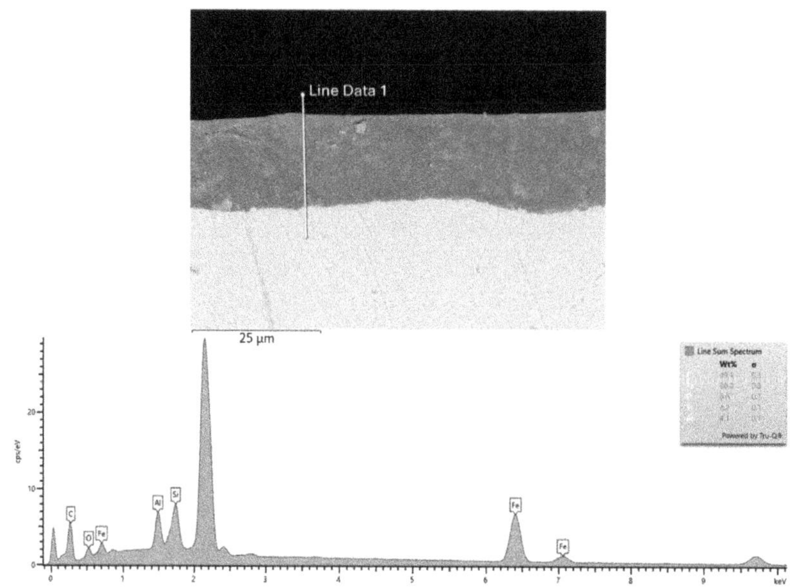

Figure 10. Energy-dispersive X-ray spectroscopy (EDX) of cataphoretic layer.

The EDX spectrum, displayed beneath the SEM image, provides information on the elemental composition along the analyzed path. The *X*-axis represents energy levels (keV), and the *Y*-axis shows counts per energy unit, with peaks indicating the presence of specific elements. The spectrum reveals that the upper layer is dominated by carbon, comprising approximately 49.1 wt.%, indicative of the organic nature of the cataphoretic paint. Iron peaks, accounting for 36 wt.%, highlight the metallic substrate, likely steel. Minor peaks for oxygen (~6.6 wt.%), silicon (~4.2 wt.%), and aluminum (~4.1 wt.%) suggest the presence of oxidized compounds, paint additives, or surface contaminants. Analysis demonstrates a clear boundary between the paint layer and the substrate, indicating the uniformity of the applied coating. The presence of oxygen at the interface points to surface oxidation, which may enhance adhesion between the organic paint and the metallic substrate.

The elemental map shown in Figure 11 for carbon (C) highlights its high concentration in the upper region, corresponding to the cataphoretic paint layer. The mapping shows a uniform distribution of carbon within the coating, with a clear demarcation at the interface between the coating and the substrate. The wt.% profile graph for carbon demonstrates consistent concentration in the coating layer, peaking above 80 wt.% near the surface. Beyond the coating–substrate interface (approximately 20–25 μm from the surface), the carbon concentration drops sharply, confirming its absence in the metallic substrate.

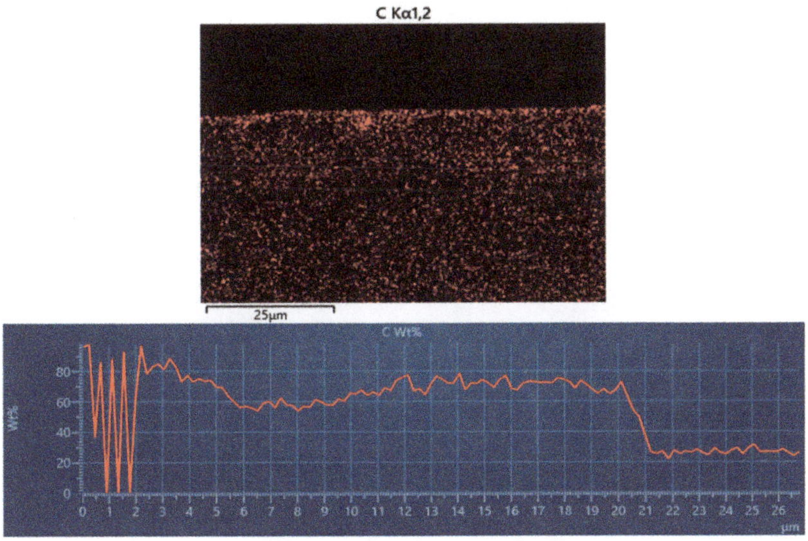

Figure 11. EDX spectrum of carbon (C).

The elemental map shown in Figure 12 for iron (Fe) shows its dominance in the lower region, representing the substrate. The mapping indicates a uniform distribution within the substrate, while confirming the lack of significant iron content in the cataphoretic paint layer. The wt.% profile graph for iron exhibits a minimal presence in the coating layer, with concentrations increasing sharply at the interface. Beyond the boundary (approximately 20–25 μm from the surface), iron stabilizes at concentrations exceeding 60 wt.%, reflecting the composition of the metallic substrate.

The oxygen (O) elemental map (Figure 13) illustrates a prominent presence in the upper region, which aligns with the cataphoretic paint layer. The distribution of oxygen appears relatively consistent throughout the coating. The wt.% profile for oxygen indicates a peak near the surface, with a concentration reaching approximately 60 wt.%. Moving towards the coating–substrate interface (about 20–25 μm from the surface), the oxygen concentration drops significantly, signifying its near absence in the underlying metal substrate.

The elemental map for silicon (Si) shown in Figure 14 clearly shows its predominant presence in the upper region of the sample, which corresponds to the coating layer. The silicon is evenly distributed within this layer, suggesting a well-formed and uniform coating. The weight percentage (wt.%) profile graph further supports this observation, displaying a peak silicon concentration of around 24 wt.% at the surface. This concentration decreases gradually, as it moves toward the coating–substrate interface. After this point, the silicon concentration becomes negligible, confirming that silicon is absent in the metallic substrate beneath the coating.

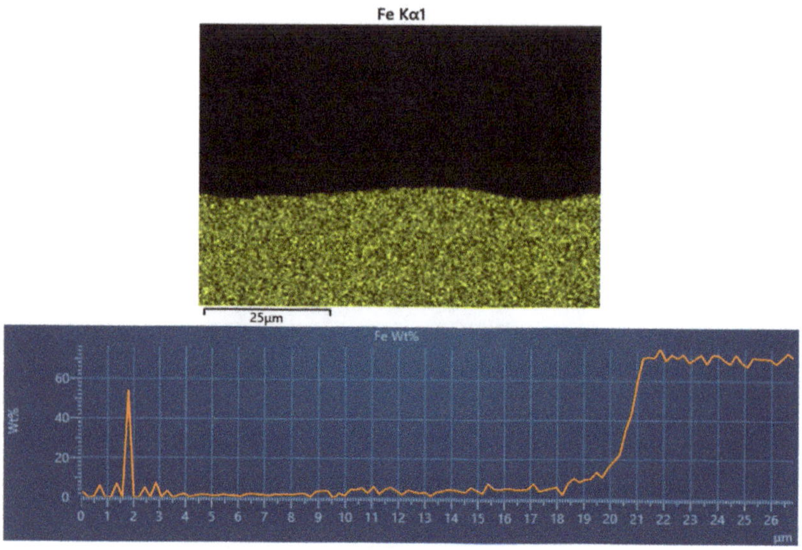

Figure 12. EDX spectrum of iron (Fe).

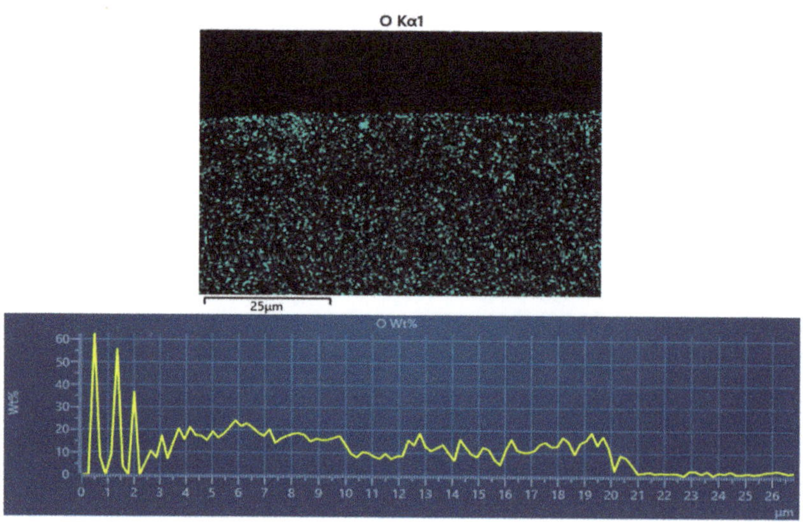

Figure 13. EDX spectrum of oxygen (O).

The elemental map for aluminum (Al) shown in Figure 15 illustrates its predominant presence in the substrate region, situated beneath the coating layer. Within the coating, aluminum is almost absent, and a sharp increase in its concentration is observed at the coating–substrate interface. This trend is further validated by the wt.% profile graph, which shows negligible aluminum content in the upper coating layer, followed by a steep rise in concentration near the interface. The concentration of aluminum remains consistently high (around 90–100 wt.%) within the substrate, indicating that aluminum is primarily confined to the metallic substrate and does not diffuse into the coating layer.

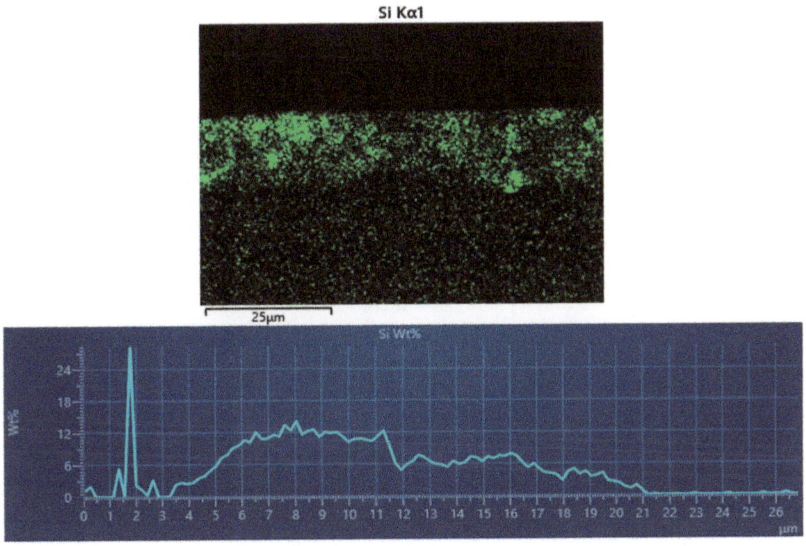

Figure 14. EDX spectrum of silicon (Si).

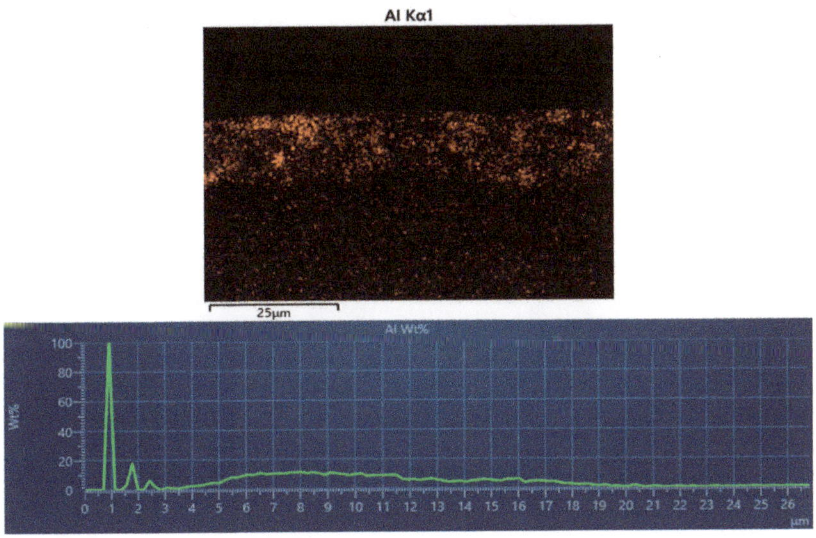

Figure 15. EDX spectrum of aluminum (Al).

In conclusion, the energy-dispersive X-ray spectroscopy (EDX) analysis offers comprehensive insights into the elemental composition and distribution across a sample subjected to a cataphoretic painting process. The combination of scanning electron microscopy (SEM) imaging and EDX spectral data reveals a clear delineation between the cataphoretic coating layer and the underlying metallic substrate, with a marked transition at the coating–substrate interface. Elemental mapping and corresponding wt.% profile graphs highlight several critical features; the cataphoretic coating is predominantly composed of carbon, confirming the organic nature of the paint, while iron is primarily concentrated in the metallic substrate, with minimal presence in the coating. Minor elements, such as oxygen, silicon, and aluminum, are distributed in both layers, with oxygen primarily localized at the coating–substrate interface, suggesting its role in surface oxidation, which may facilitate improved adhesion between the coating and substrate. Silicon is predominantly confined to the coating layer, exhibiting a uniform distribution, while aluminum shows a sharp increase in concentration near the interface, confirming its confinement to the substrate. Additionally, the EDX analysis shows that there is no significant diffusion of iron or aluminum into the coating layer, reinforcing the separation between the organic paint and the metallic substrate. The uniform distribution of silicon within the coating further highlights the consistent structure of the deposited layer. These observations substantiate the uniformity and integrity of the cataphoretic coating, and the EDX data effectively demonstrate the elemental distribution and the absence of significant diffusion between the paint and metallic substrate. This study highlights the utility of EDX in assessing coating quality and substrate interactions, providing a detailed understanding of the elemental dynamics within composite coating systems. This behavior agreed with data observed by authors Nam et al. [42], Lovejoy et al. [43], Qi et al. [44], Dalfonso et al. [45], Tromp [46], and Sanchez et al. [47].

3.5. Summary of Results

The conducted experiments underscore the critical influence of deposition time and degreasing temperature on the formation of the coating layer during cataphoretic painting (KTL). The results consistently demonstrate that, as the KTL time increases, the layer thickness rises, reaching peak values at longer deposition durations. Specifically, at a deposition time of 7 min, the maximum thickness is achieved, whereas shorter times, such as 3 min, result in thinner coatings. The impact of voltage is equally significant; at 300 V, the layer thickness increases more steeply, with minimal reduction over time, suggesting that higher voltages promote a greater accumulation of material, particularly in the early deposition stages.

The role of degreasing temperature in determining the final layer thickness is also evident. Lower degreasing temperatures, such as 40 °C, are associated with higher thicknesses, especially for shorter deposition times, whereas higher temperatures, such as 80 °C, yield reduced thicknesses under identical conditions. This trend suggests that lower temperatures enhance the coating formation process by altering surface properties or interface interactions. Furthermore, intermediate degreasing temperatures, such as 60 °C, provide an optimal balance between temperature and voltage conditions, leading to more consistent layer thicknesses across varying deposition times.

A comparison between experimental results and model predictions further validates these findings. The experimental measurements indicate a significant dependency of layer thickness on deposition time, applied voltage, solution concentration, and temperature. For instance, at 300 V and a deposition time of 7 min, the maximum thickness reaches 22.56 μm, while at 200 V under the same conditions, the thickness reduces to 18.43 μm. These results confirm that deposition time and voltage are the primary determinants of layer thickness.

The statistical model corroborates these experimental observations, explaining approximately 75% of the variability in layer thickness, with an adjusted coefficient of determination (R^2) of 70.79%. The model was found to be statistically significant ($p < 0.0001$), highlighting its reliability in describing the relationship between key process parameters and the resulting layer characteristics. Notably, while solution concentration and degreasing temperature influence layer thickness, their effects are secondary compared to deposition time and voltage. This aligns with the model's predictive accuracy, as evidenced by the minimal deviations between experimental data and model estimates.

The comparison of experimental and predicted layer thicknesses (Figure 16) shows systematic differences between model and real values. At a voltage of 200 V and a degreasing agent concentration of 15 g·L^{-1}, the measured thickness at 3 and 5 min was lower than predicted, with differences ranging from 1.2558 μm to 2.4514 μm. At 4 min the measured value was slightly higher than predicted (0.3624 μm), while at 7 min, it was again lower (2.2809 μm). At 250 V, shorter times (3–5 min) showed that the measured thickness was always lower than predicted, with differences around 2 μm. At 6 min, there was an increase in measured thickness above the predicted value (+1.6378 μm), but at 7 min, the difference was the largest (−3.0906 μm), indicating a possible inaccuracy of the prediction model at longer times. At 300 V, the measured thickness was always lower than the predicted thickness at 3–5 min, with differences ranging from 2.0351 μm to 2.7950 μm. At 6 min, there was a significant increase in thickness above the predicted value (+2.8805 μm), but at 7 min, it was again lower than predicted (−1.7219 μm).

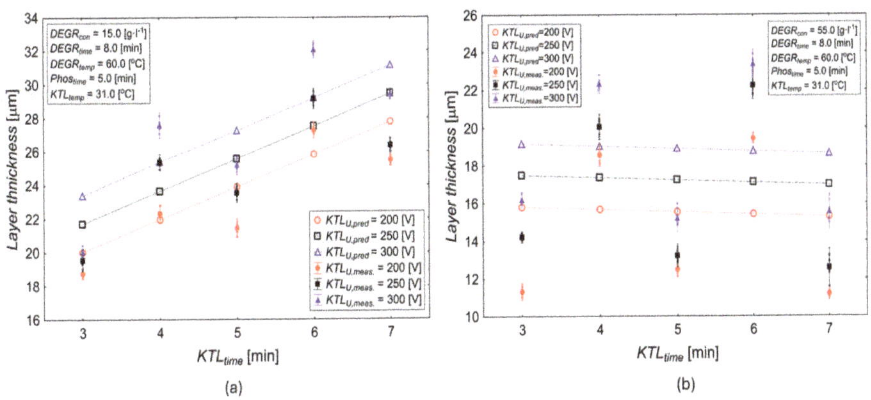

Figure 16. Comparison of measured and predicted coating thickness values depending on minimum (**a**) and maximum (**b**) degreasing concentration, KTL deposition time, and applied voltage.

At 200 V and a degreaser concentration of 55 g·L^{-1} the measured thickness at 3 and 5 min was lower than predicted, with differences ranging from 3.052 μm to 4.467 μm. At 4 min, the measured value was higher than predicted (−2.899 μm), while at 7 min, it was again lower (4.083 μm). At 250 V, shorter times (3–5 min) showed that the measured thickness was always lower than predicted, with differences around 4 μm. At 6 min, there was an increase in measured thickness above the predicted value (+5.130 μm), but at 7 min, the difference was the largest (−4.402 μm), indicating a possible inaccuracy of the prediction model at longer times. At 300 V, the measured thickness was always lower than predicted at 3–5 min, with differences ranging from 2.962 μm to 3.699 μm. At 6 min, there was a significant increase in thickness above the predicted value (4.599 μm), but at 7 min, it was again lower than predicted (3.101 μm).

In summary, the experimental findings align closely with the theoretical predictions of the statistical model. The most influential factors in determining layer thickness are deposition time and applied voltage, with solution concentration and temperature playing supplementary roles. The consistency between experimental results and model predictions supports the model's applicability in optimizing cataphoretic painting processes under varying operational conditions.

4. Conclusions

This study focused on the inter-relationship between deposition time, degreasing temperature, and applied voltage in the cataphoretic painting process, examining their cumulative effect on the thickness of the formed layers. The results showed that the optimal settings of these parameters can significantly affect the quality and uniformity of the coating, which is important for industrial surface treatment applications.

Key findings:

- The layer thickness increases with KTL deposition time, with the maximum thickness achieved at 7 min, while shorter times (e.g., 3 min) result in thinner coatings.
- The applied voltage significantly influences the layer thickness; at 300 V, the layer thickness increases more steeply, and the reduction over time is minimal, indicating a higher material accumulation in the early stages of deposition.
- Lower degreasing temperatures (e.g., 40 °C) lead to higher layer thickness, while higher temperatures (e.g., 80 °C) result in a reduction in thickness.
- The optimal combination of voltage and temperature (e.g., 60 °C) leads to more uniform layer thicknesses across different deposition times.

For future research, it would be beneficial to focus on the long-term stability and durability of cataphoretic coatings under real-world operational conditions, including corrosion resistance and mechanical property tests. This would provide a better understanding of the practical use and lifespan of coatings in industrial applications. A promising direction is also the expansion of the predictive model to include additional factors, such as electrolyte pH, solution conductivity, and surface roughness, which would increase the accuracy of layer thickness predictions and enable the better optimization of the cataphoretic painting process.

The research provided valuable insights into the impact of deposition time, voltage, and degreasing temperature on the thickness of the cataphoretic coating, but it has certain limitations. The study focused on specific values of voltage (200–300 V), deposition time (3–7 min), and degreasing temperature (40–80 °C), and expanding these parameters could offer a more comprehensive perspective. Additionally, the research mainly focused on layer thickness without a detailed analysis of factors such as electrolyte pH, solution conductivity, or surface pretreatment, whose investigation could provide a better understanding of the overall process.

The research was conducted in controlled laboratory conditions, which may limit the applicability of the results to real industrial environments. In actual operational conditions, factors such as electrolyte impurities or variations in surface pretreatment could influence the results. Additionally, while the statistical model explained about 75% of the layer thickness variability, there is room for improvement. Further refinement of factor interactions and testing additional variables could enhance the accuracy of layer thickness predictions. These limitations open opportunities for further research to expand findings and contribute to optimizing cataphoretic painting, particularly for industrial applications.

Author Contributions: P.F. and D.P. conceived and designed the experiments and wrote the paper; J.D., M.G. and P.M. analyzed and evaluated data, processed data, and performed the experiments and measurements. All authors have read and agreed to the published version of the manuscript.

Funding: This research received no external funding.

Institutional Review Board Statement: Not applicable.

Informed Consent Statement: Not applicable.

Data Availability Statement: The original contributions presented in the study are included in the article, further inquiries can be directed to the corresponding author.

Acknowledgments: This research has been designed in the framework of the projects VEGA no. 1/0488/23 and KEGA 009TUKE-4/2024, which were granted by the Ministry of Education, Science, Research and Sport of the Slovak Republic, and this publication is the result of the project "Development of excellent research capacities in the field of additive technologies for the Industry of the 21st century", ITMS: 313011BWN5, supported by the Operational Program Integrated Infrastructure funded by the ERDF.

Conflicts of Interest: The authors declare no conflicts of interest. The founding sponsors had no role in the design of the study; in the collection, analyses, or interpretation of the data; in the writing of the manuscript; or in the decision to publish the results.

References

1. Ouchi, T.; Sadoway, D.R. Positive current collector for Li||Sb-Pb liquid metal battery. *J. Power Sources* **2017**, *357*, 158–163. [CrossRef]
2. Kirchgeorg, T.; Weinberg, I.; Hoernig, M.; Baier, R.; Schmid, M.J.; Brockmeyer, B. Emissions from corrosion protection systems of offshore wind farms: Evaluation of the potential impact on the marine environment. *Mar. Pollut. Bull.* **2018**, *136*, 257–268. [CrossRef] [PubMed]
3. Flores, J.F.; Neville, A.; Kapur, N.; Gnanavelu, A. Assessing the resistance of metal matrix composites and their microstructural integrity under erosion-corrosion. *Wear* **2011**, *271*, 1331–1340. [CrossRef]
4. Babaei, N.; Yeganeh, H.; Gharibi, R. Anticorrosive and self-healing waterborne poly(urethane-triazole) coatings made through a combination of click polymerization and cathodic electrophoretic deposition. *Eur. Polym. J.* **2019**, *112*, 636–647. [CrossRef]
5. Moradi, M.; Yeganeh, H.; Pazokifard, S. Synthesis and assessment of novel anticorrosive polyurethane coatings containing an amine-functionalized nanoclay additive prepared by the cathodic electrophoretic deposition method. *RSC Adv.* **2016**, *6*, 28089–28102. [CrossRef]
6. Aghili, M.; Yazdi, M.K.; Ranjbar, Z.; Jafari, S.H. Anticorrosion performance of electro-deposited epoxy/amine functionalized graphene oxide nanocomposite coatings. *Corros. Sci.* **2021**, *179*, 109143. [CrossRef]
7. Zivkovic, L.S.; Jegdic, B.V.; Andric, V.; Rhee, K.Y.; Bajat, J.B.; Miskovic-Stankovic, V.B. The effect of ceria and zirconia nanoparticles on the corrosion behaviour of cataphoretic epoxy coatings on AA6060 alloy. *Prog. Org. Coat.* **2019**, *136*, 105219. [CrossRef]
8. Oliviera, M.G.; Romano, A.P.; Vandermiers, C.; Mathieu, X.; Poelman, M. Influence of the stress generated during an ageing cycle on the barrier properties of cataphoretic coatings. *Prog. Org. Coat.* **2008**, *63*, 323–329. [CrossRef]
9. Garcia, S.J.; Rodriguez, M.T.; Izquierdo, R.; Suay, J. Evaluation of cure temperature effects in cataphoretic automotive primers by electrochemical techniques. *Prog. Org. Coat.* **2007**, *60*, 303–311. [CrossRef]
10. Zanella, C.; Pedrotti, A.; Fedel, M.; Deflorian, F. Influence of the electrochemical behavior of metal substrates on the properties of cataphoretic clearcoat. *Prog. Org. Coat.* **2014**, *77*, 1987–1992. [CrossRef]
11. Calovi, M.; Russo, F.; Rossi, S. Synergic behavior of graphene-based filler and thermochromic pigments in cataphoretic coatings. *Prog. Org. Coat.* **2021**, *2021*, 105978.
12. Darowicki, K.; Slepski, P.; Szocinski, M. Novel application of dynamic electrochemical impedance monitoring to a cataphoretic coating process. *Prog. Org. Coat.* **2020**, *149*, 105906.
13. Sternadelova, K.; Krupova, H.; Matysek, D.; Mohyla, P. Quality assessment of the vitreous enamel coating applied to the weld joint. *MM Sci. J.* **2023**, *2023*, 6333–6338.
14. Brüggemann, M.; Rach, A. *Electrocoat: Formulation and Technology*; Vincentz Network: Hannover, Germany, 2020.
15. Goldschmidt, A.; Streitberger, H.J. *BASF Handbuch Lackiertechnik*; Vincentz Network: Hannover, Germany, 2014.
16. Brock, T.; Groteklaes, M.; Mischke, P. *European Coatings Handbook*, 2nd ed.; Vincentz Network: Hannover, Germany, 2010.
17. Ovsik, M.; Manas, M.; Stanek, M.; Dockal, A.; Mizera, A.; Fluxa, P.; Bednarik, M.; Adamek, M. Nano-mechanical properties of surface layers of polyethylene modified by irradiation. *Materials* **2020**, *13*, 929. [CrossRef]
18. Akafuah, N.K.; Poozesh, S.; Salaimeh, A.; Patrick, G.; Lawler, K.; Saito, K. Evolution of the automotive body coating process-A Review. *Coatings* **2016**, *6*, 24. [CrossRef]
19. Skotnicki, W.; Jedrzejczyk, D. The comparative analysis of the coatings deposited on the automotive parts by the cataphoresis method. *Materials* **2021**, *14*, 6155. [CrossRef]
20. Jaczewski, M. Industrial Paint Coatings for Metal Surfaces. In *Tikkurila Protective Coating Guide*, 2nd ed.; Tikkurila: Vantaa, Finland, 2008.
21. Javidi, A.; Rieger, U.; Eichlseder, W. The effect of machining on the surface integrity and fatigue life. *Int. J. Fatigue* **2008**, *30*, 2050–2055. [CrossRef]
22. Rivera, J.L.; Reyes-Carrillo, T. A framework for environmental and energy analysis of the automobile painting process. *Procedia CIRP* **2014**, *15*, 171–175.
23. Goeke, S.; Biermann, D.; Stickel, D.; Stemmer, P.; Fischer, A.; Geenen, K.; Huth, S.; Theisen, W. Enhancing the surface integrity of tribologically stressed contacting surfaces by an adjusted surface topography. *Procedia CIRP* **2014**, *13*, 214–218.
24. *VDA 239-100*; Spray and Micro-Dosage Application—Technical Cleanliness of Components in Contact with Fluids. Normservis, 2016. Available online: https://eshop.normservis.sk/norma/vda-239-100-1.5.2016.html (accessed on 15 January 2025).
25. Marine Steels. DIN EN 10130 DC01 Steel—Cold Rolled Low Carbon Steel Plate. Available online: https://sk.marinesteels.com/stainless-steel/stainless-steel-plate/din-en-10130-dc01-steel.html (accessed on 18 January 2025).
26. Dean, A.; Voss, D.; Draguljic, D. *Design and Analysis of Experiments*; Springer International Publishing AG: Cham, Switzerland, 2017; ISBN 978-3-319-52250-0.

27. KTL s.r.o. Princip Činnosti. KTL s.r.o. Available online: https://www.ktlpresov.sk/princip-cinnosti (accessed on 24 January 2025).
28. *ISO 16232:2018*; Road Vehicles—Cleanliness of Components and Systems. International Organization for Standardization: London, UK, 2018. Available online: https://www.iso.org/standard/71583.html (accessed on 25 January 2025).
29. Built In. Shapiro-Wilk Test: A Guide to Normality Testing. Available online: https://builtin.com/data-science/shapiro-wilk-test (accessed on 25 January 2025).
30. Garcia, S.J.; Suay, J. Optimization of deposition voltage of cataphoretic automotive primers assessed by EIS and AC/DC/AC. *Prog. Org. Coat.* **2009**, *66*, 306–313.
31. Kamas, T.; Tekkalmaz, M. Coupled field modeling of E/M impedance of piezoelectric wafer active sensor for cataphoretic coating thickness measurement. *Smart Mater. Struct.* **2017**, *26*, 045035.
32. Krenicky, T.; Goncharov, O.Y.; Kuchar, J.; Sapegina, I.V.; Kudlacek, J.; Faizullin, R.R.; Korshunov, A.I.; Cerny, D. Chemical vapor deposition of tantalum carbide in the TaBr5-CCl4-Cd system. *Coatings* **2024**, *14*, 547. [CrossRef]
33. Kania, H.; Saternus, M.; Kudlacek, J. Impact of Bi and Sn on microstructure and corrosion resistance of zinc coatings obtained in Zn-AlNi bath. *Materials* **2020**, *13*, 3788. [CrossRef] [PubMed]
34. Hamera, P.; Bennison, A.D. TIB Clean A300—lowering the temperature of alkaline degreasing. *Ochr. Przed Koroz.* **2021**, *64*, 144–147.
35. Wang, C.S.; Cai, N.; Zhang, D.V.; Zhang, J.X.; Chang, T.H.; Li, S.M.; Chao, Y.Q.; Hu, J.Q. Preparation of optimum degreasing-sintering process for metal-polymer blending low temperature 3D printing. *Rapid Prototyp. J.* **2019**, *25*, 515–529. [CrossRef]
36. Swierczynski, T. Leraclen® ATR and Lerabilt® 268—low-temperature, alkaline products for degreasing prior to hot-dip galvanizing and electroplating. *Ochr. Przed Koroz.* **2022**, *65*, 324–326.
37. Rossi, S.; Calovi, M.; Fedel, M. Corrosion protection of aluminum foams by cataphoretic deposition of organic coatings. *Prog. Org. Coat.* **2017**, *109*, 144–151.
38. Calovi, M.; Dire, S.; Ceccato, R.; Deflorian, F.; Rossi, S. Corrosion protection properties of functionalised graphene-acrylate coatings produced via cataphoretic deposition. *Prog. Org. Coat.* **2019**, *136*, 105261.
39. Su, Y.; Zhitomirsky, I. Cataphoretic assembly of cationic dyes and deposition of carbon nanotube and graphene films. *J. Colloid Interf. Sci.* **2013**, *399*, 46–53.
40. Li, C.H.; He, Y.; Zhao, Y.; Li, Z.J.; Sun, D.; Li, H.J.; Chen, W.; Yan, J.; Wu, G.Y.; Yuan, X. Cataphoretic deposition of an epoxy coating with the incorporation of Ti3C2Tx@Mg-Al layered double hydroxide for long-term active corrosion protection effect. *Prog. Org. Coat.* **2023**, *175*, 107333. [CrossRef]
41. Kania, H.; Saternus, M.; Kudlacek, J.; Svoboda, J. Microstructure Characterization and Corrosion Resistance of Zinc Coating Obtained in a Zn-AlNiBi Galvanizing Bath. *Coatings* **2020**, *10*, 758. [CrossRef]
42. Nam, D.; Opanasyuk, A.S.; Koval, P.V.; Ponomarev, A.G.; Jeong, A.R.; Kim, G.Y.; Jo, W.; Cheong, H. Composition variations in $Cu_2ZnSnSe_4$ thin films analyzed by X-ray diffraction, energy dispersive X-ray spectroscopy, particle induced X-ray emission, photoluminescence, and Raman spectroscopy. *Thin Solid Films* **2014**, *562*, 109–113. [CrossRef]
43. Lovejoy, T.C.; Ramasse, Q.M.; Falke, M.; Kaeppel, A.; Terborg, R.; Zan, R.; Dellby, N.; Krivanek, O.L. Single atom identification by energy dispersive x-ray spectroscopy. *Appl. Phys. Lett.* **2012**, *100*, 154101. [CrossRef]
44. Qi, P.; Samadi, N.; Martinson, M.; Ponomarenko, O.; Bassey, B.; Gomez, A.; George, G.N.; Pickering, I.J.; Chapman, L.D. Wide field imaging energy dispersive X-ray absorption spectroscopy. *Sci. Rep.* **2019**, *9*, 17734.
45. D'Alfonso, A.J.; Freitag, B.; Klenov, D.; Allen, L.J. Atomic-resolution chemical mapping using energy-dispersive x-ray spectroscopy. *Phys. Rev. B* **2010**, *81*, 100101. [CrossRef]
46. Tromp, R.M. Energy-dispersive X-ray spectroscopy in a low energy electron microscope. *Ultramicroscopy* **2024**, *259*, 113935. [CrossRef]
47. Sanchez, D.F.; Simionovici, A.S.; Lemelle, L.; Cuartero, V.; Mathon, O.; Pascarelli, S.; Bonnin, A.; Shapiro, R.; Konhauser, K.; Grolimund, D.; et al. 2D/3D Microanalysis by Energy Dispersive X-ray Absorption Spectroscopy Tomography. *Sci. Rep.* **2017**, *7*, 16453. [CrossRef]

applied sciences

MDPI

Article

Methodology for Verification of Geometrically Complex Components Through Reverse Engineering

Peter Gabštur *, Marek Kočiško, Jakub Kaščak and Martin Pollák

Department of Computer Aided Manufacturing Technologies, Faculty of Manufacturing Technologies with a Seat in Presov, Technical University of Kosice, Bayerova 1, 08001 Presov, Slovakia; marek.kocisko@tuke.sk (M.K.); martin.pollak@tuke.sk (M.P.)

* Correspondence: peter.gabstur@tuke.sk

Abstract: Verification of geometrically complex components is a crucial step in ensuring the quality and functionality of industrial products. This article presents a comprehensive methodological approach to their verification using reverse engineering techniques. The study highlights the limitations of traditional measurement methods for complex geometries and proposes a framework that integrates advanced 3D scanning and digital modeling. The methodology involves obtaining accurate geometric data, with deviations during repeated measurements being well within the tolerance range. With a tolerance of ±1 mm, the maximum deviation was −0.66 mm, while the maximum deviation with the highest tolerance of 4.5 mm was 1.19 mm. These data were compared with predefined 3D models to analyze deviations and identify process errors. By applying this detailed methodology, more precise fixtures were created, resulting in measurements with deviations well below the required tolerances, compared to measurements without this methodology. A practical example of the design and verification of a welding fixture for cabin manufacturing illustrates the effectiveness of reverse engineering in detecting inconsistencies and optimizing manufacturing processes. The findings emphasize the importance of geometric accuracy in quality control and demonstrate how reverse engineering can streamline verification processes and enhance product reliability. This approach is particularly valuable for the manufacturing industry, which requires high precision in the production of geometrically complex components.

Keywords: reverse engineering; verification methodology; Kreon ACE-7-30; PolyWorks Inspector; welding; welding fixtures

1. Introduction

Currently, the verification of geometric components in industrial applications primarily relies on traditional measurement methods, such as coordinate measuring machines (CMMs) and manual measurements. While these techniques are effective for assessing basic geometric shapes, they encounter limitations when applied to complex geometries with minor deviations. Traditional methods can be time-consuming and prone to human error, ultimately affecting measurement accuracy and the overall quality of the components [1–4]. With advancements in reverse engineering technologies, more efficient and precise verification of complex components, including welded structures, has become possible. Reverse engineering involves analyzing an existing product or component to understand its design, function, and parameters. It is widely used in manufacturing, quality control, digitalization, and metrology. The process consists of digitizing an object through 3D scanning, processing the acquired data, reconstructing a CAD model, and validating its accuracy. Three-dimensional scanners enable the capture of highly detailed geometric data, facilitating the creation of digital models that can be compared against predefined CAD specifications. This approach provides precise measurements and an in-depth deviation analysis between scanned data and original models, allowing for the identification of potential manufacturing defects, welding-induced deformations, and optimization of production processes. Consequently, reverse engineering enhances the ability to swiftly detect inconsistencies, contributing to improved quality and reliability of welded components.

Similar challenges in component verification and the application of reverse engineering have been explored in previous research. The study conducted by Li. et al. examines a reverse engineering approach to facilitate the remanufacturing of worn mechanical components with uncontrolled failure conditions and complex geometries. The research outlines a methodology that includes 3D surface data acquisition, nominal model reconstruction, precise registration, defect extraction, tool path generation, and machining processes. A Prominent Cross-Section algorithm with curvature constraints is employed to identify damaged areas and exclude defective point clouds during reconstruction. Additionally, an improved Iterative Closest Point (ICP) algorithm integrates curvature and distance constraints to achieve optimal alignment between worn and nominal models. The study concludes with the successful remanufacturing of mechanical components [5].

R. Helle and his team also addressed this topic by examining the application of reverse-engineering-based 3D scanning technologies for quality control. Their study involved scanning a 3D-printed component with a complex geometry using a laser scanner. Two surface modeling techniques were tested: the primitive-based method, suitable for manufacturing, and NURBS modeling, which provided a more accurate surface representation. The research confirmed the suitability of 3D scanning for complex shapes while emphasizing the need to improve the efficiency and accuracy of reconstruction methods [6]. Furthermore, the study by Liu X. et al. focuses on the analysis of residual deformations occurring during the resistance spot welding (RSW) process on a basic structural panel (BSP), which was designed using reverse engineering and finite element analysis (FEA) simulations. Here, reverse engineering was employed to compare the original and final geometries of the panel to determine the overall residual deformation, highlighting the interaction of deformations across multiple welding processes [7].

P. Ruediger et al. investigated the use of visualization and machine learning for analyzing parameters influencing manufacturing quality. Their reverse-engineering-based approach identifies key parameters and their impact on parts using dimensionality reduction. The prototype system enables the determination of optimal settings for quality control, which the authors validated through a sheet metal deformation example [8].

Katsuhiro Saiga and colleagues present a sustainable approach to reverse engineering that minimizes the need for sophisticated technologies. Their five-step methodology—from geometric modeling to 3D printing—accelerates the process while maintaining accuracy. This approach supports sustainable product development and production optimization [9].

The reason for conducting this research was the need to address the issue of cabin deformations that occurred during the initial welding phase when the cabin was welded. This approach led to deformations caused by the weight of the structure and handling of components, which negatively affected the accuracy of the final products. Therefore, the welding process was divided into several parts, with the key factor being the dimensional stability of the smaller parts of the cabin, aimed at eliminating inaccuracies during assembly and ensuring the accuracy of the overall product.

This research addresses the limited accuracy of traditional measurement methods for inspecting complex components and the lack of standardized procedures for verifying welding fixtures using reverse engineering. While existing literature covers 3D scanning and digital reconstruction, it lacks a systematic methodology for welding processes to quantify accuracy and effectiveness. The study aims to develop a methodology for verifying welding fixtures with 3D scanning and deviation analysis, compare it with traditional methods, and identify key factors affecting measurement accuracy and quality control. This work fills a gap in the existing literature and highlights the practical benefits of reverse engineering in manufacturing optimization.

2. Materials and Methods

The research focused on designing a geometrically complex welding fixture intended for the assembly of a specific cabin structure (front panel). The primary objective was to develop a fixture that would ensure stability and precision during the welding of sheet metal components, minimizing geometric inaccuracies and deformations caused by heat input during the process. The methodological approach emphasized the detailed design of the fixture, which underwent thorough verification and analysis to ensure its effectiveness in real manufacturing conditions.

The fixture design was carried out using advanced CAD software tools, considering all factors affecting welding accuracy as well as the specific characteristics of individual cabin components. Following the conceptual design, a verification process was conducted, which included testing and measuring geometric deviations occurring during welding. This research thus aimed to optimize manufacturing processes and eliminate potential errors in geometric parameters, thereby enhancing the overall quality and reliability of the final product.

During the assembly of the cabin, which consists of welded complex components, geometric inaccuracies, and deformations arose due to heat input during welding, improper handling of semi-finished products, and material processing errors such as bending and curvature. These inaccuracies affected the final quality and functionality of the cabin, as deviations from dimensional accuracy requirements led to assembly issues or incorrect positioning of components. The solution to this problem was the design of an effective welding fixture that ensured stability and precision during welding, which was verified using reverse engineering. Until now, quality control has mainly relied on visual assessment and manual measurements, which do not provide sufficient accuracy for detecting minor yet critical geometric defects. This situation underscores the necessity for a new, more precise system for monitoring geometric errors during welding.

This research was based on a comprehensive methodology that incorporated several key steps and the use of specific technologies and tools to ensure the precision and quality of welding processes:

1. Identification of the primary issue, which was ensuring geometric accuracy during the welding of cabin sheet metal components.
2. Review of the current state of welding fixtures and verification of geometric parameters.
3. Consideration of the specific geometric shapes of the cabin and the design of a fixture tailored to these components.
4. Development of a geometrically complex welding fixture aimed at minimizing deformations and deviations of components during welding, utilizing Creo PTC software (version 11.0.0.0).
5. Design of a welding fixture prototype incorporating input parameters, including material selection (steel S235JR–1.0038) and the chosen welding technology (MAG).
6. Manufacturing the fixture with a focus on dimensional accuracy in positioning semi-finished products (locks, engraving).

7. Application of the 3D scanner Kreon ACE-7-30 (Kreon company, Limoges, France) and the PolyWorks Inspector Standard software (version 2022) for data collection, measurement of geometric deviations of welded components (fixture), and analysis against digital reference data. Measurement results were compared with predefined tolerances to assess the fixture's effectiveness in minimizing errors.
8. Evaluation of results in terms of deviation acceptability within established tolerances, allowing for an objective assessment of the fixture's efficiency.
9. Presentation of test results and evaluation of the design's performance.
10. Discussion of potential process improvements and methods for achieving more precise component positioning and optimization of the welding process.
11. Formulation of conclusions regarding the effectiveness of the proposed welding fixture in optimizing the production process. Recommendations for further improvements in fixture design and measurement technologies, which could enhance manufacturing accuracy in the future, along with continued research on additional components.

2.1. Identification of Critical Factors Affecting Accuracy

The dimensional accuracy of cabin weldments is a crucial factor influencing their functionality, assembly, and overall quality. Dimensional inconsistencies arise due to multiple critical factors that impact various stages of the manufacturing process. The most common causes include inconsistent quality of input semi-finished products, thermal deformations during welding, inaccuracies in welding fixtures, and improperly configured welding parameters. In the case of complex structures, these errors may become irreparable, meaning that inaccuracies in individual components can have a significant impact on the final product and lead to costly rework or rejection. Even minor geometric deviations can cause substantial issues during assembly, such as misalignment of parts, restricted functionality of moving components, or a reduction in the aesthetic quality of the final product [10,11].

In the production of complex structures such as cabins, even small dimensional and geometric deviations can accumulate over time, leading to significant overall defects and affecting the quality of the final product. These deviations can occur during various stages of production and material processing, significantly impacting the precision and proper functioning of the final assembly process.

Figure 1 illustrates a welded cabin structure where dimensional deviations arise due to several potential factors. One of the main causes is errors in the processing technology, such as laser cutting or material bending. Although these technologies are highly precise, they can sometimes cause minor material deformations, which later manifest as deviations in the final dimensions of the parts. Additionally, defects in the semi-finished products, such as waves, bumps, or incorrectly cut sections, may also cause deviations that accumulate and affect the quality of the final structure. A visible misalignment in the joint precision, which appears as an evident deviation between the joined parts, causes improper fitting of individual components. These misaligned joints later reflect in incorrect assembly, where components that should fit together do not match precisely, complicating the subsequent assembly phase. This leads to problems when joining individual cabin parts, which may result in irregular seating or improper connections between different parts of the structure. These issues often require additional adjustments or repairs, which can significantly extend production time and increase costs.

One of the primary challenges in geometrically complex components is thermal deformation, which occurs due to uneven heating and cooling of the material. This phenomenon causes shrinkage, displacement, and warping of parts, leading to deviations from the required dimensions. Another critical factor is the quality of semi-finished materials, where material inhomogeneity and residual stress contribute to deformations even in the initial phases of the process [12].

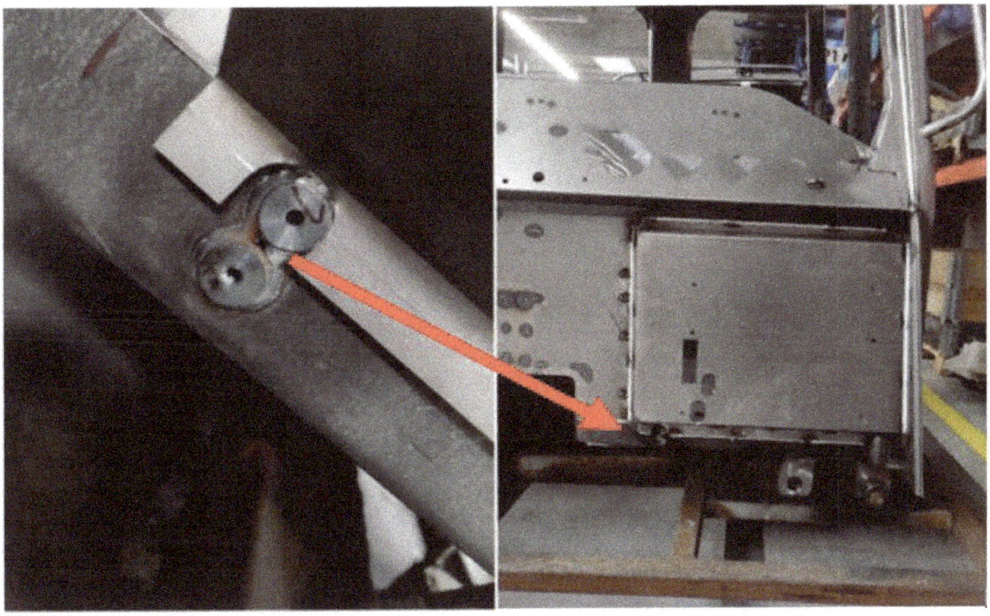

Figure 1. Shows dimensional and shape deviations in a cabin structure due to processing errors, material defects, and thermal deformations during welding, causing assembly issues.

Improperly designed or worn-out welding fixtures can negatively affect the correct positioning of components, increasing the likelihood of dimensional errors. Similarly, incorrect welding parameter settings—such as welding speed, current, or voltage—can result in poor-quality weld joints and uneven stress distribution within the material. To mitigate these issues, thorough process planning, optimization, and the implementation of appropriate control mechanisms are essential. The use of simulation tools allows for the prediction and minimization of thermal deformations already in the design phase. Additionally, regular quality control of input materials, calibration of fixtures, and personnel training play a crucial role. The proper identification and resolution of these critical factors are fundamental to achieving the required accuracy of cabin weldments and ensuring long-term reliability [13,14].

To verify the deficiency illustrated in Figure 1, a reverse engineering method was employed, providing a detailed and precise assessment of geometric deviations between the actual weldment and its digital 3D model. Following an initial visual inspection, which can reveal fundamental defects, the weldment was measured using advanced reverse engineering technology. These devices enable accurate geometry capture, allowing the obtained values to be compared against the original CAD model. In this study, a Kreon ACE-7–30 articulated-arm 3D scanner was selected, capable of detecting even the smallest geometric deviations with high precision, ensuring that any discrepancies affecting assembly quality and accuracy are identified [15,16].

The scanner was certified for a reach of 3000 mm, meaning it can perform measurements at distances up to this value with high precision. This certification is essential as it ensures that the device meets the quality and measurement reliability requirements set by the relevant standards and regulations. The verification results of its reliability can be seen in Table 1 [17].

Table 1. Technical specifications of the Kreon ACE-7-30 scanner.

Tests	Product Specification Testing Methodology	Specs	Measured [mm]	Result
Psize.Sph. 1 × 25:Tact.Aarm (±)	Maximum deviation in the measurement of sphere diameter (element measurement accuracy)	0.024	0.010	OK
Pform.Sph. 1 × 25:Tact.Aarm	Maximum measurement error in the determination of sphere shape (dispersion accuracy)	0.041	0.017	OK
Ldia. 5 × 5:Art:Tact.Aarm	Maximum deviation in sphere center measurement caused by probe orientation change	0.095	0.058	OK
Euni:0:Tact.Aarm (±)	Maximum permissible measurement error between two points (length measurement error)	0.079	0.069	OK
SPOT	Repeatability of point measurement	0.054	0.030	OK

This process of comparing the measured parameters with the 3D model provides detailed information on any deviations that may result from improper welding fixture settings, thermal deformations of the material, or insufficiently precise welding parameter adjustments. These deviations may manifest as dimensional errors, which could affect subsequent assembly or the functional integrity of the weldment within the assembly. By identifying and quantifying these deviations, it is possible to determine precisely which aspects of the manufacturing process require adjustment, whether it involves modifying welding parameters, optimizing fixtures, or refining production processes. Reverse engineering thus not only confirms the existence of deficiencies but also provides specific data to improve the manufacturing process and minimize errors [18,19].

To improve access to the measured surfaces of the cabin weldment and optimize scanning angles, an elevated platform, as shown in Figure 2, was used. This setup ensured better visibility and reduced obstacles during the measurement process. Positioning the weldment at a more accessible height allowed for easier and more precise scanning of complex areas. The elevated platform also helped minimize potential errors from poor scanning angles, ultimately improving the accuracy and reliability of the measurements. This approach contributed to more efficient data collection and enhanced the overall quality of the inspection process.

As shown in Figure 3, which displays the color deviation map, the inspection process involves selecting input values—specifically, the permissible tolerances based on the technical drawing documentation (in our case, ±6 mm). These values are represented in green on the map, indicating that the measured deviations fall within acceptable limits. Blue areas on the map denote negative deviations, meaning that the measured point is below the required value. Conversely, red areas indicate positive deviations, signifying that the measured point exceeds the specified dimensional requirements.

If the overall model exhibits deviations in red or blue areas that exceed the allowed tolerances, it indicates that the model is defective and requires adjustments or corrections in the manufacturing process to achieve the required precision and quality.

Figure 2. Illustration of the measured object and scanning equipment setup for optimal access to the surfaces of the cabin weldment.

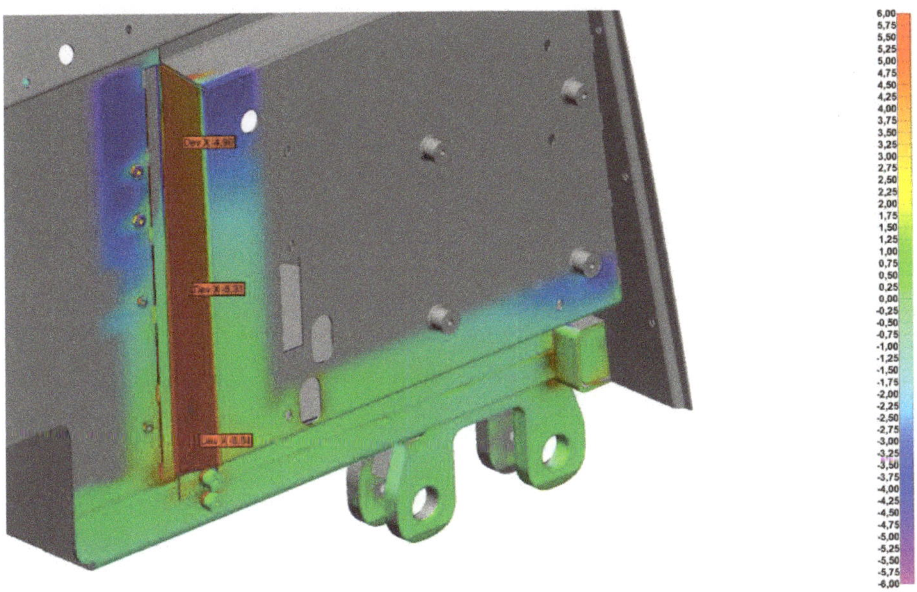

Figure 3. The graphical representation shows weldment deviations, with green indicating acceptable limits, blue for negative deviations, and red for exceedances, helping to identify areas needing adjustment.

2.2. Factors Influencing Geometric Accuracy and Quality

Deficiencies in technological processes can arise for various reasons, with the most common factors including inappropriate production parameters, material properties, the quality of input semi-finished products, and the influence of external environmental conditions. These errors can lead to a reduction in the quality of final products, affect their mechanical properties, and ultimately cause non-conformities in assembly processes. When processing metal components, particularly sheet metal parts, it is essential to precisely monitor each step of the manufacturing process to minimize deviations and defects that may occur during bending, cutting, or welding.

One of the key areas where defects occur is the welding process, as highlighted by the red circles in Figure 4. These defects can arise due to several factors, such as improper welding parameters (e.g., incorrect heat settings, inadequate welding speed, or unsuitable filler materials), insufficient control over the welding environment (e.g., drafts or fluctuating ambient temperature), or even operator errors. Such defects can result in issues like warping, cracking, or porosity in the weld, all of which compromise the structural integrity and dimensional accuracy of the welded components. These welding-related deficiencies often require rework or corrections, increasing production time and costs [20,21].

Figure 4. The front panel of the cabin showing thermal zones where welding-induced temperature variations cause dimensional deviations.

During welding, significant thermal changes take place in the material, leading to the formation of the heat-affected zone (HAZ). This zone is critical because its properties differ from those of the unwelded material due to localized heating and subsequent cooling. Incorrect welding parameter settings, such as welding current or welding speed, can cause undesirable structural changes in the material, resulting in cracking or warping [22].

In the detailed view in Figure 5, the heat-affected zone forms in different locations depending on the welding process, joint type, and applied parameters. This zone is situated near the weld metal itself, where significant microstructural changes occur due to thermal exposure. This phenomenon is particularly critical in thin sheet metals, where excessive heat input can lead to overheating and unwanted deformation after cooling [23].

In the processing of sheet metal components, errors may also arise due to incorrect bending, where improper machine settings or inappropriate tool selection lead to deviations in angles and dimensions. These errors can significantly affect subsequent welding operations, as improperly prepared parts may not fit together precisely, resulting in increased weld gaps and reduced joint strength [24].

Additionally, improper material handling can cause surface defects, which later impact the final quality of the welded joint. As shown in Figure 6, properly bent components are correctly positioned, ensuring precise alignment. The correct bend position guarantees perfect contact between parts, minimizing the occurrence of thermal deformations in the welding area and ensuring high weld quality.

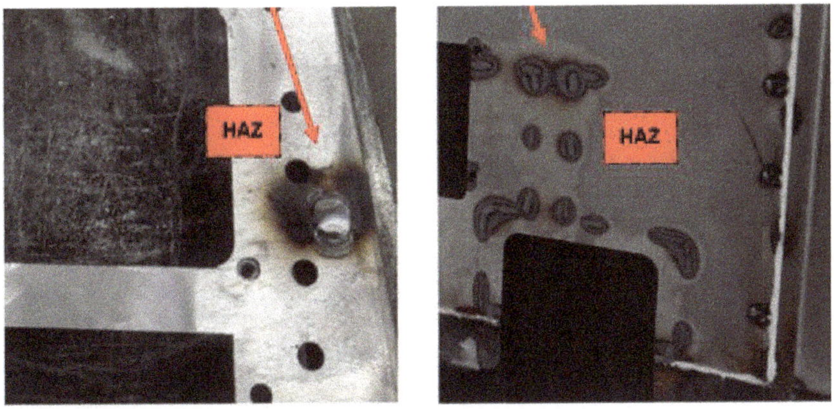

Figure 5. Heat-affected zones near the weld can alter material properties, impacting the structural integrity of the cabin weldment.

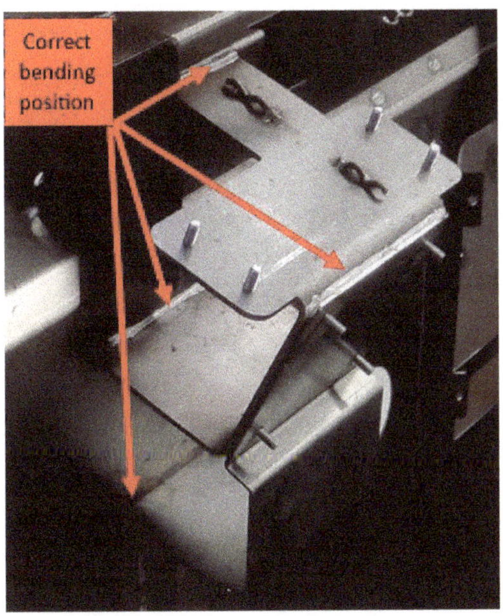

Figure 6. Detail of the correct bending and positioning of sheet metal components on the cabin.

2.3. Errors in the Design and Wear of Welding Fixtures and Their Impact on the Accuracy of Component Positioning

Welding fixtures are devices or tools used to hold and position parts during welding to ensure their correct orientation and alignment. The design of a welding fixture involves various steps and considerations to guarantee precise and efficient welding. Improperly designed or worn-out welding fixtures can cause significant issues affecting the overall quality and accuracy of welds. If the fixture does not provide adequate clamping of the components, they may experience undesired movement during welding. This can lead to displacement, rotation, or oscillation of the parts, increasing the probability of defects such as misalignment, deformations, or incorrect weld geometry.

A key aspect of fixture design is the proper placement of clamping points. Incorrect positioning can cause uneven material stress during welding, leading to distortions. For example, if the clamping points are placed too close together, excessive stress may be concentrated in certain areas, potentially causing positional shifts, which are critical for complex structures. Worn-out or damaged clamping mechanisms can result in improper component fixation, allowing movement during welding. If the contact surfaces of the fixture are worn or contaminated, stability and positioning may be compromised. Dirt and wear can reduce positioning accuracy and lead to instability during welding. Over prolonged use, geometric deviations or reduced stiffness may develop in the fixture. If a welding fixture consists of multiple components, ensuring proper assembly is crucial to keeping parts in the correct position throughout the welding process. To facilitate accurate component placement, the use of locks or engraved markings is essential, aiding in the proper alignment of individual fixture elements [25,26]. Minimizing welding processes and utilizing cold joints is an effective strategy to reduce deformations and enhance part accuracy, particularly for thinner materials. In thin sheet metal components, welding is more prone to thermal deformations, which can affect shape and geometric accuracy. On the other hand, welding thicker plates is generally easier, as the material withstands high temperatures better and results in stronger welds. However, thicker plates increase the overall weight of the fixture and make handling more challenging [27].

One effective approach to minimizing welding-induced deformations is to divide larger welded assemblies into smaller sections. This process offers several advantages. Smaller parts welded separately are less prone to deformation, as thermal effects act on a smaller volume of material. By gradually joining these parts into larger assemblies, geometric accuracy can be better controlled, preventing major shape alterations. This approach also simplifies fixture design, reducing both costs and setup time. Additionally, each subassembly can be individually inspected for quality and accuracy, allowing for the early detection and correction of potential issues [28].

An overview of the design factors of the welding fixture is presented in Table 2, which highlights the key categories and their respective elements. These factors are critical for ensuring the proper functioning, accuracy, and safety of the welding fixture, including aspects such as geometric and dimensional requirements, material selection, type of welding, fixture adjustability, and safety measures. Proper consideration of these factors during the design phase ensures the fixture meets the necessary technical specifications and performs reliably across different applications [29,30].

Table 2. Comprehensive view of the design factors of the welding fixture.

Category	Key Factors
Geometric and dimensional requirements	Shape determination (simple/complex), dimensional parameters (length, width, height, diameter)
Material of the welded part/fixture	Material type: steel, aluminum, stainless steel, etc.
Type of welding	Welding method: MIG/MAG, TIG, robotic, other types of welding
Type of weld	Joint type: corner, butt, spot, straight, etc.
Functional requirements	Welding position: horizontal, vertical
Orientation and fixation	Workpiece fixation method, stability, and alignment during welding (clamps, stops)
Technical parameters	Positioning accuracy: how precisely the part must be placed in the fixtureFixture adjustability: flexibility for different types of parts
Safety assurance	Protection against scratches and damage to the workpiece (rubber, covers)
Fixture geometry tolerance	Tolerance definition for geometric shapes of the fixture, such as flatness, perpendicularity, and parallelism of fixture elements
Fixture repeatability	Ensuring that the fixture provides consistent results with each use, with minimal deviations in positioning accuracy
Fixture dimensional tolerance	Defining dimensional tolerances of the fixture (e.g., length, width, height ±)

The prototype of the rotary welding fixture, shown in Figure 7, was designed based on the digital model of the front panel of the cabin (yellow model) and is engineered to ensure precise and stable fixation of components during the welding process. The parts are secured within the fixture using various clamping elements, such as mechanical clamps, toggle clamps, and stops, which keep the individual components in the required position throughout the entire welding operation. This fixture is equipped with a rotary mechanism that allows rotation around its axis, providing flexibility for welding from different angles. Rotational functionality eliminates the need for repositioning and excessive handling, thereby increasing the efficiency and accuracy of the welding process.

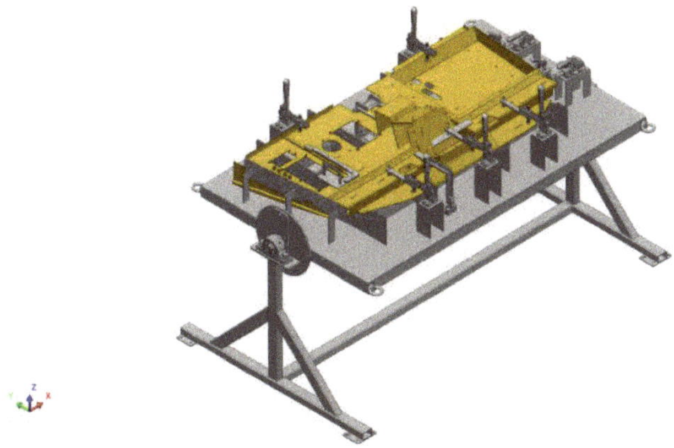

Figure 7. Prototype of the rotating welding fixture structure.

The fixture's structure was built using various types of profiles and sheet metal components, which were processed using laser cutting technology. This method ensures high precision and machining quality, which is essential for achieving the required technical parameters and geometric accuracy of the components. To guarantee the precise positioning of individual fixture elements, locks integrated into the base plate were implemented. These locks enable consistent and repeatable positioning, eliminating potential assembly errors. The design also included engraving (mechanical marking), which serves to identify positions or key reference points, facilitating easier orientation during fixture assembly and improving the overall efficiency of the work process. The combination of locks and engravings ensured high positioning accuracy while minimizing assembly deviations. For certain fastened components, such as clamps, a mechanical joint method was chosen. This approach was selected to minimize the risk of deformations that could occur during the welding process while also allowing for easy disassembly in case of a mechanism failure. The use of bolted joints enables simple replacement or repair of damaged parts without requiring complex technological interventions.

Finally, casters were mounted on the bottom section of the fixture, facilitating easier movement and handling during transportation.

2.4. Manufacturing of the Prototype and Processes Related to Welding, Assembly, and Surface Treatment

During the manufacturing of the prototype welding fixture, several key phases were followed to ensure the precision and functionality of the final product. To prevent potential bending errors and achieve the required accuracy in the production of the welding fixture, a specific methodology was chosen, focusing on precise processing and component assembly.

No materials requiring subsequent bending were used in the manufacturing process; instead, laser-cut parts were employed. This production method enabled high accuracy, minimizing geometric deviations and ensuring a quick alignment of individual components during assembly. The components that were later assembled into the welding fixture were designed with interlocking features, ensuring precise geometric accuracy during assembly while reducing the risk of assembly errors. These interlocks acted as positioning guides, ensuring the correct placement of components, thereby eliminating the risk of misalignment.

After welding, toggle clamps and positioning stops were added to the fixture to provide stability, correct positioning, and accuracy throughout the entire welding process. For quick assembly and potential adjustments after prolonged use, these elements were fastened using bolted joints. This fastening method was selected for its flexibility and precision, allowing for the fine-tuned positioning of individual components to achieve the stability and accuracy necessary for proper part alignment.

Furthermore, during the fixture design phase, particular emphasis was placed on minimizing thermal defects in the heat-affected zone (HAZ), which could later impact the quality of the weld. By ensuring precise and stable fixation of components, the risk of material deformation during welding—caused by improper clamping or excessive heat input—was effectively eliminated. As can be seen in Figure 8, this approach resulted in a high-quality and precise outcome, fully prepared for the subsequent production stages.

Figure 8. Manufactured prototype of the welding fixture.

During verification, sheets of various thicknesses, ranging from 4 to 5 mm, were used, and welding was carried out using the MAG (Metal Active Gas) method. During the process, it was necessary to minimize thermal effects that could lead to deformation or excessive internal stress.

The first step was to properly clamp the sheets to prevent movement during welding. Initially, tack welding was performed at regular intervals, which fixed the material in the desired position and reduced twisting during the later application of the main weld.

When welding larger parts or joining sheets of different thicknesses, we selected procedures that ensured an even distribution of heat. Instead of welding in a single line, we applied a backstep welding procedure, where welds were placed in the opposite direction to prevent heat from concentrating in one area. This prevented excessive elongation of the material in one direction and minimized twisting.

An important factor was also controlling the heat input. We used a welding wire with a diameter of 1.0 mm, suitable for sheet thicknesses in the range of 4 to 5 mm, and worked with a welding current of 140 to 180 A and a voltage of 18 to 22 V. The average wire feed speed was 6 to 9 m/min, ensuring a stable melting process and optimal penetration of the weld metal into the base material.

To protect the welding area from oxidation and contamination, a mixture of protective gases, CO_2 and Ar, was used, with a higher CO_2 content (approximately 18 to 25%) supporting good penetration of the weld metal and a stable electric arc. Short breaks were made between individual welds to allow the material to cool naturally and prevent excessive heat buildup.

Finally, the fixture underwent surface treatment, specifically coating and painting, to protect it from external influences such as corrosion, scratches, and chemical exposure. The paint coating not only improved the fixture's aesthetic appearance but also significantly increased its durability, which was essential for its long-term use in real industrial conditions.

2.5. Verification of the Welding Fixture Using the Kreon ACE-7-30 Scanner in PolyWorks Inspector

Before the welding fixture was introduced into serial production, a comprehensive verification process was conducted to ensure its accuracy and functionality. This step was essential to confirm that the fixture correctly positions individual components during manufacturing and meets the precision and tolerance requirements specified in the technical drawings. The primary objective was to identify any deviations that could affect manufacturing quality, particularly concerning locking mechanisms, clamps, and other key fixture components. To achieve this, reverse engineering techniques were employed, allowing for a detailed geometric analysis of the fixture. In this case, reverse engineering was used to verify whether the shape and dimensions of the fixture corresponded to the original CAD models. This process involved 3D scanning, which provided a digital representation of the fixture's actual state. Subsequent analysis helped identify potential inaccuracies in the fixture's construction, which could cause issues in serial production. This approach effectively mitigated risks such as incorrect part positioning or misalignment, which could lead to increased production costs, reduced product quality, or even fixture failure in operation.

During the scanning process, environmental factors that could affect measurement accuracy were considered. The lighting was stabilized to minimize its impact on scanning accuracy, as different lighting conditions can influence the quality of the captured data. Additionally, the surface properties of the object were considered during measurement. Surfaces with high gloss or reflectivity can cause inaccuracies, so specific scanner settings were applied to compensate for these factors and improve the scanning quality. The use of a calibrated scanner also played a crucial role.

To ensure the reliability and repeatability of the scanning process, the measurement was conducted in the same environment, where external influences such as temperature changes and vibrations were minimized. A stable table was used for the measurement, ensuring stability and eliminating any external movements that could affect accuracy. The repeatability of the measurement was verified by performing multiple scans of the same object, and the results were compared with reference values.

The entire verification process of the welding fixture was carried out using PolyWorks Inspector software, version 2022 (Figure 9), which is equipped with advanced tools for 3D scanning and geometric deviation analysis. This software is ideal for detecting inaccuracies and comparing the actual condition of the fixture with the specifications outlined in the technical documentation.

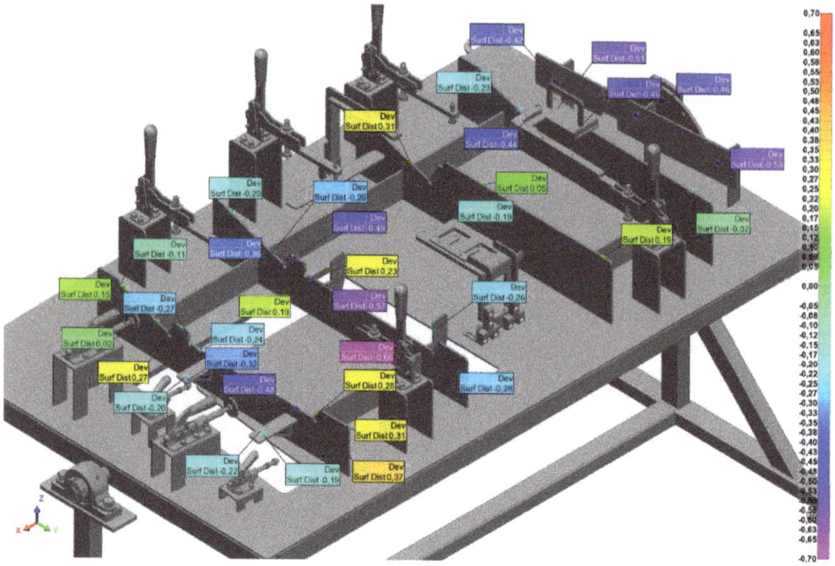

Figure 9. Imported 3D model of the fixture with a tolerance color map showing deviations from nominal dimensions and a color scale.

Figure 9 illustrates the selection of all analyzed parts of the fixture, with clearly marked areas where deviations or critical geometric features were identified. The analysis results are visualized using tolerance color maps, allowing for a quick interpretation of the differences between the actual and nominal models. Green represents values within the nominal dimensions, meaning no significant deviations. Yellow and red indicate positive values, where the actual dimension exceeds the nominal value. Blue and purple signal negative values, where the actual dimension is smaller than the nominal value.

Detailed step-by-step description of the verification process of the welding fixture using the Kreon ACE-7-30 Scanner in PolyWorks Inspector:

(a) Preliminary inspection—before starting the verification process, a thorough inspection of the physical condition of the welding fixture was conducted. All fastenings, locks, and screws securing the correct positioning of the parts were checked to prevent any unwanted movement during scanning.

(b) Scanner setup and calibration—the Kreon ACE-7-30 scanner was configured and calibrated to achieve the required scanning accuracy. This scanner collects data in the form of 3D point clouds, and its high precision makes it ideal for measuring geometric deviations.

(c) Scanner and software integration—once the scanner was set up, it was connected to a computer running PolyWorks Inspector. This software was used to process 3D data and compare them with the CAD model of the fixture.

(d) CAD model import—the CAD model of the fixture, created in Creo PTC 11.0.0, was imported into PolyWorks Inspector. This model contained all the geometric tolerance requirements, component positioning data, and dimensional specifications.

(e) Data acquisition process—the Kreon ACE-7-30 scanner was placed on the fixture, and 3D point data were collected from all critical areas, including locking points, fastenings, and other key structural details.

(f) Sequential scanning procedure—the scanning process was performed in multiple steps, focusing on each section of the fixture to ensure a complete geometric record.

Each scanned point was added to the overall point cloud, forming a comprehensive digital representation of the fixture.

(g) Deviation analysis in PolyWorks Inspector—the PolyWorks Inspector software calculated deviations between the actual state of the fixture (obtained from the scanner) and the requirements in the CAD model. These deviations were visualized using color-coded maps, highlighting areas that exceeded tolerance limits.

(h) Evaluation of measurement results—once the results were displayed in PolyWorks Inspector, areas where deviations occurred were analyzed in detail. These deviations were assessed to determine whether they were within acceptable tolerances or if they could potentially cause problems in serial production. Each deviation was categorized based on its magnitude and severity, allowing for precise identification of which parts of the fixture required adjustments or modifications. This analysis provided a comprehensive overview of the fixture's condition.

(i) Final adjustments and reporting—based on the verification data, necessary adjustments to the fixture were made as required. After a successful verification, a detailed report was generated containing all analysis results, detected deviations, implemented modifications, and the final state of the fixture.

The comprehensive analysis was crucial in ensuring that the welding fixture met all the initial project requirements and was fully prepared for efficient and seamless integration into serial production. By using the Kreon ACE-7-30 3D scanner and PolyWorks Inspector software, the verification method allowed for a detailed analysis of the fixture's geometry, ensuring that the welding fixture met the precise specifications and tolerances required for optimal performance in a production environment. The final report generated after the verification process served as a detailed documentation of the fixture's condition, changes made, and verification results. This report not only confirmed that the fixture met all design specifications but also provided valuable information for future audits and continuous improvement of the manufacturing process. It demonstrated a commitment to accuracy and quality, which is in line with industry standards, and helped build confidence in the fixture's performance in the production environment.

In summary, this step-by-step verification process ensured that the welding fixture was fully optimized and ready for serial production, with the certainty that it would consistently provide the required accuracy and performance within the manufacturing process. The use of advanced 3D scanning added an additional layer of quality assurance, which directly contributed to overall efficiency.

3. Results

In terms of the acceptability of deviations within the specified tolerances for verifying the effectiveness of the designed fixture, it was assessed whether the fixture met all accuracy and functionality requirements outlined in the technical drawings and technological specifications. After performing 3D measurements and analyzing the fixture's geometry, the actual values were compared with the CAD model specifications.

As shown in Figure 10, all obtained results were subjected to a detailed analysis, during which deviations between the actual and expected conditions of the fixture were evaluated. Based on this analysis, the extent of these deviations was determined, and it was assessed whether they fell within the predefined tolerances set during the fixture development. These tolerances were essential for ensuring the proper functionality of the fixture and guaranteeing the quality of the final products. Figure 10 shows the measurement of flatness in nine planes, with tolerances specified in the range of 1 mm to 4.5 mm in each plane, as indicated in the technical drawing documentation. This allowed for an accurate evaluation of whether the fixture met the requirements for geometric accuracy.

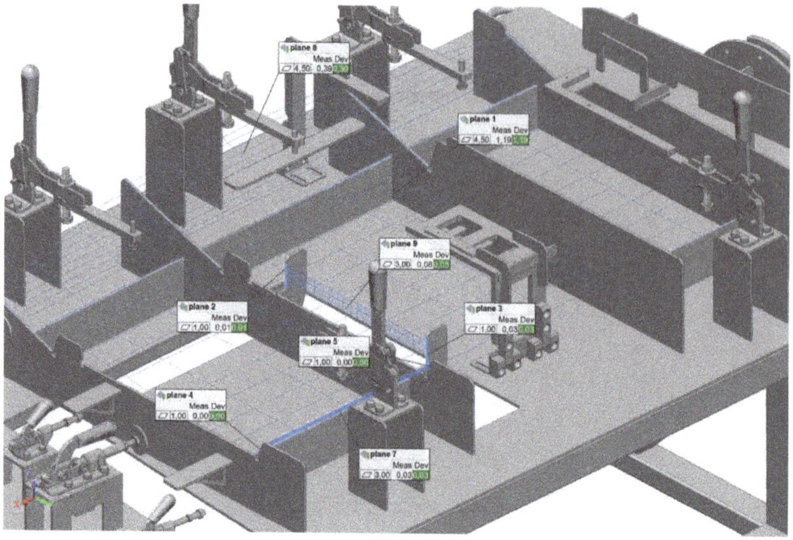

Figure 10. Illustration of the verification process of the fixture's flatness.

By utilizing scanning or CAD models, it was possible to obtain complete surface representations of the object and subsequently compare them with reference values. The PolyWorks software allowed the definition of entire surfaces and their accuracy analysis in comparison with reference CAD models or ideal surfaces. If specific inspection of a certain surface area was needed, as shown in Figure 11, it was possible to select and define only that section, allowing for more detailed analysis in areas where a higher risk of errors or issues was anticipated. Additionally, it was possible to select the weldment display for better illustration and apply the elements of the fixture that are movable or adjustable during the process.

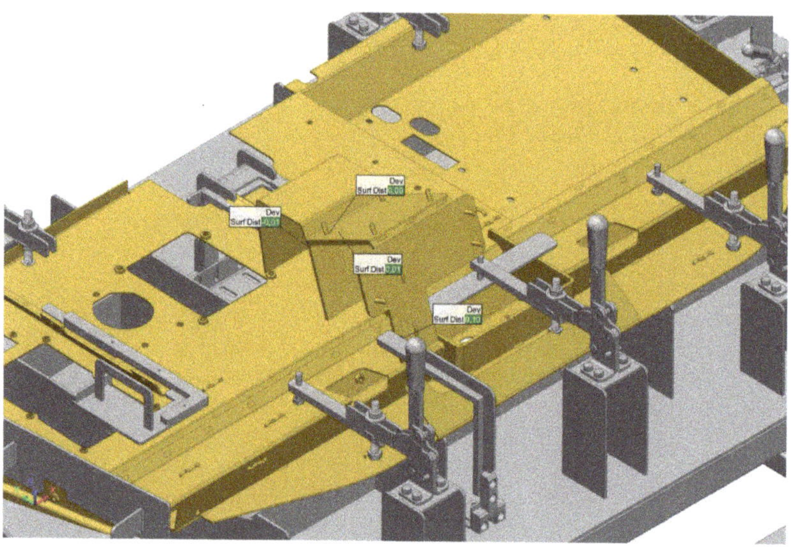

Figure 11. Verification of the geometric parameters of the welding fixture, including the weldment display and application to selected parts.

In Figure 12, a sample of 13 points was selected within the tolerance range of ±0.7 mm for better representativeness of the results. After evaluating these points, it was determined that all measured values fell within the required tolerances, with no significant deviations detected that could negatively affect the effectiveness of the fixture or the quality of the final product. This result confirms that the designed fixture is fully functional, meets all specified technical parameters, and was manufactured with high precision. These findings also confirm that the fixture is ready for implementation in serial production without the risk of quality issues.

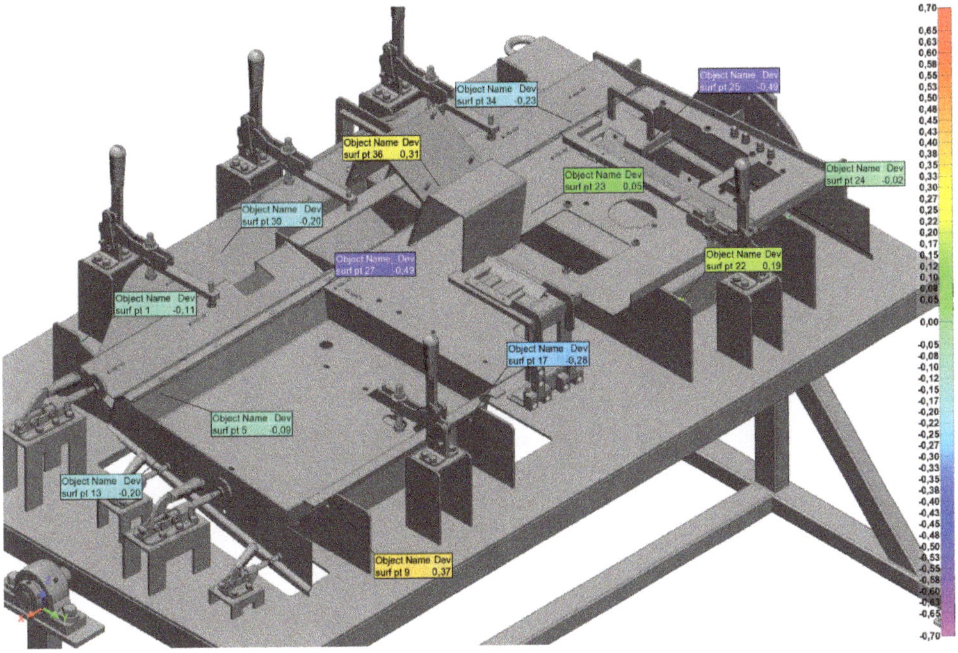

Figure 12. Illustration of a selected representative sample from the total surface area of the fixture.

Reverse engineering played a crucial role in the verification of the fixture by enabling precise analysis of the physically manufactured fixture and its comparison with the original CAD model. The geometric data obtained through this process allowed for a detailed assessment of deviations between the actual and expected state.

4. Discussion

When verifying the suitability of the welding fixture and its subsequent application in serial production, verification plays a key role. Traditional measurement methods, such as dimension measurement using micrometers or other measuring instruments, may be insufficient for accurately assessing all geometric deviations that can affect part positioning and welding accuracy. In this process, it is essential to consider not only basic dimensional tolerances but also more complex factors, such as proper component positioning, their mutual alignment, and potential deformations during welding. Reverse engineering combined with 3D scanning enabled a detailed analysis of the fixture and an evaluation of all these factors.

The results obtained from PolyWorks Inspector confirmed that deviations from the CAD model were within acceptable tolerances, which would have been difficult to assess using conventional measuring tools.

Table 3, referencing Figure 10, displays the results of flatness measurements of the fixture in individual planes. For each plane, the measured values, required tolerances, and actual deviations are provided. Based on these data, all measurements are within the specified tolerances (marked as "Pass"), confirming that the fixture meets the requirements for flatness and geometric accuracy.

Table 3. Flatness measurement results.

Name	Control (Flatness)	Measurement [mm]	Tolerances [mm]	Deviation [mm]	Test
Plane 1	⏥	1.186	4.500	1.186	Pass
Plane 2	⏥	0.010	1.000	0.010	Pass
Plane 3	⏥	0.030	1.000	0.030	Pass
Plane 4	⏥	0.002	1.000	0.002	Pass
Plane 5	⏥	0.002	1.000	0.002	Pass
Plane 7	⏥	0.035	3.000	0.035	Pass
Plane 8	⏥	0.390	4.500	0.390	Pass
Plane 9	⏥	0.075	3.000	0.075	Pass

Table 4 presents the results of surface distance measurements during the verification of the welding fixture. Each surface was measured to assess flatness and deviation from the nominal value in comparison with the tolerance range of ±1.000 mm. All measured surfaces, including Surf 1, Surf 5, Surf 9, and others, showed deviations within the permissible tolerances, indicated by the "Pass" status, confirming that the surface distances of the fixture met the requirements for implementation into production.

Table 4. Surface distance measurement results.

Name	Control (Flatness)	Measurement [mm]	Tolerances [mm]	Deviation [mm]	Test
Surf 1	Surface Distance	−0.109	±1.000	−0.109	Pass
Surf 5	Surface Distance	−0.092	±1.000	−0.092	Pass
Surf 9	Surface Distance	0.373	±1.000	0.373	Pass
Surf 13	Surface Distance	−0.202	±1.000	−0.202	Pass
Surf 17	Surface Distance	−0.282	±1.000	−0.282	Pass
Surf 22	Surface Distance	0.187	±1.000	0.187	Pass
Surf 23	Surface Distance	0.048	±1.000	0.048	Pass
Surf 24	Surface Distance	−0.024	±1.000	−0.024	Pass
Surf 25	Surface Distance	−0.486	±1.000	−0.486	Pass
Surf 27	Surface Distance	−0.487	±1.000	−0.487	Pass
Surf 30	Surface Distance	−0.203	±1.000	−0.203	Pass
Surf 34	Surface Distance	−0.232	±1.000	−0.232	Pass
Surf 36	Surface Distance	0.315	±1.000	0.315	Pass

Without these advanced methods, it would be extremely challenging to determine whether the fixture is truly suitable for serial production and whether it ensures the required quality of welded components. Given the accuracy and functionality demands of the manufacturing process, verification using reverse engineering is essential for ensuring the required quality and eliminating the risk of errors that could affect the entire production process.

Based on the verification and testing of the welding fixture, although the results meet the required tolerances, there are several areas where the process could be further improved and optimized. The first step is to enhance the accuracy of component positioning. Automation of positioning systems, for example, through robotic arms or automatic clamping mechanisms, could improve the speed and precision of part placement, thereby reducing the risk of human errors.

Another possible improvement is to enhance the mechanical and thermal properties of the fixture. The use of materials with better thermal properties or greater resistance to deformation could help prevent thermal stresses during welding, increasing the fixture's precision and lifespan. However, this approach would involve higher financial costs.

Advanced simulation tools could predict the behavior of materials during welding, allowing for the optimization of component positioning and clamping even before physical production begins. This would reduce the need for physical testing and increase efficiency. Implementing online monitoring systems during welding would enable real-time parameter adjustments, ensuring that the welding process remains within optimal conditions, thereby minimizing potential errors.

5. Conclusions

Reverse engineering has proven to be a highly effective tool in optimizing manufacturing processes, especially in reducing geometric deviations and improving the quality of final products. The use of 3D scanning to collect detailed geometric data allowed for the identification of inconsistencies in design and manufacturing during the early stages of production. This approach enabled the timely implementation of necessary adjustments, leading to significant improvements in the precision of manufacturing processes.

Unlike traditional measurement methods, which can be time-consuming and prone to human error, reverse engineering offers a faster, more accurate, and automated alternative. Three-dimensional scanning allows for the acquisition of complex geometric data, which can be immediately analyzed using specialized software, eliminating the need for manual measurements and reducing the risk of errors. As a result, not only is accuracy improved, but the overall manufacturing process becomes more efficient, minimizing the need for rework and reducing the costs of rework and claims.

The research focused on developing a methodology for verifying geometric components through 3D scanning and deviation analysis, contributing to improved geometric accuracy and product quality. The proposed methodology, implemented in real-world manufacturing processes, allowed for the early detection of inconsistencies and errors, achieving higher compliance with quality requirements and minimizing the need for rework, thus reducing repair costs. The aim was to enhance the reliability and quality of products during the early stages of production, which had a positive impact on the overall efficiency of the manufacturing process. Tests showed that the implementation of this methodology led to a measurable reduction in geometric and welding errors, as well as an increase in the quality of the final products, particularly in the case of welding fixtures and other critical components. The proposed methodology combined theoretical approaches with practical experiments, emphasizing the iterative process of fixture efficiency.

Measurement results indicate that we were able to approach zero deviation. With a tolerance of ± 1 mm, the lowest deviation was almost zero, specifically 0.002 mm, which demonstrates the effectiveness of the applied methodology. With the highest tolerance of ± 4.5 mm, the maximum deviation was 1.19 mm, and with a tolerance of ± 1 mm, we also recorded a deviation of -0.66 mm. This progress is the result of a precise process of verifying geometric components through 3D scanning and deviation analysis.

However, it is important to note that although we aim to ensure consistent conditions during repeated measurements, these processes may be influenced to some extent by various factors, such as small variations in material properties, environmental conditions, or technical settings during individual measurements. These factors may slightly affect the results, although the methodology overall showed significant improvement in accuracy and quality. Despite these potential influences, we still achieve values that are below the requirements set for production, significantly reducing the need for rework and repair costs.

The results suggest that further improvements in the design of welding fixtures are needed to achieve even better performance. Future efforts should focus on more precise adaptation of fixtures to different material properties and changes that occur during the manufacturing process. However, these materials may be more expensive and require additional modifications in production. Such a step could be advantageous in applications where the highest precision is required, but economic considerations and manufacturing capabilities will need to be considered.

A crucial step in process optimization is improving measurement technologies and integrating them with IoT elements directly into welding fixtures. Integrating intelligent sensors and measuring devices that allow real-time parameter adjustments could significantly improve the stability and quality of the manufacturing process. These technologies would enable continuous monitoring of key parameters such as temperature fluctuations, material stress, and deviations from the desired geometry, thus minimizing the likelihood of errors and increasing the overall reliability of the process.

In conclusion, it is also important to note that the method brought measurable improvements in the geometric accuracy of products, with deviations well below the set tolerances, confirming the effectiveness of this methodology in real-world conditions. The implementation of the methodology contributed to higher compliance with quality requirements, which is particularly important in the production of critical components such as welding fixtures. Monitoring and analyzing geometric deviations using 3D scanning allowed for high accuracy in the early stages of production, eliminating the need for further interventions and reducing repair costs.

Although the results showed a positive impact on the quality and reliability of products, further research is recommended, focusing on the broader use of advanced analytical methods and simulations, as well as testing the methodology with different materials and manufacturing conditions. This approach could bring even better results in optimizing welding processes and further reducing geometric deviations.

The results of applying the methodology to other parts of the cabin confirmed its effectiveness in other areas of production, opening the possibility for its broader application in various industrial sectors and processes. This provided a valuable foundation for future research and the implementation of technology that could significantly improve the quality and efficiency of production in various fields.

Author Contributions: Conceptualization, P.G. and J.K.; methodology, P.G.; software, P.G.; validation, P.G., J.K. and M.K.; formal analysis, J.K.; investigation, M.P.; resources, P.G.; data curation, J.K.; writing—original draft preparation, P.G.; writing—review and editing, J.K.; visualization, P.G.; supervision, J.K.; project administration, M.K.; funding acquisition, M.K. and M.P All authors have read and agreed to the published version of the manuscript.

Funding: This research was funded by project KEGA 002TUKE-4/2023 and VEGA 1/0121/23, granted by the Ministry of Education, Research, Development and Youth of the Slovak Republic, and by the EU NextGenerationEU through the Recovery and Resilience Plan for Slovakia under the project No. 09I03-03-V05-00015, and No. 09I05-03-V02-00042.

Institutional Review Board Statement: Not applicable.

Informed Consent Statement: Not applicable.

Data Availability Statement: The original contributions presented in this study are included in the article; further inquiries can be directed to the corresponding author.

Conflicts of Interest: The authors declare no conflicts of interest.

Abbreviation

The following abbreviations are used in this manuscript:

CMM	Coordinate Measuring Machine
ICP	Iterative Closest Point
NURBS	Non-Uniform Rational B-Splines
RSW	Resistance Spot Welding
BSP	Base Sheet Panel
FEA	Finite Element Analysis
HAZ	Heat Affected Zone

References

1. Mizera, O.; Čepová, L.; Sadílek, M.; Čep, R.; Hrubý, R.; Zelinka, J. The Problems of Measuring Selected Geometric Deviations on a CMM After Machining. In *Advances in Manufacturing II: Volume 5-Metrology and Measurement Systems*; Springer International Publishing: Cham, Switzerland, 2019; pp. 158–171. [CrossRef]
2. Pan, F.; Nie, L.; Bai, Y.; Wang, X.; Wu, X. Geometric errors measurement for coordinate measuring machines. In *IOP Conference Series: Earth and Environmental Science*; IOP Publishing: Bristol, UK, 2017; Volume 81, p. 012117. [CrossRef]
3. Page, D.; Koschan, A.; Abidi, M. Methodologies and techniques for reverse engineering—The potential for automation with 3-d laser scanners. In *Reverse Engineering: An Industrial Perspective*; Springer: London, UK, 2008; pp. 11–32. [CrossRef]
4. Swornowski, P.J. A critical look at the coordinate measuring technique. *Mechatronics* **2013**, *23*, 80–93. [CrossRef]
5. Li, L.; Li, C.; Tang, Y.; Du, Y. An integrated approach of reverse engineering aided remanufacturing process for worn components. *Robot. Comput.-Integr. Manuf.* **2017**, *48*, 39–50. [CrossRef]
6. Helle, R.H.; Lemu, H.G. A case study on use of 3D scanning for reverse engineering and quality control. *Mater. Today Proc.* **2021**, *45*, 5255–5262. [CrossRef]
7. Liu, X.; Wei, Y.; Wu, H.; Zhang, T. Factor analysis of deformation in resistance spot welding of complex steel sheets based on reverse engineering technology and direct finite element analysis. *J. Manuf. Process.* **2020**, *57*, 72–90. [CrossRef]
8. Ruediger, P.; Claus, F.; Hamann, B.; Hagen, H.; Leitte, H. Combining visual analytics and machine learning for reverse engineering in assembly quality control. *Electron. Imaging* **2020**, *33*, jist0956. [CrossRef]
9. Saiga, K.; Ullah, A.S.; Kubo, A. A sustainable reverse engineering process. *Procedia CIRP* **2021**, *98*, 517–522. [CrossRef]
10. Guo, Z.; Liu, X.; Rao, X.; Yuan, C. Effect of defects on the structural integrity of ship piping welds under simulated piping conditions. *Ocean. Eng.* **2024**, *308*, 118372. [CrossRef]
11. Serrati, D.S.; Machado, M.A.; Oliveira, J.P.; Santos, T.G. Non-destructive testing inspection for metal components produced using wire and arc additive manufacturing. *Metals* **2023**, *13*, 648. [CrossRef]
12. Shravan, C.; Radhika, N.N.; Deepak Kumar, N.H.; Sivasailam, B. A review on welding techniques: Properties, characterisations and engineering applications. *Adv. Mater. Process. Technol.* **2023**, *10*, 1126–1181. [CrossRef]
13. Naksri, C.; Chuchom, S.; Chaiprapat, S. Design of Fixture for Welding. In *IOP Conference Series: Materials Science and Engineering*; IOP Publishing: Bristol, UK, 2021; Volume 1163, p. 012007. [CrossRef]
14. Arnaoutis, V.; Rosić, B.V.; Lutters, E. Parametric design evolution for production setups; a case study for welding fixtures. *Procedia CIRP* **2023**, *120*, 1522–1527. [CrossRef]
15. Javaid, M.; Haleem, A.; Kumar, L. Dimensional errors during scanning of product using 3D scanner. In *Advances in Engineering Design: Select Proceedings of FLAME, Uttar Pradesh, Noida, India 3 October 2018*; Springer: Singapore, 2019; pp. 727–736. [CrossRef]
16. Haleem, A.; Javaid, M.; Singh, R.P.; Rab, S.; Suman, R.; Kumar, L.; Khan, I.H. Exploring the potential of 3D scanning in Industry 4.0: An overview. *Int. J. Cogn. Comput. Eng.* **2022**, *3*, 161–171. [CrossRef]
17. ACE Measuring Arm Range Precisely for You, n.d. Available online: https://www.kreon3d.com/products/ace-measuring-arm (accessed on 7 February 2025).
18. Jędrych, M.; Gorzkiewicz, D.; Deja, M.; Chodnicki, M. Application of 3D scanning and computer simulation techniques to assess the shape accuracy of welded components. *Int. J. Adv. Manuf. Technol.* **2024**, 1–9. [CrossRef]
19. Hegedűs-Kuti, J.; Szőlősi, J.; Varga, D.; Abonyi, J.; Andó, M.; Ruppert, T. 3d scanner-based identification of welding defects—Clustering the results of point cloud alignment. *Sensors* **2023**, *23*, 2503. [CrossRef] [PubMed]
20. Meng, W.; Li, Z.; Huang, J.; Wu, Y.; Chen, J.; Katayama, S. The influence of various factors on the geometric profile of laser lap welded T-joints. *Int. J. Adv. Manuf. Technol.* **2014**, *74*, 1625–1636. [CrossRef]
21. Liu, S.; Wu, Z.; Liu, H.; Zhou, H.; Deng, K.; Wang, C.; Liu, L.; Li, E. Optimization of welding parameters on welding distortion and stress in S690 high-strength steel thin-plate structures. *J. Mater. Res. Technol.* **2023**, *25*, 382–397. [CrossRef]
22. Neuvonen, R.; Peltoniemi, T.; Skriko, T.; Ghafouri, M.; Amraei, M.; Ahola, A.; Björk, T. The effects of geometric and metallurgical constraints on ultra-high strength steel weldments. *Structures* **2023**, *56*, 104900. [CrossRef]

23. Singh, I.J.; Murtaza, Q. A comprehensive review on effect of cold metal transfer welding parameters on dissimilar and similar metal welding. *J. Eng. Res.* **2023**. [CrossRef]
24. Ikumapayi, O.M.; Akinlabi, E.T.; Madushele, N.; Fatoba, S.O. A brief overview of bending operation in sheet metal forming. In *Advances in Manufacturing Engineering: Selected Articles from ICMMPE 2020*; Springer: Berlin/Heidelberg, Germany, 2019; pp. 149–159. [CrossRef]
25. Seloane, W.T.; Mpofu, K.; Ramatsetse, B.I.; Modungwa, D.J.P.C. Conceptual design of intelligent reconfigurable welding fixture for rail car manufacturing industry. *Procedia CIRP* **2020**, *91*, 583–593. [CrossRef]
26. Kadam, G.; Nagarajan, B.; Mirji, V.; Sawant, S. Design of Rigid-Locking Variable-Angular-Span Welding Fixture. *Eng. Proc.* **2024**, *66*, 35. [CrossRef]
27. Kang, S.; Yun, W.; Kim, H.; Kim, J.; Ji, C.; Lee, K.; Kim, J.; Jang, H.-L.; Chun, K. A Study on Minimizing Welding Deformation of Joints for the Sealing of Emission After-Treatment Structure. *Materials* **2021**, *14*, 6982. [CrossRef]
28. Liu, S.; Wu, Z.; Zhou, W.; Zhou, H.; Zhang, K.; Yin, D.; Lei, Y.; Qiu, Y. A review of welding simulation methods for large components. *Prog. Nat. Sci. Mater. Int.* **2023**, *33*, 551–568. [CrossRef]
29. Jha, N.K. Design and analysis of welding fixture for elementary weld joints. *CVR J. Sci. Technol.* **2018**, *15*, 90–95. [CrossRef]
30. Viňáš, J.; Brezinová, J.; Guzanová, A.; Svetlík, J. Degradation of renovation layers deposited on continuous steel casting rollers by submerged arc welding. *Proc. Inst. Mech. Eng. Part B J. Eng. Manuf.* **2013**, *227*, 1841–1848. [CrossRef]

applied sciences

MDPI

Article

Controlling Product Properties in Forming Processes Using Reinforcement Learning—An Application to V-Die Bending

Ciarán-Victor Veitenheimer *, Dirk Alexander Molitor, Viktor Arne and Peter Groche

Institute for Production Engineering and Forming Machines, Technical University of Darmstadt, Otto-Berndt-Straße 2, 64287 Darmstadt, Germany
* Correspondence: ciaran.veitenheimer@ptu.tu-darmstadt.de

Abstract: Uncertainty is unavoidable in forming processes due to fluctuating properties in the semi-finished product, the tool system and the environment. For this reason, numerous scientists have addressed this issue by developing control approaches like self-optimizing machine tools or the control of product properties. Machine learning algorithms, in particular reinforcement learning (RL) methods, show promising results for controlling production processes in this way. In this paper, the application of RL is demonstrated on an industrially commonly used process, V-die bending. For this purpose, first a flexible tool system is developed that allows the bending angle to be adjusted continuously between 80 and 110°. The developed tool is initially simulated through an FEM model in order to create a sufficient database for the training of an RL agent for springback compensation. The pre-trained agent is then used to control the springback in the real process. To close the resulting sim-to-real gap, it is then retrained on the experimentally generated data. It is shown that the springback can be significantly reduced compared to the uncontrolled case in both the simulative and the experimental process.

Keywords: die bending; property control; reinforcement learning; metal forming; closed-loop control

1. Introduction

Uncertainty is omnipresent in forming processes and is reflected in fluctuating product properties. Fluctuating influences from the semi-finished product, the tool, the machine and the environment prevent reliable manufacturing of components. For this reason, scientists have been researching approaches that contribute to the control of uncertainty for several years. Self-optimizing machine tools [1] and closed-loop control of product properties [2] are seen as promising higher-level approaches. Both approaches have in common that information about product properties of previous production steps is used to manipulate the product properties of subsequent workpieces. For this purpose, models are often used that draw on relationships between the control variables of the process and the product properties. In most cases, the relationships have high complexities, interdependencies and non-linearities, cannot be modelled by analytical models or are unsuitable for control due to long computing times [3]. Therefore, machine learning algorithms are increasingly used that allow a data-driven description of the relationship and whose model parameters are fitted in such a way that manually parameterized cost functions are minimized.

Reinforcement Learning (RL) algorithms are particularly suitable for deriving optimal control variables in forming processes. Using the example of a Finite-Element Method (FEM) deep drawing simulation, Dornheim et al. [4] show that at discrete time steps, the blankholder force can be optimized with respect to sheet thinning, cracking and stress peaks. Idzik et al. [5] show that coupling a RL algorithm with an analytical model in a rolling process leads to pass schedules that optimize both the mechanical properties of the components and the energy efficiency of the process. A similar approach is presented by Reinisch et al. [6] in a multi-stage open-die forging process, where the RL algorithm is rewarded for low geometry deviations of the components as well as for a high utilization of the available press force. Deng et al. [7] demonstrate the suitability of RL algorithms for controlling the sheet thickness in sheet metal strip rolling by adjusting the roll forces and tilts, resulting in reductions in sheet thickness variations.

The first applications of RL algorithms for the design and control of forming processes can be found in the literature. However, their application is largely limited to simulative investigations. This is due to implementation barriers that make it difficult to transfer policies generated from simulation data to real processes. Reasons for this include sim-to-real gaps [8] that often cannot be closed due to a lack of availability of experimental data and latencies between reference variables and manipulated variables that are not simulated [9]. Thus, the full potential of RL for forming process has not yet been exploited. Therefore, in this paper, we present an approach for how RL agents trained on FEM simulation data can be transferred to experimental operation and contribute to the control of product properties. As an example process, a flexible die bending process is used that allows the production of different bending angles and materials. On the one hand, the process is modelled using an FEM simulation, which is used to generate synthetic data, and on the other hand, a flexible die bending tool is designed and manufactured to carry out the experimental tests.

The paper is organized as follows. Section 2 introduces the basic principles of the die bending process and RL algorithms, which are fundamental for the understanding of the paper. Section 3 presents the methodology for RL-based control of the die bending process. The subject of the section is the setup of the FEM simulation, the experimental setup and the procedure for the application and training of the applied RL algorithm. Section 4.1 first presents simulative results on the performance of the RL algorithm under different levels of uncertainty and then its performance in the real process. The paper concludes with a summary of the results in Section 5.

2. State of the Art

2.1. Die Bending

Die bending is a bending process with a straight tool movement where the tool system consists of a punch and a die. Depending on the tool shape, a distinction is made between V and U die bending. It is often used industrially due to its increased robustness, compared to other bending processes. As long as the bending part has not established full contact with the punch and die, it is referred to as free bending. As soon as the workpiece establishes form closure in the tool, i.e., the bending angle is equal to the die angle, free bending is complete and the coining process begins. This is characterized by a sharp increase in the process force (F_P) which is required to press the workpiece into the die. This significantly increases the dimensional accuracy of the finished component and makes precision die bending the most accurate bending process. An exact calculation to quantify the punch force ($F_{P,req}$) required for coining is not known from the literature and can be almost arbitrarily large. However, it can be estimated approximately, e.g., according to [10] using Equation (1). The die angle is represented by α, r_p indicates the punch radius, b describes the sheet width and s_0 the initial sheet thickness of the semi-finished product. Frictional influences are taken into account by the friction coefficient μ, the tensile strength R_m, 0.2%, yield strength $R_{p0.2}$ and uniform elongation A_g represent material parameters.

$$F_{P,req} = \frac{bs_0}{2}\left\{\frac{R_m - R_{p0.2}}{3A_g}\left(\frac{s_0}{r_P + \frac{s_0}{2}}\right)^2 + R_{p0.2}\left(\frac{s_0}{r_P + \frac{s_0}{2}}\right)\right\}\frac{\cos\frac{\alpha}{2} + \mu\sin\frac{\alpha}{2}}{\frac{\alpha\pi}{2\cdot 180}} \quad (1)$$

Although die bending results in higher dimensional accuracy compared to free bending processes, it usually requires a corresponding tool for each bending angle. In order to reach high accuracies, the process parameters usually have to be determined empirically. If process conditions fluctuate, such as varying materials or sheet thicknesses, this process often has to be repeated individually. A large number of tools, lengthy process setup and high forces can quickly result in high process costs. Moreover, it is often not possible to react appropriately to the uncertainty in forming processes, induced by the semi-finished product, tool, machine and environment, due to the rigidity of the process.

2.1.1. Springback

Dimensional inaccuracies occur during bending in the form of springback, which is caused by the state of stress in the bent component.

Figure 1 schematically shows two characteristic stress curves in the sheet metal during a die bending process. The dotted line represents the neutral fiber, which is characterized by a stress-free state. In the overlying (underlying) area of the sheet metal, compressive stresses (tensile stresses) prevail, which are highlighted in red (blue) in Figure 1. These stresses increase in magnitude with the distance from the neutral fiber. This becomes clear in the effective stresses during the bending process (blue line). The elastic parts of the bending can regress after the component has been unloaded and the residual stresses (red line) thus cause a springback in the component. In order to be able to clearly distinguish between the different bending angles, they are defined as follows:

- $\alpha_{des}^{(unl)}$: Desired bending angle after unloading the component;
- $\alpha_{act}^{(unl)}$: Actual bending angle after unloading the component;
- $\alpha_{des}^{(loa)}$: Desired bending angle with loaded component (in process);
- $\alpha_{act}^{(loa)}$: Actual bending angle with loaded component (in process).

Figure 2 shows this graphically. The schematically depicted probability density functions imply that variables describing the actual state are subject to uncertainty. The springback of the workpiece can thus be defined as $\Delta\alpha = \alpha_{act}^{(unl)} - \alpha_{act}^{(loa)}$.

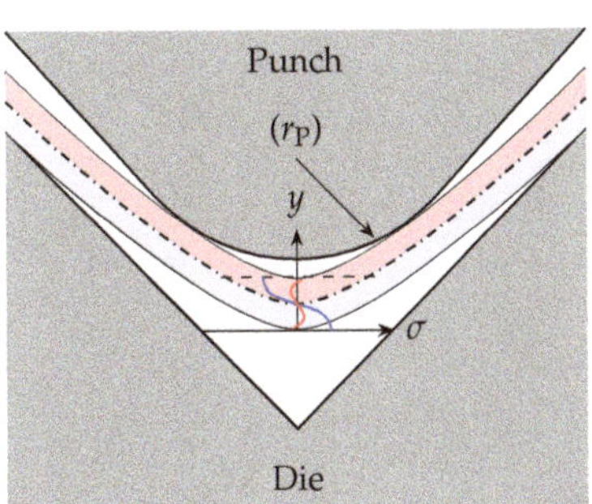

Figure 1. (Residual) Stresses.

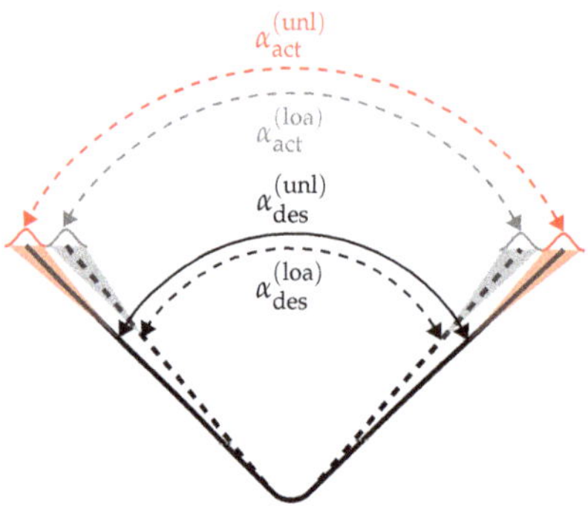

Figure 2. Stresses and bending angles in flexible die bending.

2.1.2. Control Approaches

There are numerous examples in the literature of how to compensate for springback during sheet metal bending. Numerical or experimental tests are usually carried out to investigate the springback for specific combinations of material, sheet thickness and bending radius. However, compensation usually refers to (iterative) design adjustments to the tool system in order to match the desired product geometry. References [11,12] both propose approaches based on the displacement adjustment method. Here, the shape deviation is iteratively reduced based on the displacement of simulation nodes and can therefore be time consuming. Due to the higher flexibility of air bending, multiple approaches can be found in the literature for controlling the springback inline. One possible approach is incremental bending. Reference [13] presents press brake bending, where the punch stroke is divided into incremental steps, after which the workpiece is unloaded multiple times. From the multiple collected pairs of loaded and unloaded bending angles, the actual thickness and material properties of the workpiece are predicted based on an analytical model. This information is then used to predict the springback using the plane strain theory. Similarly, reference [14] uses a semi-analytical process simulation in combination with online measurements to compensate for springback during air bending In order to compensate for the effects of fluctuating material properties on a bending angle, reference [15] uses force signals from an adjacent forming stage of a progressive die for material characterization. It has been demonstrated that a LASSO regression may serve to reduce the deviation by 24%. More recent studies focus on the use of artificial neural networks for springback prediction. Reference [16] shows an example where neural networks are trained with FEM simulations to predict springback. The predictions of the network align well and thus represent an inline-capable alternative to FEM. Reference [17] developed a tool path planning strategy based on RL and supervised learning. The efficacy of this approach is evaluated through a case study, which involves an FEM simulation of a rubber tool forming process. It has been demonstrated that a substantial reduction in geometry deviations can be achieved. Reference [18] presents a method for predicting springback in air bending of high-strength sheet metal using a neural network. The springback is predicted by the model based on process parameters like sheet thickness, punch radius and material coefficients. The model was then validated by numerical and experimental tests. Reference [19] develops a neural network control system along with a stepped binder force trajectory to control the springback in a steel channel forming process. In order to demonstrate the generalizability of the model, various configurations of material and lubrication conditions that had not been used for training were employed. Although significant improvements in accuracy were achieved, the study also highlights the need for more advanced sensing technologies and the implementation of real-time closed-loop control to further enhance process stability and performance. Reference [20] addresses this issue by presenting an online springback compensation model for an air bending process. The study demonstrates that in-line control of the bending process can be achieved by integrating camera-based bending angle detection with neural networks and the press controller. The results show that the control strategy significantly improves the accuracy of the achieved bending angle.

2.2. *Reinforcement Learning*

Like supervised and unsupervised learning, RL is a machine learning class and a subfield of artificial intelligence. It describes sequential decision making problems that an agent learns through feedback. Since the agent is initialized with no information about the environment and the influence of actions, it has to collect information via trial and error. At each time step, the agent chooses an action $a_t \in \mathcal{A}$ and subsequently receives the new state of the environment $s_{t+1} \in \mathcal{S}$ along with a scalar reward signal $r_{t+1} \in \mathcal{R}$. Based on this observation, the agent adapts his action to maximize the reward. It is therefore a promising approach to control (partially) unknown systems. A visual representation of the described structure is given in Figure 3. The theoretical foundations of RL and decision making are described in the following sections. Like supervised and unsupervised learning, RL is a machine learning class and a subfield of artificial intelligence. It describes sequential decision-making problems that an agent learns through feedback. Since the agent is initialized with no information about the environment and the influence of actions, it has to collect information via trial and error. At each time step, the agent chooses an action $a_t \in \mathcal{A}$ and subsequently receives the new state of the environment $s_{t+1} \in \mathcal{S}$ along with a scalar reward signal $r_{t+1} \in \mathcal{R}$. Based on this observation, the agent adapts his action to maximize the reward. It is therefore a promising approach to control (partially) unknown systems. A visual representation of the described structure is given in Figure 3. The theoretical foundations of RL and decision making are described in the following sections.

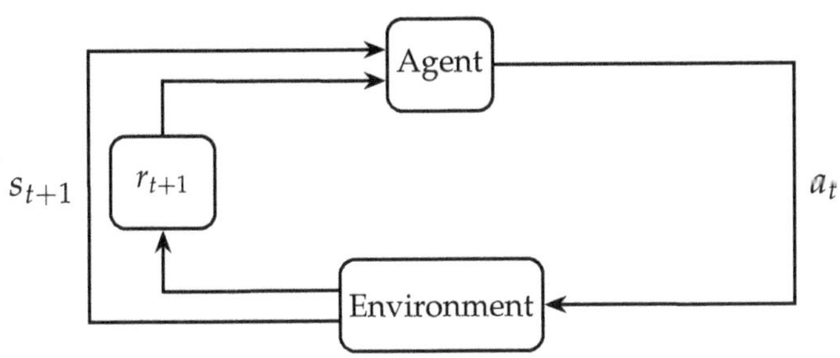

Figure 3. Structure of reinforcement learning.

2.2.1. Markov Decision Processes

A Markov Decision Process (MDP) is a framework used to model optimization problems and forms the basis of RL. The state of an MDP is defined by the fact that the probability of occurrence of a future state is independent of past conditions as long as the current state is known. This can be defined mathematically as follows:

$$\mathbb{P}[\mathcal{S}_{t+1} \mid \mathcal{S}_t] = \mathbb{P}[\mathcal{S}_{t+1} \mid \mathcal{S}_1, \ldots, \mathcal{S}_t] \tag{2}$$

$\mathbb{P}$ describes the state transition probability matrix, whereas $\mathcal{S}$ represents the states of the environment. If $\mathbb{P}$ is known, it is fully described through a mathematical model. If this model is used to learn an optimal policy to fulfill a task, RL is considered as model-based.

On the contrary, the term model-free RL is used if $\mathbb{P}$ is not known. The approaches presented in this paper are model-free.

2.2.2. Optimal Control

The behavior of an agent is determined by its policy π, which is either deterministic (Equation (3)) or stochastic (Equation (4)). The policy maps each state to a single action (deterministic) or multiple actions with different probabilities (stochastic).

$$\pi(s) = a \tag{3}$$

$$\pi(a|s) = \mathbb{P}[a_t = a|s_t = s] \tag{4}$$

The return J describes the cumulative, discounted reward of an MDP. Given a policy π, the state-value function $V^{\pi}(s)$ represents the expected return J starting from state s and following policy π. It therefore evaluates how good it is for the agent to be in a certain state. It is formally described as:

$$V^{\pi}(s) = \sum_{a\in\mathcal{A}} \pi(a|s)\left(\mathcal{R}(s,a) + \gamma \sum_{s'\in\mathcal{S}} \mathbf{P}(s'|s,a)V^{\pi}(s')\right) \tag{5}$$

Similar to the state-value function, an action-value function can be formulated:

$$Q^{\pi}(s,a) = \mathcal{R}(s,a) + \gamma \sum_{s'\in\mathcal{S}} \mathbf{P}(s'|s,a) \sum_{a'\in\mathcal{A}} \pi(a'|s')Q^{\pi}(s',a') \tag{6}$$

As the state-value function, it describes the expected return J, starting from state s and following strategy π, while also considering action a. In order to solve the RL-task, the agent has to maximize the return and therefore find the optimal policy π^*.

$$\pi^* = \arg\max_{a\in\mathcal{A}} Q^*(s,a) \tag{7}$$

This can either be achieved by directly optimizing the policy or by learning the state-value function Q. One possible approach of policy optimization is to use the policy gradient method. The policy is parameterized with θ in order to be able to adjust the policy iteratively. This is equivalent to the optimization of the parameterized expected return J_θ. Policy Gradient (PG) methods use gradient ascent to maximize the objective J_θ in the direction of the gradient $\nabla_\theta J_\theta$. The update rule for the parameters θ is then carried out as follows, where α denotes the learning rate.

$$\theta_{t+1} = \theta_t + \alpha\nabla_\theta J_\theta \tag{8}$$

As mentioned, another approach to solving the RL problem is to learn the action-value function. In environments with discrete action and state spaces, this task can, e.g., be performed with tables. The complexity increases drastically in continuous or higher-dimensional state and action spaces. Q-Learning is then usually performed with neural networks. Those value estimation approaches are in general computational more efficient and therefore easier to implement. But these are often less suitable for large and continuous spaces. Policy estimation approaches, on the other hand, usually deliver good results in high-dimensional spaces with continuous action selection. Still, they are difficult to implement because of their underlying intractable mathematics.

2.2.3. Actor-Critic Algorithms

Actor-Critic (AC) algorithms represent a class of RL algorithms that combine elements of policy optimization and value estimation. As presented by [21], their main advantage lies in efficiently computing policies and value functions, leading to more stable learning processes. AC networks typically consist of two main components: an actor network and a critic network. The actor network is responsible for learning the policy, while the critic network estimates the value function. By iteratively improving the policy based on feedback from the value estimator, AC networks enable effective exploration and exploitation in the environment. The flexibility and adaptability of neural networks allow the application of AC networks in a variety of RL scenarios, both in continuous and discrete action spaces. These networks also support asynchronous learning, which potentially increases learning speed. AC networks therefore provide a robust method for solving complex RL problems as they combine the benefits of value estimation and policy optimization. The Soft Actor-Critic (SAC) algorithm used in this work is a variant of AC proposed by [22] which additionally incorporates entropy maximization as a regularization term during policy optimization. This encourages exploration by discouraging premature convergence to suboptimal policies and can therefore lead to a more efficient and stable learning process.

3. Methodology

In order to compensate for the springback, the component must be overbent to the angle $\alpha_{\mathrm{des}}^{(\mathrm{loa})}$. As previously mentioned, the angle necessary for springback compensation depends on many process parameters and must be determined by the process controller. In this paper, it is intended to control the springback of three different materials in different thicknesses (see Table 1). Thus, a tool design that allows for continuous adjustment of the angle is necessary. Subsequently, a product controller needs to be developed that can predict the springback of different combinations of material, sheet thickness and target angle $\alpha_{\mathrm{des}}^{(\mathrm{unl})}$. The use of RL models as product controllers poses challenges in terms of data generation due to the associated data inefficiency. For this reason, a FEM-model of the process is also being developed which can be used to pretrain RL agents. Differences to the real process (sim-to-real gap) should then be closed by retraining on experimental data.

Table 1. Material and sheet thickness combinations.

Sheet Thickness [mm]	0.5	0.75	1
DC01	x	x	x
EN AW 6082T6	x	-	x
Copper	x	-	x

3.1. Design of Variable Die Bending Tool

In order to make a V-die bending process variable regarding its bending angle and thus allow the angle to be controlled, additional Degrees Of Freedom (DOF) need to be introduced in the tooling system. The concept developed for a flexible die is shown in Figure 4 in the form of a principle sketch. It shows that the die is divided into two halves, maintaining a common pivot point (MD). It has a vertical DOF and is guided horizontally in the die center. It is furthermore supported against the floor by a spring of stiffness c_{d}. Moreover, each die half is guided at its other end through a floating bearing. The angle β_{g}, in which the floating bearing is arranged, ensures that the die angle (α_{d}) is reduced when the center point is lowered ($x_{\mathrm{d}} > 0$). The die angle can therefore be set depending on the vertical displacement of MD.

Apart from the die, the punch is designed to be flexible, too. Figure 5 shows the principle sketch of a punch with variable punch angle. As with the die, the punch is divided into two halves with a common pivot point (MP). The ends of the punch halves (C) are connected to a floating bearing (D) via springs. It must also be ensured that D is always vertically aligned above C. The floating bearing allows for the compensation of the horizontal offset that occurs when the punch angle is reduced. Similar to the die, springs (c_P) ensure that the appropriate coining forces can be applied for varying punch angles. Viewed individually, the punch has two DOF (x_{P1}, x_{P2}), as the two halves of the punch can move independently. In combination with the die, these DOF are reduced to the displacement of the die center point x_d and thus the stroke height.

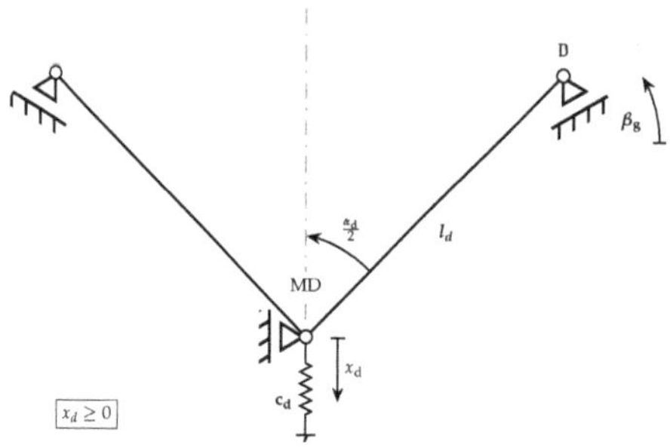

Figure 4. Principle sketch of the die.

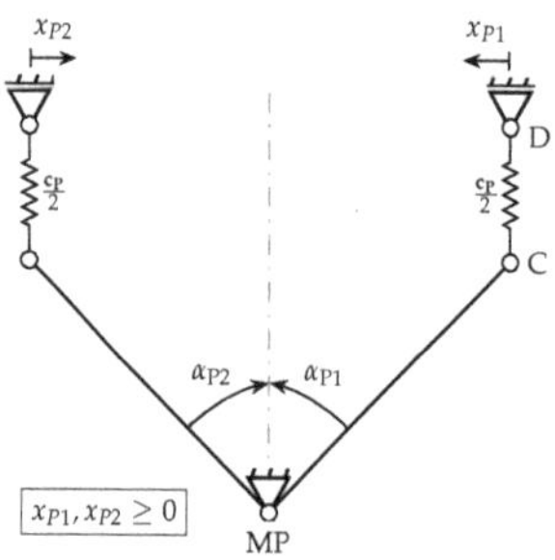

Figure 5. Principle sketch of the punch.

The tooling system is dimensioned in a way that sheet materials with a length of 160 mm and a width of 50 mm can be processed. The corresponding materials and sheet thicknesses are shown in Table 1. It should also be possible to realize angles in a range from 80 to 110°. The required force for bending is estimated according to Equation (1) for each material and sheet thickness combination. The maximum estimated stamping force is then used to select the springs.

$$F_{P,max} = \max\left(F_{P,req_i}\right), i \in \mathbf{m} \tag{9}$$

The vector **m** denotes all possible combinations of material and sheet thickness. By integrating the additional DOF into the tool, the product property, i.e., the bending angle, can be controlled. Derived from [2], a cascaded controller for springback compensation is shown in Figure 6. The RL model is referred to as RL-based product-controller in the offline control loop and represents a SAC RL agent. The observable state contains the target angle $\alpha_{des}^{(unl)}$ and vector **m**. The material vector **m** includes the elastic modulus, the Poisson ratio and the tensile strength of the selected material and sheet thickness. Based on the inputs, the agent selects a compensated angle $\alpha_{des}^{(loa)}$, which is forwarded to the online controller. The online control ensures that machine-side and environmental influences such as gear backlash and actuator inaccuracies can be eliminated. This is achieved by a camera-based inline measurement of the bending angle (50 Hz sampling frequency) and a direct feedback of the measured bending angle into the machine control (PLC). Reference [20] provides a more detailed description of the similar measurement system and information flows in an air bending process. In this work, the camera (acA1600-60hm, Basler AG, Ahrensburg, Germany) of the measuring system is mounted on the moving part of the die so that no distortions occur due to perspective shifts. Subsequent to the execution of the forming operation, the actual unloaded bending angle is transmitted to the product controller. Thereafter, a scalar reward is calculated in accordance with Equation (10). As the agent tries to maximize the reward, the deviation $\Delta\alpha_{\text{ctrl}}$ is minimized.

$$r = 10 - \Delta\alpha_{\text{ctrl}} \text{ with } \Delta\alpha_{\text{ctrl}} = \alpha_{\text{des}}^{(\text{unl})} - \alpha_{\text{act}}^{(\text{unl})} \tag{10}$$

Due to the data inefficiency of RL methods, the variable die bending process is also represented numerically in the form of an FEM simulation. In Figure 6, the interaction of the simulation with the product property controller is represented by dashed lines. Due to the deterministic nature of a simulation, the online control of $\alpha_{\text{des}}^{(\text{loa})}$ is not necessary. The numerical mapping of the process otherwise has the same input and output variables as the real process. Inputs are the material vector and the target angle $\alpha_{\text{des}}^{(\text{unl})}$. The product property controller can therefore be used to control the springback in the simulation and in the real process. To increase efficiency, the RL-based controller should first interact with the simulation and then be transferred to the real process. Differences between the real process and the simulation are to be overcome by transfer learning.

A more detailed overview of the development process of the RL controller is shown in Figure 7. It can be seen that the basic structure of the RL structure remains the same between the simulation and real process and that the use of knowledge from simulations is enabled by retraining the agent on real process data. As previously mentioned, the state $s^{\text{Sim/Exp}}$ consists of the vector $\mathbf{m}$ and the observed unloaded angle $\alpha_{\text{act}}^{(\text{unl})}$. Based on this, the compensated target angle $\alpha_{\text{des}}^{(\text{loa})}$ is predicted and used as input for the simulation or the process controller. The selected hyperparameters are presented in Table 2 and were determined through a grid search, with the explored parameter space also detailed in the table.

Table 2. Overview of selected hyperparameters and search space for grid search.

	Learning Rate	Batchsize	ϵ	τ
Selected Hyperparameters	1×10^{-2}	128	0.05	0.1
Search Space	1×10^{-1}–1×10^{-4}	8–256	0.01–0.2	0.01–0.2

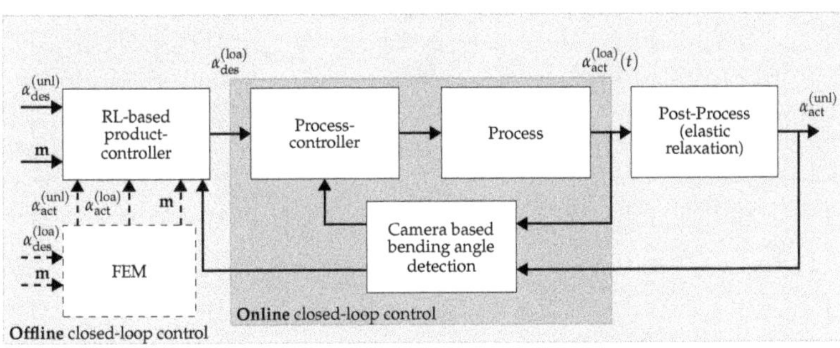

Figure 6. Block diagram of the property closed-loop controlled die bending process.

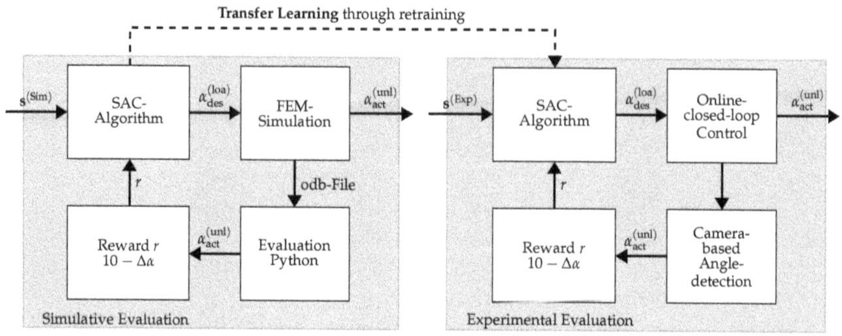

Figure 7. Structure of the AI-supported product controller when applied to the die bending process.

3.2. Finite Element Simulation

The simulation model is set up using the software SIMULIA Abaqus CAE/2023 as a two-dimensional implicit model. All tool parts are defined as rigid bodies based on the described principle (Figures 4 and 5). A plane strain state is assumed for the sheet metal since the width of the sheet is significantly larger than the thickness of the sheet [13,23]. Due to the shorter calculation time, an x-symmetry is also applied to the center of the die. The time sequence of the simulation process is divided into three steps. In the first step, all individual parts are fixed and the sheet metal is placed in the center of the die. In the second step, all springs are preloaded. In the third step, forming takes place by moving the punch downwards. The displacement of the punch required to achieve the target angle $\alpha_{des}^{(unl)}$ is calculated using the trigonometric relationship:

$$s_P(\alpha_{des}^{(loa)}) = \left(\cos\left(\frac{\alpha_{des}^{(loa)}}{2}\right) + \cos\left(\frac{\alpha_{max}}{2}\right)\right) \cdot l_d + s_0 \tag{11}$$

It consists of the cosine of the target angle in the process and the angle of the die at the uppermost point ($x_d = 0$), the length of a die half and the sheet thickness. A constant friction coefficient of $\mu = 0.1$ is applied to model the contact between the tool and the sheet metal, which is a widely accepted approximation for metallic contacts in metal forming simulations without detailed surface characterization (see [24]). After the third step, the deformed component is loaded into a new model in its final state of the previous simulation and the applied loads are removed. This causes the springback to form and the unloaded bending angle ($\alpha_{act}^{(unl)}$) can be measured. Figure 8 shows the described process and the simulation model schematically. In order to map the influence of uncertainty occurring in the semi-finished product, multiple data sets are created in which uncertainty is applied to two material parameters (v, E). An overview of the magnitude of the maximum uncertainty applied and the sample size of the different data sets are shown in Table 3. The applied uncertainty is chosen stochastically (uniformly distributed) at the start of every simulation. The data sets are labeled as follows: no uncertainty (nu), low uncertainty level (lu), medium uncertainty level (mu) and high uncertainty level (hu).

(a) (b) (c) (d) (e) (f)

Figure 8. Schematic sequence of the simulative forming process with springback for a target angle of 80°and DC01 with a sheet thickness of 1 mm: (**a**) Initial state (**b**) Preloading springs (**c**) Begin of forming operation (**d**) Completed forming operation (**e**) Simulation of springback (**f**) Final geometry.

Table 3. Applied uncertainties of the four data sets.

Uncertainty	nu	lu	mu	hu
δE	±0	±2500	±5000	±10,000
δv	±0	±0.01	±0.02	±0.04
N	3094	1725	3310	2069

3.3. Experimental Setup

To validate the concept experimentally, a tool is designed based on the principle sketches shown above. The real tool system, consisting of die and punch, is shown in Figure 9. It is used in combination with the prototype of the 3D Servo Press [25], where the die is mounted on the press table and the punch is centered above it on the ram. The press is controlled in the task space using model-based control concepts presented by [26]. The available, additional DOF of the 3D Servo Press are not used in this case such that the tilting DOFs are locked and only vertical displacements of the tool are performed. The guiding elements shown in the principle sketches were implemented in the tool using linear guides, plain bearings and slots. In order to apply the spring force evenly and without jamming to the die, the individual spring shown in the principle is replaced by a parallel connection of four springs. The arrangement of the springs in the punch remains as shown in the principle sketch (Figure 5). A camera can be attached to the moving parts of the die, so it is moved along with it. This prevents the measured bending angle from being influenced by perspective distortions induced by a relative displacement between the camera and the part. After setting up the angle detection, which is based on an edge detection algorithm and the live image of the camera, it is therefore sufficient to calibrate the detected angle just once. Further information about the camera setup can be found in [20].

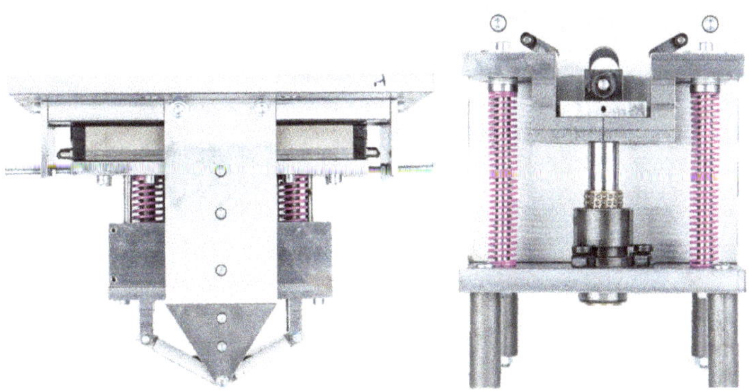

(a) (b)

Figure 9. Die bending tool consisting of flexible punch (**a**) and die (**b**).

4. Results

In the following, the springback behavior of the different materials and sheet thicknesses of the simulative and real process is shown. This is followed by a discussion of springback using the control system presented.

4.1. Simulative Results

The springback behavior of the simulated process is shown in Figure 10. All figures show the occurring springback $\Delta\alpha$ for the different configurations (**m**) plotted against the target angle ($\alpha_{act}^{(loa)}$). The top row refers to data set nu (no uncertainty) and the bottom row to data set mu (medium uncertainty).

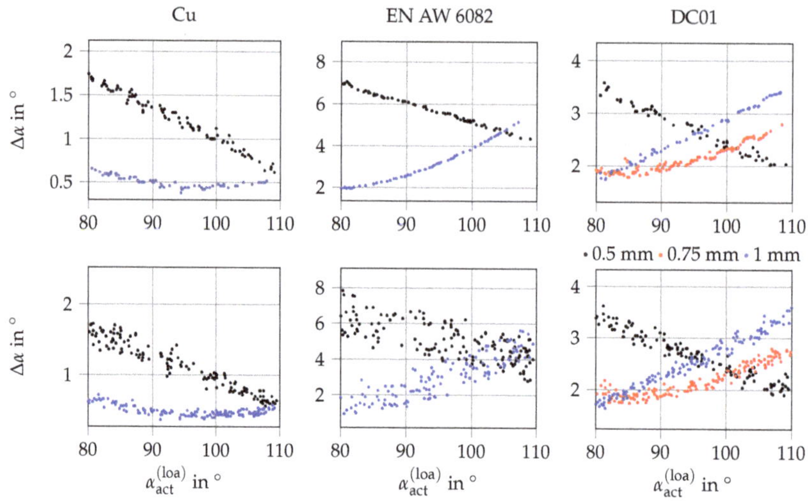

Figure 10. Simulated springback behavior with constant material parameters (dataset nu, **top**) and material parameters subject to uncertainty (dataset mu, **bottom**).

It can be clearly seen that the highest springback occurs in the aluminum sheets with a thickness of 0.5 mm. This is expected, since in general, the springback increases with decreasing sheet thickness. It also increases with decreasing modulus of elasticity at similar tensile strength. For some material and sheet thickness combinations, springback increases with increasing forming angle ($\alpha_{act}^{(loa)}$), while it decreases for others. Since this behavior only occurs for materials of higher tensile strength, this leads to the following assumption. The applied stamping force is directly determined by the springs and their deflection, which remains equal for all materials and sheet thicknesses. This may mean that the applied force is not sufficiently high for thicker sheets of stiffer materials, like the used aluminum alloy and steel, at large angles. This assumption was confirmed by simulations with doubled spring stiffnesses, which can be seen in Figure 11.

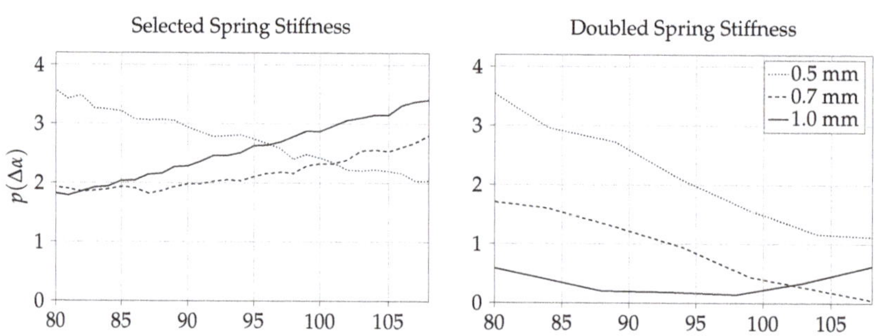

Figure 11. Simulated springback for different spring stiffnesses of punch and die for DC01.

When comparing data set nu with mu, it is obvious that the variance of the springback increases with rising uncertainty.

With the four data sets presented in Section 3, RL models are trained to learn the material-, sheet thickness- and angle-dependent springback behavior. The trained agents then predict the springback and compensate for it using the control loop. The results are summarized for each data set in Figure 12 in the form of a Probability Density Function (PDF). An overview over the average and maximum deviations of the individual data sets is also shown in Table 4. The PDF of the agent belonging to data set nu shows that the mean deviation could be significantly reduced ($|\overline{\Delta\alpha_{ctrl}}| = 0.16°$). It can also be seen that both the mean and the maximum deviation increase with increasing uncertainty. To test whether the performance of the trained agents can be improved with knowledge of the applied uncertainty, a further series of tests is carried out. In these, the actual material properties ($E + \delta E$ resp. $v + \delta v$ instead of E resp. v) were fed to the agents, to see if the springback behavior of untrained configurations is learned. In a real environment, the feedback of the uncertainty or the actual material parameters could be achieved, for example, by measuring the forming forces in the process [27]. However, given the considerable difference between synthetically generated force signals and real measured signals, especially in the case of a flexible tool system, the approach described above was chosen. The results of these tests are also shown in Table 4 (gray values) and in the form of PDFs (Figure 13), separated by data set. The results presented in this study are derived from 30 training runs, with both the mean and maximum deviations being displayed. It shows clearly that with knowledge of the applied uncertainty, $\Delta\alpha_{ctrl}$ can be reduced significantly. Table 4 shows that both the mean and maximum deviation can be significantly improved for all datasets. The difference between agents with knowledge about the applied uncertainty and those without becomes clearer as the level of applied uncertainty increases.

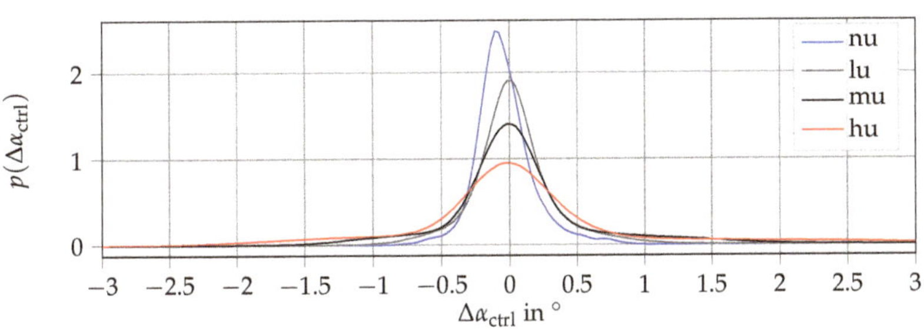

Figure 12. Bending angle deviations using the trained models on simulative generated datasets with different levels of uncertainty.

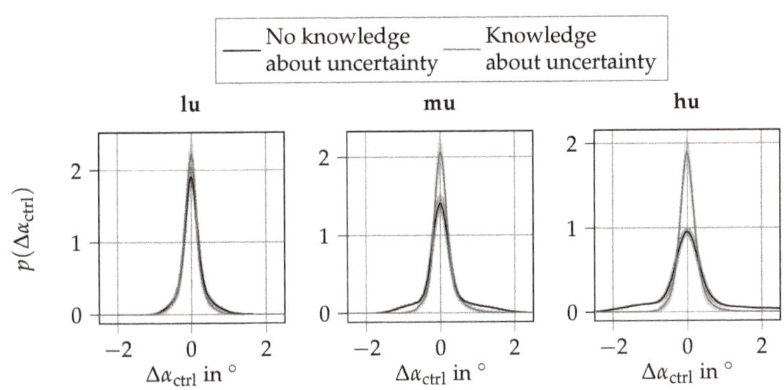

Figure 13. Comparison of PDF of the bending angle deviations with and without knowledge of artificially applied uncertainty.

Table 4. Maximum and average occurring bending angle deviations using the SAC algorithm as an AI-supported product controller on the different simulation data sets without (black) and with knowledge (gray) about uncertainty of the material parameters (mean values based on 30 trained models each).

Metric	nu	lu	mu	hu
$\max(\lvert\Delta\alpha_{ctrl}\rvert)$ in °	0.88	2.28	2.58	4.52
$\max(\lvert\Delta\alpha_{ctrl}\rvert)$ in °	-	1.59	1.96	1.72
$\overline{\lvert\Delta\alpha_{ctrl}\rvert}$ in °	0.16	0.21	0.33	0.51
$\overline{\lvert\Delta\alpha_{ctrl}\rvert}$ in °	-	0.17	0.19	0.19

4.2. Experimental Results

The following section examines the use of RL models for bending angle control in an experimental environment. As shown in Section 3, simulatively trained agents are initially used to predict the springback. Subsequently, the models are retrained based on experimental data and validated on an additional experimental dataset.

For the experimental validation of the agent trained with simulatively generated data, the angular range is traversed with a stepsize of 2.5° (see Table 5). Figure 14 shows the springback $\Delta\alpha$ plotted against the nominal angle $\alpha_{des}^{(unl)}$. The data points represent the experimentally measured values and the dashed lines the springback predicted by the agent. The used agent was trained on the nu data set, as information on material fluctuations cannot be fed back in the test setup shown. The agent can therefore only learn the mean springback of datasets subject to uncertain material parameters, which corresponds to training on the dataset without applied uncertainty. Figure 14 shows the springback behavior broken down by material. For copper sheets, it can be seen for both thicknesses that the material springs back significantly more in the experimental case than predicted by the agent. Since it could be shown that the agent reduces the deviation in the simulated process to an average of 0.16° (max. 0.88°), this means that the material springs back significantly more in the real process. This may be due to the fact that the material parameters of copper used for the simulation were not determined experimentally with the material used, but represent literature values. As shown by [28], the prediction quality in V-die bending is also highly dependent on the rolling direction of the sheet and the underlying criteria and laws for different materials. For 1 mm thick aluminum sheets, the graph shows a good match between predicted and actual springback. Small deviations in the springback behavior can be seen, but these should be largely due to fluctuations in the material or process. Even for 0.5 mm thick aluminum sheets, there is good agreement between prediction and experiment for desired angles $\alpha_{des}^{(unl)}$ greater than 95°. Below this, the springback increases in the predicted and thus simulative case compared to the experimental investigation as the desired angle decreases. As already discussed in Section 4.1, this could also be due to the coining component during die bending, which behaves slightly differently in the real process than in the simulation. For sheets of the two thinner thicknesses of DC01, higher springbacks occur in the simulation than in the real test. For 1 mm thick sheets, on the other hand, the predicted and the experimentally determined springback agree well.

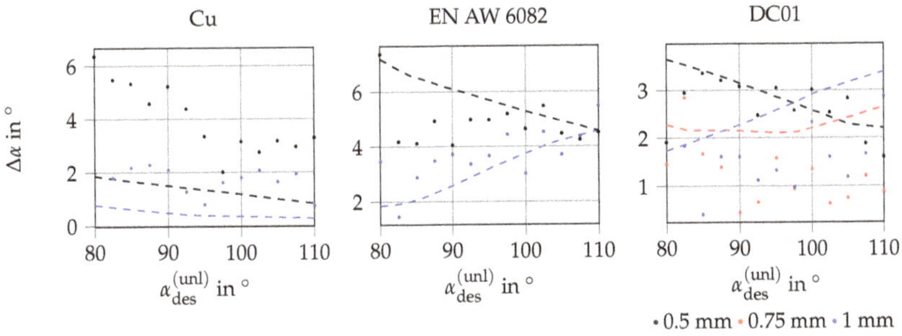

Figure 14. Comparison of experimental springback on *validation dataset* and the predicted springbacks of the SAC algorithm trained on simulation data (dashed line).

In summary, it can be said that the deviation $\Delta\alpha_{ctrl}$ is already improved by the simulative training of the agents. However, as simulation and reality do not align exactly, there is a sim-to-real gap that limits the prediction quality of the agent. In order to obtain more accurate results, the gap must be closed through transfer learning. To accomplish this, agents that were originally trained on synthetic data can be further retrained using experimental datasets. As shown in Table 5, a total of an additional 330 tests were carried out to generate an experimental database. The agent was then retrained *off-policy* on the generated training dataset. The plots in Figure 15 show this dataset (scatter) with the predictions of the final trained agents (dashed lines). The predictions of the agent now align very well with the average springback of the according material and thickness configuration. However, the experimental process underlies uncertainty which, in contrast to simulation, cannot be eliminated. As a result, the springback fluctuates visibly even under similar conditions. Thus, the average and maximum deviation are reduced compared to the sim-to-real data set, but remain high compared to the simulation datasets (see Table 6). The above results are based on the training data set. In order to rule out overfitting, the best retrained agent is revalidated according to the test matrix shown in Table 5. Again, the springback ($\Delta\alpha$) of the different materials and sheet thicknesses is plotted over the target angle ($\alpha_{des}^{(unl)}$) in Figure 16. Since the agent is not retrained during validation, the predictions (dashed lines) remain the same. Finally, the PDFs of each dataset are compared with each other in Figure 17. The corresponding characteristic values are summarized in Table 6. It can be seen that the retrained agent achieves similar performance for the training and validation data set. The mean deviation $|\overline{\Delta\alpha_{ctrl}}|$ is within 0.1°, while the maximum occurring deviation is higher in the training data set. Due to the size of the datasets, this is likely caused by stochastic uncertainty in the process.

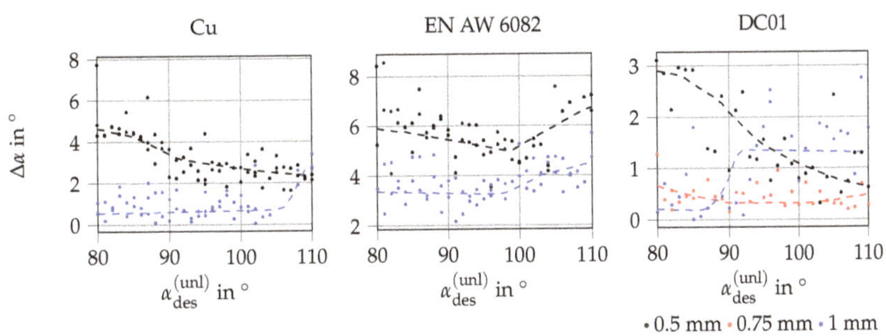

Figure 15. Comparison of the experimental springback on the *training dataset* and the predicted springback of the SAC algorithm retrained on experimental data (dashed line).

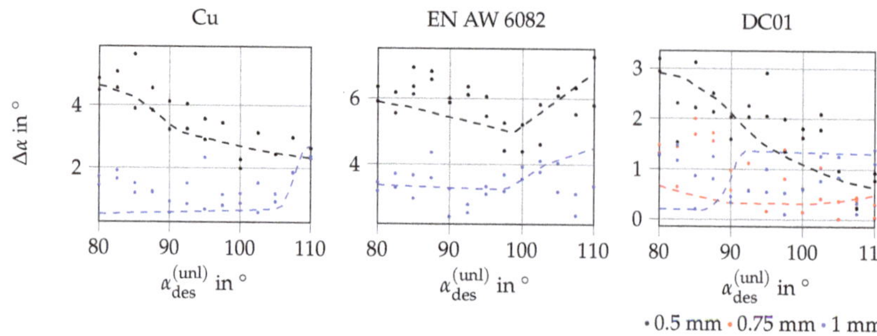

Figure 16. Comparison of the experimental springback of the *validation dataset* and the predicted springback of the SAC algorithm retrained on experimental data (dashed line).

Table 5. Overview of experimentally conducted test series and corresponding parameters.

Dataset	N	Stepsize	$\alpha_{des}^{(unl)} \in \ldots^\circ$
Sim-to-Real	95	2.5°	[80, 110]
Training	330	1°	[80, 110]
Validation	175	2.5°	[80, 110]

Table 6. Error metrics for bending angle deviations.

Data Set	$\lvert\overline{\Delta\alpha_{ctrl}}\rvert$	max $\lvert\overline{\Delta\alpha_{ctrl}}\rvert$	R^2
Sim-to-Real	1.28	4.5	−0.03
Training	0.48	3.13	0.88
Validation	0.57	1.86	0.86

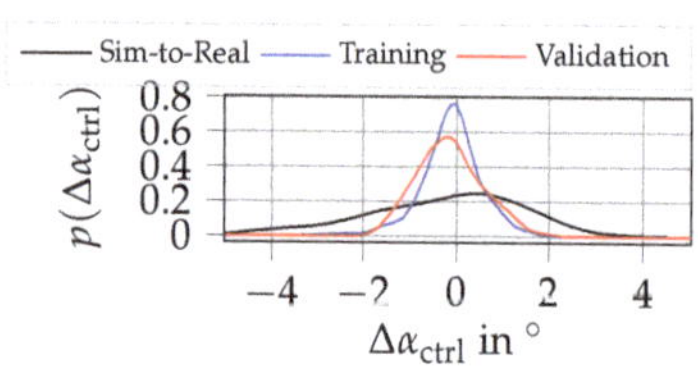

Figure 17. PDF of bending angle deviations.

5. Conclusions

In accordance with the prevailing trend towards mass customization [29], product life cycles are becoming increasingly shorter and batch sizes are decreasing. Therefore, as illustrated by [30], forming processes and production processes in general need to be flexibilized, which is accompanied by the requirement for process control. RL algorithms are a promising approach for this purpose. In this paper, a novel concept for flexible V-die bending is presented which allows the bending angle to be adjusted and hence allows springback compensation. As springback is mutually dependent on numerous parameters, this results in a complex control task. At first, a parameterized FEM model is developed to accommodate variable angles, sheet thicknesses, and material properties. The general effectiveness of RL as a product property controller could be proven, as the mean bending angle deviation is reduced to 0.16° across all data sets. Additionally, the maximum occurring deviations could be reduced significantly as well. However, when applied to the real-world process, a noticeable sim-to-real gap was observed due to dissimilarities in springback behavior. To address this gap, retraining of the agent on the experimental data set was conducted. This improves the accuracy noticeably, but the process is subject to uncertainty that the agent cannot detect in the real environment. In the simulative environment, it was shown that the prediction quality of the agent could be improved when the agent possesses knowledge of the actual material properties. Earlier studies by [20,31] show that force contains valuable information about the material properties and therefore springback behavior. Deriving from this, the integration of force sensors as additional input for the agent seems to be a promising approach to take process fluctuations into account and further reduce the deviation. The use of RL as a product property controller improves the dimensional accuracy of the bending workpieces in the shown process, which is why its application to other forming processes should be further investigated. It is of particular interest to consider more complex forming processes with multiple DOF, which allows for advanced manipulation.

Author Contributions: Conceptualization, C.-V.V., D.A.M. and V.A.; methodology, D.A.M. and V.A.; investigation, C.-V.V.; writing—original draft preparation, C.-V.V.; writing—review and editing, D.A.M., V.A. and P.G.; visualization, C.-V.V. and D.A.M.; supervision, P.G.; project administration, P.G.; funding acquisition, P.G. All authors have read and agreed to the published version of the manuscript.

Funding: The results of this paper are achieved within the project "ProKI—Demonstrations und Transfernetzwerk KI für die Umformtechnik ProKI-Darmstadt" funded by the German Federal Ministry of Education and Research (BMBF). The authors wish to thank for funding and supporting this project.

Institutional Review Board Statement: Not applicable.

Informed Consent Statement: Not applicable.

Data Availability Statement: The data presented in this study are available on request from the corresponding author.

Conflicts of Interest: The authors declare no conflicts of interest.

Abbreviations

The following abbreviations are used in this manuscript:

AC	Actor-Critic
DOAJ	Directory of Open Access Journals
DOF	Degrees Of Freedom
FEM	Finite-Element Method
LD	Linear Dichroism
MDPI	Multidisciplinary Digital Publishing Institute
PDF	Probability Density Function
RL	Reinforcement Learning
SAC	Soft Actor-Critic
TLA	Three Letter Acronym

References

1. Möhring, H.; Wiederkehr, P.; Erkorkmaz, K.; Kakinuma, Y. Self-optimizing machining systems. *CIRP Ann.* **2020**, *69*, 740–763. [CrossRef]
2. Allwood, J.; Duncan, S.; Cao, J.; Groche, P.; Hirt, G.; Kinsey, B.; Kuboki, T.; Liewald, M.; Sterzing, A.; Tekkaya, A. Closed-loop control of product properties in metal forming. *CIRP Ann.* **2016**, *65*, 573–596. [CrossRef]
3. Volk, W.; Groche, P.; Brosius, A.; Ghiotti, A.; Kinsey, B.; Liewald, M.; Madej, L.; Min, J.; Yanagimoto, J. Models and modelling for process limits in metal forming. *CIRP Ann.* **2019**, *68*, 775–798. [CrossRef]
4. Dornheim, J.; Link, N.; Gumbsch, P. Model-free adaptive optimal control of episodic fixed-horizon manufacturing processes using reinforcement learning. *Int. J. Control Autom. Syst.* **2020**, *18*, 1593–1604. [CrossRef]
5. Idzik, C.; Krämer, A.; Hirt, G.; Lohmar, J. Coupling of an analytical rolling model and reinforcement learning to design pass schedules: Towards properties controlled hot rolling. *J. Intell. Manuf.* **2023**, *35*, 1469–1490. [CrossRef]

6. Reinisch, N.; Rudolph, F.; Günther, S.; Bailly, D.; Hirt, G. Successful pass schedule design in open-die forging using double deep Q-learning. *Processes* **2021**, *9*, 1084. [CrossRef]
7. Deng, J.; Sierla, S.; Sun, J.; Vyatkin, V. Reinforcement learning for industrial process control: A case study in flatness control in steel industry. *Comput. Ind.* **2022**, *143*, 103748. [CrossRef]
8. Zhao, W.; Queralta, J.; Qingqing, L.; Westerlund, T. Towards Closing the Sim-to-Real Gap in Collaborative Multi-Robot Deep Reinforcement Learning. In Proceedings of the 2020 5th International Conference on Robotics and Automation Engineering (ICRAE), Singapore, 20–22 November 2020; pp. 7–12.
9. Dulac-Arnold, G.; Levine, N.; Mankowitz, D.; Li, J.; Paduraru, C.; Gowal, S.; Hester, T. An empirical investigation of the challenges of real-world reinforcement learning. *arXiv* **2020**, arXiv:2003.11881.
10. Zünkler, B. *Untersuchung des überelastischen Blechbiegens, von einem Einfachen Ansatz Ausgehend*; Hanser Verlag: München, Germany, 1965; Volume 6.
11. Gan, W.; Wagoner, R. Die design method for sheet springback. *Int. J. Mech. Sci.* **2004**, *46*, 1097–1113. [CrossRef]
12. Zhang, Z.; Wu, J.; Zhang, S.; Wang, M.; Guo, R.; Guo, S. A new iterative method for springback control based on theory analysis and displacement adjustment. *Int. J. Mech. Sci.* **2016**, *105*, 330–339. [CrossRef]
13. Wang, J.; Verma, S.; Alexander, R.; Gau, J. Springback control of sheet metal air bending process. *J. Manuf. Process.* **2008**, *10*, 21–27. [CrossRef]
14. Heller, B.; Chatti, S.; Ridane, N.; Kleiner, M. Online-Process Control of Air Bending for Thin and Thick Sheet Metal. *J. Mech. Behav. Mater.* **2004**, *15*, 455–462. [CrossRef]
15. Havinga, J.; van den Boogaard, T.; Dallinger, F.; Hora, P. Feedforward control of sheet bending based on force measurements. *J. Manuf. Process.* **2018**, *31*, 260–272. [CrossRef]
16. Sharad, G.; Nandedkar, V. Springback in Sheet Metal U Bending-Fea and Neural Network Approach. *Procedia Mater. Sci.* **2014**, *6*, 835–839. [CrossRef]
17. Liu, S.; Shi, Z.; Lin, J.; Yu, H. A generalisable tool path planning strategy for free-form sheet metal stamping through deep reinforcement and supervised learning. *J. Intell. Manuf.* **2025**, *36*, 2601–2627. [CrossRef]
18. Fu, Z.; Mo, J. Springback prediction of high-strength sheet metal under air bending forming and tool design based on GA–BPNN. *Int. J. Adv. Manuf. Technol.* **2011**, *53*, 473–483. [CrossRef]
19. Viswanathan, V.; Kinsey, B.; Cao, J. Experimental Implementation of Neural Network Springback Control for Sheet Metal Forming. *J. Eng. Mater. Technol.* **2003**, *125*, 141–147. [CrossRef]
20. Molitor, D.; Arne, V.; Kubik, C.; Noemark, G.; Groche, P. Inline closed-loop control of bending angles with machine learning supported springback compensation. *Int. J. Mater. Form.* **2024**, *17*, 8. [CrossRef]
21. Konda, V.; Tsitsiklis, J. Actor-Critic Algorithms. *Adv. Neural Inf. Process. Syst.* **1999**, *12*, 1–7.
22. Haarnoja, T.; Zhou, A.; Abbeel, P.; Levine, S. Soft Actor-Critic: Off-Policy Maximum Entropy Deep Reinforcement Learning with a Stochastic Actor. *arXiv* **2018**, arXiv:1801.01290.
23. Shahzamanian, M.; Lloyd, D.; Partovi, A.; Wu, P. Study of influence of width to thickness ratio in sheet metals on bendability under ambient and superimposed hydrostatic pressure. *Appl. Mech.* **2021**, *2*, 542–558. [CrossRef]
24. Doege, E.; Behrens, B.-A. *Handbuch Umformtechnik: Grundlagen, Technologien, Maschinen*; Springer: Berlin/Heidelberg, Germany, 2016; pp. 1–948.
25. Groche, P.; Scheitza, M.; Kraft, M.; Schmitt, S. Increased total flexibility by 3D Servo Presses. *CIRP Ann.* **2010**, *59*, 267–270. [CrossRef]
26. Molitor, D.; Arne, V.; Spies, D.; Hoppe, F.; Groche, P. Task space control of ram poses of multipoint Servo Presses. *J. Process Control* **2023**, *129*, 103057. [CrossRef]
27. Unterberg, M.; Niemietz, P.; Trauth, D.; Wehrle, K.; Bergs, T. In-situ material classification in sheet-metal blanking using deep convolutional neural networks. *Prod. Eng. Res. Dev.* **2019**, *13*, 743–749. [CrossRef]
28. Mulidrán, P.; Spišák, E.; Tomáš, M.; Rohal, V.; Stachowicz, F. The Springback Prediction of Deep-Drawing Quality Steel used in V-Bending Process. *Acta Mech. Slovaca* **2020**, *23*, 14–18. [CrossRef]
29. Tseng, M.M.; Wang, Y.; Jiao, R.J. Mass Customization. In *CIRP Encyclopedia of Production Engineering*; Laperrière, L., Reinhart, G., Eds.; Springer: Berlin/Heidelberg, Germany, 2017. [CrossRef]
30. Yang, D.; Bambach, M.; Cao, J.; Duflou, J.; Groche, P.; Kuboki, T.; Sterzing, A.; Tekkaya, A.; Lee, C. Flexibility in metal forming. *CIRP Ann.* **2018**, *67*, 743–765. [CrossRef]
31. Schenek, A.; Görz, M.; Liewald, M.; Riedmüller, K.R. Data-Driven Derivation of Sheet Metal Properties Gained from Punching Forces Using an Artificial Neural Network. *Key Eng. Mater.* **2022**, *926*, 2174–2182. [CrossRef]

applied sciences

MDPI

Article

Development of a Small CNC Machining Center for Physical Implementation and a Digital Twin

Claudiu-Damian Petru, Fineas Morariu, Radu-Eugen Breaz *, Mihai Crenganiș, Sever-Gabriel Racz, Claudia-Emilia Gîrjob, Alexandru Bârsan and Cristina-Maria Biriș

Engineering Faculty, Lucian Blaga University of Sibiu, 550024 Sibiu, Romania; claudiudp@yahoo.com (C.-D.P.); fineas.morariu@ulbsibiu.ro (F.M.); mihai.crenganis@ulbsibiu.ro (M.C.); gabriel.racz@ulbsibiu.ro (S.-G.R.); claudia.girjob@ulbsibiu.ro (C.-E.G.); alexandru.barsan@ulbsibiu.ro (A.B.); cristina.biris@ulbsibiu.ro (C.-M.B.)

* Correspondence: radu.breaz@ulbsibiu.ro

Featured Application: The results of this research can be used to develop low-cost, open-architecture CNC (computer numerical control) systems and digital twins that make them easier to use and continually improve their performance. These CNC systems can be used for applications such as the engraving or milling of soft materials.

Abstract: This work aimed to develop both a real implementation and a digital twin for a small CNC machining center. The X-, Y-, and Z-axes feed systems were realized as closed-loop motion loops with DC servo motors and encoders. Motion control was provided by Arduino boards and Pololu motor drivers. A simulation study of the step response parameters was carried out, and then the positioning regime was studied, followed by the two-axis simultaneous motion regime (circular interpolation). This study, based on a hybrid simulation diagram realized in Simulink–Simscape, allowed a preliminary tuning of the PID (proportional integral derivative) controllers. Next, the CAE (computer-aided engineering) simulation diagram was complemented with the CAM (computer-aided manufacturing) simulation interface, the two together forming an integrated digital twin system. To validate the contouring performance of the proposed CNC system, a circular groove with an outer diameter of 31 mm and an inner diameter of 29 mm was machined using a 1 mm cylindrical end mill. The trajectory followed the simulated 30 mm circular path. Two sets of controller parameters were applied. Dimensional accuracy was verified using a GOM Atos Core 200 optical scanner and evaluated in GOM Inspect Suite 2020. The results demonstrated good agreement between simulation and physical execution, validating the PID tuning and system accuracy.

Keywords: Arduino; CNC; digital twins; circular interpolation; motion control; PID controller; simulation

1. Introduction

CNC equipment allows workpieces to be machined by various processes (e.g., milling) automatically by executing a numerical code (program). This is made possible using high-performance motion control systems in several motion directions (axes). The motion control systems used are called feed drive systems.

Automated part machining and numerical control code generation is currently performed using computer-aided techniques (CAD, CAM). However, the correlation of simultaneous multi-axis movements and the complexity of CNC equipment and machining

 https://doi.org/10.3390/app15105549

operations mean that the simple use of CAD/CAM techniques is no longer sufficient when it comes to the most efficient use of CNC equipment.

One of the current approaches involves the development of digital twins of CNC machines, which mirror real systems in the virtual environment and allow, among other things, a more realistic simulation study of the processes taking place on them.

On the other hand, the emergence on the market of increasingly advanced variants of industrial CNC equipment has led to the ever-increasing price of such equipment. There is thus a need to develop CNC equipment solutions with a better price/performance ratio, which can be achieved by using open-architecture CNC equipment.

This work aims to design and build a compact CNC system and develop a simulation-based digital twin for the purpose of validating motion control and testing PID tuning strategies prior to physical implementation.

2. Literature Review

CNC machines use feed drive systems realized in the form of closed-loop motion control systems and use PID (proportional–integral–derivative) controllers as common control strategies.

The majority of control systems in use today, including motion control systems, use the PID controller due to its simple structure, ease of implementation, and the existing knowledge base of PID control over a long time [1–3].

A model reference adaptive control (MRAC) based on PID control was proposed in [4], which was also successfully validated experimentally. However, experimental validation was carried out by studying the tracking error on a single axis, without studying the contouring regime (simultaneous displacements on two axes).

The work in [5] presented a new method for tuning a nonlinear PID (NLPID) controller using a particle swarm optimization based on a generalized opposition-based learning (G-PSO) algorithm. The results showed an improvement in the system's step response (reduced settling time and suppression of overshoot) but did not test the system's behavior in the fast single-axis positioning regime or in the simultaneous multi-axis motion (contouring) regime.

The work presented in [6] proposed a fuzzy PID controller based on a sparse fuzzy rule base (S-FPID). It was demonstrated by means of simulation and experiments that the proposed controller can effectively control the position of a nonlinear CNC feed drive. Also in this case, the validation of the proposed method was based only on the study of the step response, without taking into account either the one-axis positioning regime or the two-axis simultaneous displacement regime with working feed-forward algorithms (contouring regime).

Open-architecture CNC systems are built modularly, with each module fulfilling specific tasks for CNC equipment [7]. The open nature of these systems gives the user the possibility to intervene on component modules, either hardware or software, which can lead to performance improvements [8,9]. Open-architecture CNC systems are used not only in machine tools but also in industrial robots [10].

The study in [11] presented the development of a three-axis CNC machining center with open architecture. The motion control systems were based on a master–slave combination between Raspberry Pi 3B and Arduino MEGA 2560 boards. The implementation of linear and circular interpolation algorithms was reported, but the concrete implementation of the motion control was not specified, nor were experimental results related to the system performance presented.

In [12], the development of a laser engraving system with open architecture was presented. The motion control was realized based on Arduino MEGA boards (Arduino

S.R.L., Monza, Italy). The study and improvement of the system behavior were based on the step response, without taking into account the positioning or contouring work regimes. Another implementation of the open architecture of a CNC system, also a laser engraver, was presented in [13]. Motion control was implemented using Arduino boards, but this time, open loop control with stepper motors was used.

Another approach showing the design and development of a writing machine in the form of a CNC equipment using Arduino boards, stepper motors for the X- and Y-axes, and a servo motor for the Z-axis was presented in [14]. The performance of the machine was demonstrated experimentally, but it should be kept in mind that due to the purpose of the machine, writing and drawing, the technological resisting forces had negligible values.

For machining processes run on CNC equipment, digital twins technology can be used for intelligent monitoring and analysis by mirroring the actual machining process in a virtual environment [15–19]. A study that used digital twins technology to synchronize the load exerted on the main spindle and NC code data and to uniformize the machining load by analyzing the stored data was presented in [20]. Digital twins technology is also used to simulate machining processes on CNC machines. This helps to avoid various abnormal situations that can occur in the process, such as collisions, and to improve machining efficiency and quality [21–24]. Digital twins can also be used for error compensation and fault diagnosis purposes [25–27]. Specific instruments were also used to develop intelligence-driven digital twins of the CNC system [28].

3. Main Contribution

The present approach primarily focused on the development of a compact CNC machine that is both cost-effective and suitable for the designated purpose, namely the machining of soft materials.

Cost-effectiveness was achieved by using established yet efficient solutions, such as closed-loop motion control systems based on DC motors and PID control as tuning strategies. Although starting from the use of known solutions, the proposed approach also had innovative aspects, which will be presented in the conclusion section.

A simulation study was conducted utilizing MATLAB, Simulink, and Simscape, encompassing the dynamics of the system (step response), the rapid single-axis positioning regime, and the contouring regime involving simultaneous two-axis motions.

The use of the Simscape module, together with Simulink, allowed the realization of a hybrid simulation, which emphasized the phenomena taking place in the system, instead of the mathematical relationships, thus favoring a better understanding of the functioning of the system under study.

The control strategy involved the implementation of numerical PID controllers on each motion axis. The physical implementation of the control strategies was intended to demonstrate the efficacy of a low-cost solution based on Arduino development boards for a CNC system, particularly in the context of closed-loop control systems.

Also, the implementation of the control with Arduino development boards was performed using a specialized module from Simulink (Simulink Support Package for Arduino Hardware), which allowed a way of working that did not require the direct realization of a program (code). The Arduino-based control system was realized in the form of functional blocks, and the code was generated automatically. In this way, the focus was on efficiency and functionality and not on the correctness of a program syntax.

The final objective was to realize an integrated digital twin CAE/CAM simulation system by integrating simulation diagrams from Simulink and Simscape (MATLAB v. 2016–v. 2024, The MathWorks Inc, Natick, MA, USA) with a kinematic model of the system realized in a CAM software (ENCY v.1, Limassol, Cyprus) package.

This digital twin integrated the CAD/CAM approach, allowing an advanced simulation of the processes taking place on the CNC equipment with a CAE approach, which allowed the estimation of the technological forces and moments acting on the equipment. A summary of the particularities of the present approach, compared to the results presented in literature, is presented in Table 1.

A graphical abstract of the approach proposed in this paper is presented in Figure 1.

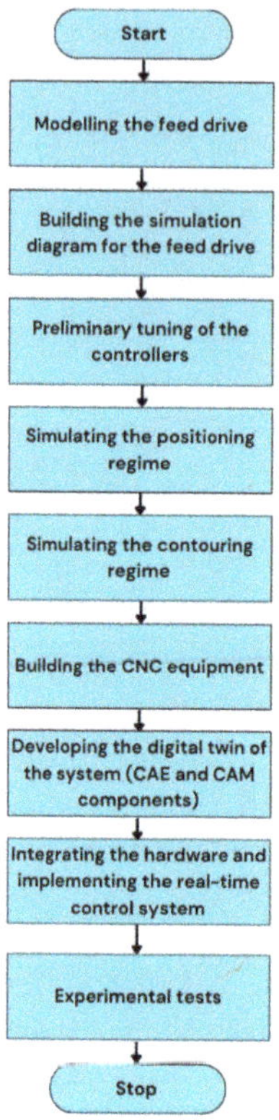

Figure 1. Workflow diagram of the proposed approach.

Table 1. Comparison between literature references and the present approach.

Open-Architecture CNC Systems	Literature References	Present Approach
Control strategy	- PID (continuous and numerical) [1–3] - Advanced PID [2,4,5] - Fuzzy PID [6] - Other advanced control strategies [7]	- Numerical PID
Working regimes studied by means of simulation	- Step response (mostly) [1,4–6,12] - Positioning regime (none) - Contouring regime (few to none) [11]	- Step response - Positioning regime - Contouring regime
Working regimes studied by means of experimental tests	- Positioning regime (few to none) [13,14] - Contouring regime (few) [9,10]	- Contouring regime (circular interpolation)

Table 1. *Cont.*

Open-Architecture CNC Systems	Literature References	Present Approach
Simulation approach	- MATLAB and Simulink (v. 2016–v. 2024) or other software packages based on numerical integration (CAE approach) [4,6,12]	- MATLAB and Simulink and Simscape (v. 2016–v. 2024) (CAE approach) - CAM approach - integrated CAE + CAM approach (digital twin)
Motion control platform	- Arduino + Raspberry PI (medium cost) [11–14] - Yaskawa servo driver (medium to high cost) [4,6] - GALIL motion control card (medium to high cost) [5,6]	- Arduino + low cost Pololu DC motor drivers

4. Materials and Methods

4.1. Modeling the Feed Drive Systems

The feed systems on the three axes of motion (X, Y, and Z) were modeled as closed-loop motion control systems, using the DC servo motor as the actuation device and the incremental encoder as the feedback device. A ball screw—nut mechanism was used as the transmission element. A block diagram of the system is shown in Figure 2.

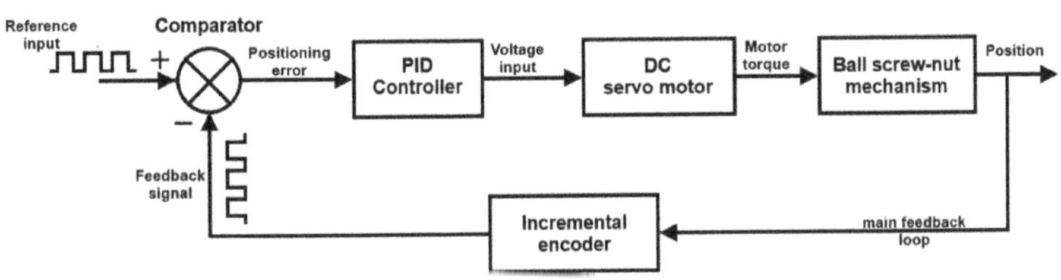

Figure 2. Block diagram of the feed system.

It should be noted that although all three feed axes (X, Y, and Z) were numerically controlled, at this stage, the machine was a 2.5D machine. This meant that while the X- and Y-axes could perform simultaneous movements (in contouring mode), the Z-axis could only operate in single-axis mode. (It could not operate simultaneously with either the X- or Y-axis.) This was because the Z-axis drive system, including the servo motor and ball screw, was different from the X- and Y-axes systems.

Normally, the mathematical modeling of such a system is based on the transfer function in the relation below, considering the voltage applied on the armature as the input signal and the angular velocity of the motor as the output signal, which can be expressed as:

$$\frac{\Omega_m(s)}{U(s)} = \frac{K_t}{(J_{red}s + B_{red})(Ls + R) + K_t K_v}. \tag{1}$$

The notations employed in the aforementioned relation were defined as follows:

- Voltage applied to the armature) u [V] (the term *U(s)* represents the Laplace transform of the voltage input);

- Angular speed of the motor shaft ωm [rad/s] (the term Ωm (s) represents the Laplace transform of the angular speed output;
- Armature moment of inertia J_m [kgm^2];
- Coefficient of viscous friction B_m [Nms/rad];
- Back electromotive force (emf) voltage constant K_v [V/rad/s];
- Electromagnetic torque constant K_t [Nm/A];
- Armature resistance R [Ω];
- Armature inductance L [H].

In order to take into account all components of the motion control system, not just the motor, it was necessary to modify the above transfer function by relating the inertia and friction of moving parts other than the motor to the motor shaft. This was achieved by substituting J_m and B_m with J_{red} and B_{red}, respectively. This substitution signified reductions in the moment of inertia and viscous friction constant at the motor shaft. It is important to note that the ball screw was the sole rotating moving element considered in this system, with the coupling between the motor and the ball screw being negligible.

The values for J_{red} and B_{red} were calculated using the following relations:

$$J_{red} = J_m + \frac{J_s}{i^2}. \tag{2}$$

$$B_{red} = B_m + \frac{B_s}{i^2}. \tag{3}$$

The notations employed in Equations (2) and (3) are defined as follows:

- Reduced moment of inertia at the motor shaft J_m [kgm^2];
- Coefficient of viscous friction reduced at the motor shaft B_{red} [Nms/rad];
- Transfer ratio of the transmission between motor and screw i.

Since the motor and the lead screw were coupled directly, i = 1; therefore, Equations (2) and (3) can be rewritten as:

$$J_{red} = J_m + J_s. \tag{4}$$

$$B_{red} = B_m. \tag{5}$$

Finally, the transfer function between the voltage applied on the armature as the input signal and the angular velocity of the motor as the output signal can be expressed as:

$$\frac{\Omega(s)}{U(s)} = \frac{K_t}{(J_{red}s + B_{red})(Ls + R) + K_t K_v}. \tag{6}$$

where Ω(s) is the Laplace transform of the angular velocity of the motor shaft input.

The simulation study of such systems characterized by transfer functions is usually performed in Simulink using specific blocks. This type of simulation is based on the representation of systems by equations and mathematical relations.

In this study, an alternative approach to simulation studies was adopted, utilizing the Simscape software (MATLAB v. 2016–v. 2024) module. The simulation was based on the physical phenomena occurring in the system, as opposed to the previous approach of using mathematical equations and relationships. The conventional transfer function blocks were substituted by blocks that emulated the behavior of physical systems, such as the "DC motor" block. The diagram for the simulation of an X-axis motion control system realized in Simscape is shown in Figure 3.

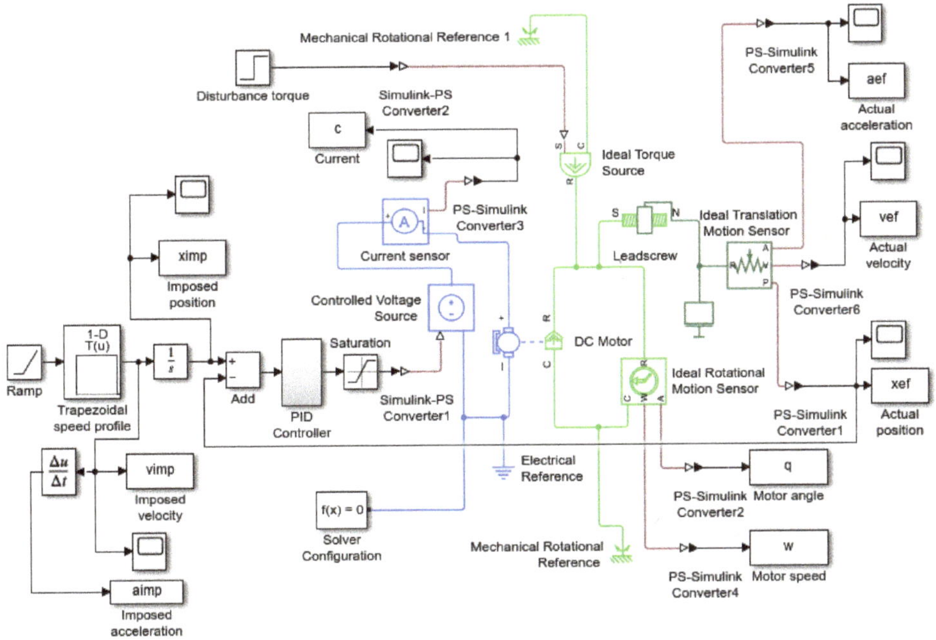

Figure 3. Simulation diagram for the motion control system made in Simscape.

The feed drive system for the X- and Y-axes was characterized by the parameters shown in Table 2.

Table 2. Parameters of the motion control system (X- and Y-axes).

Parameter	Meaning	Values (for X- and Y-Axes)
M_{mn}	Nominal torque developed by the motor	0.14 Nm
K_t	Electromagnetic torque constant	0.06 Nm/A
K_v	Back electromotive voltage constant	0.06 Vs/rad
B_m	Coefficient of viscous friction of the motor	0 Nms/rad
L	Armature inductance	0.029 H
R	Armature resistance	1.33 Ohm
J_m	Armature moment of inertia	0.00004 kgm^2.
p_s	Pitch of the lead screw	0.006 m
d	Ball screw diameter	0.01 m
L	Ball screw length	0.15 m
ρ	Screw material density (steel)	7800 kg/m^3
J_s	Lead screw moment of inertia	$1.1486 \cdot 10^{-6}$ kgm^2
J_{red}	Reduced moment of inertia at the motor shaft	$4.1149 \cdot 10^{-5}$ kgm^2
m	Overall mass of the system	m = 20 kg

4.2. System Dynamics

The control strategy used for the motion control systems involved the use of PID controllers.

One of the main objectives of the research was to provide users with a practical, cost-effective control solution for an open-architecture CNC equipment.

The classical PID controller was found to fulfill these requirements. As experimental research has shown, acceptable results have been obtained using this type of controller.

Also, industrial practice indicates that even in commercial industrial CNC industrial equipment, the classical PID control algorithm is the most widely used. Even if in the settings of such equipment there is a possibility to activate more advanced control facilities (e.g., feed-forward algorithms), these are often not activated.

The numerical implementation of a more advanced PID algorithm would have required more computational power than what is available on Arduino development boards, which are fitted with only 8-bit microcontrollers.

It should be kept in mind here that the Arduino boards need to run both the PID controller algorithm and handle the input quantities generated by the linear or circular interpolation algorithm.

Since the actual system is controlled numerically, the numerical form of the transfer function, shown in the equation below, was used for the PID controller:

$$H_{PID}(z) = K_P + K_I T_s \frac{1}{z-1} + K_D \frac{N}{1 + N T_s \frac{1}{z-1}}. \quad (7)$$

where:

K_P is the proportional constant;

K_I is the integration constant;

K_D is the derivative constant;

T_s is the sampling time of the system.

The interactive "Tune" facility, available for the "PID Controller" block in Simulink, was used to tune the controller. It is imperative to consider that in the context of automatic position or displacement feedback systems, the overshoot should ideally be zero, or as close to this value as possible. Additionally, the physical limits of variation of the controller parameters for the experimental system must be taken into account. Preliminary experimental tests established that for the proportionality constant K_P, the limit was 200, and for the derivative constant K_D, the limit was 1. No limit was established for the integration constant K_I.

For the preliminary tuning of the controller using the specific tuning interface in MATLAB/Simulink, the control objective pursued a balanced approach between reference tracking (which would result in a better contouring accuracy of the CNC equipment) and input disturbance rejection (which would lead to a high robustness of the system to the action of technological resistive forces due to the machining process).

Figure 4 shows some forms of the step response and the values of the controller parameters for which they were obtained.

A summary of the PID controller parameter values, as well as system characteristics (rise time, settling time, overshoot) for the step responses shown in Figure 4a–d, is presented in Table 3.

At this stage of studying the motion control system, the following can be stated:

- It was determined that acceptable behavior was obtained for values of the proportional constant K_P of approximately 150;
- The integral constant K_I exerted a relatively minor influence on the behavior of the system; that is to say, the smaller the integration constant, the better the system's behavior;
- Furthermore, it was not possible to draw any useful conclusions about the influence of the derivative constant and the filter coefficient N. The practical implementation of the effect of the derivative in the operation of the experimental system differed from the implementation used in the interactive tuning interface in Simulink.

In order to ensure the fidelity of the simulation process, a modified transfer function of the numerical PID controller was employed, thereby ensuring the simulation's proximity to the behavior of the real system:

$$H_{PID}(z) = K_P + K_I \frac{T_s}{z-1} + K_D \frac{z-1}{T_s z} \tag{8}$$

Table 3. Parameters of the controller and system characteristics.

Step Response	Controller Parameters	System Characteristics
Figure 4a	$K_P = 367.9$, $K_I = 18.73$, $K_D = -20.25$, $N = 12.12$	Rise time = 0.155 s Settling time = 1.03 s Overshoot = 6.24%
Figure 4b	$K_P = 255.6$, $K_I = 8.604$, $D = -24.94$, $N = 8.05$	Rise time = 0.334 s Settling time = 1.06 s Overshoot = 4.4%
Figure 4c	$K_P = 357.3$, $K_I = 17$, $K_D = -30.36$, $N = 11.58$	Rise time = 0.16 s Settling time = 0.768 s Overshoot = 12.2%
Figure 4d	$K_P = 155.2$, $K_I = 3.294$, $K_D = -17.59$, $N = 8.717$	Rise time = 0.561 s Settling time = 0.966 s Overshoot = 0.938%

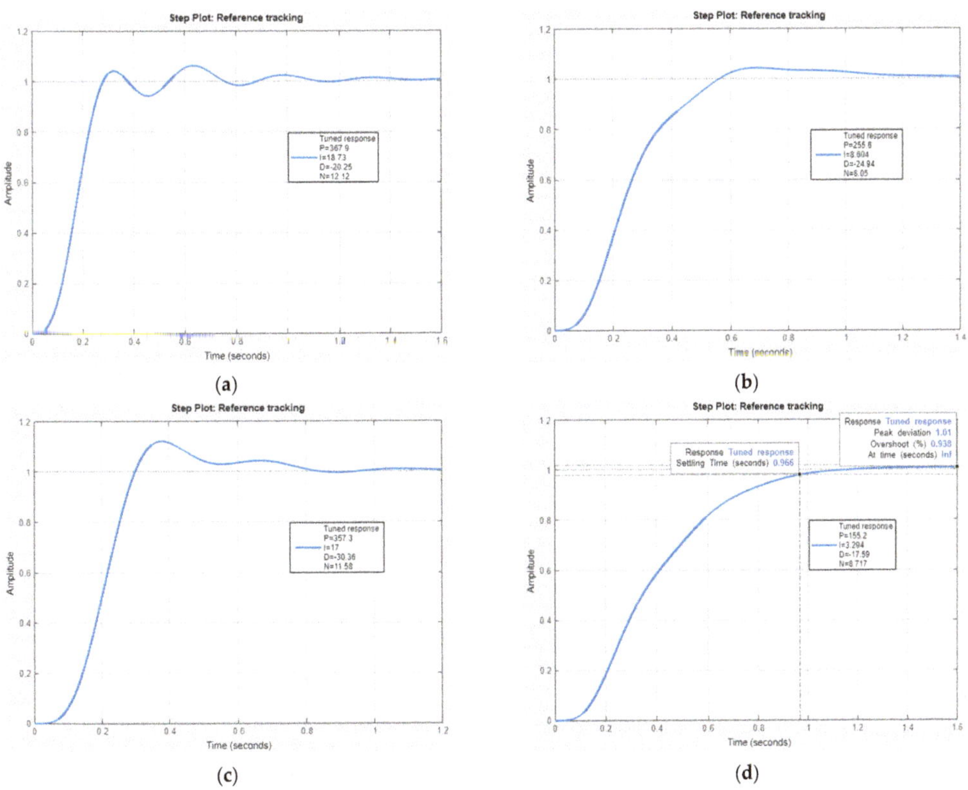

Figure 4. Step responses for different PID controller parameter values. for (**a**–**c**) only the controller parameters are shown. For the step response in (**d**), the values of the overshoot (0.938%) and settling time (0.966 s) are also shown.

To implement this form of transfer function, the Simulink blocks shown in Figure 5a were used. The "Saturation" block with ±24 V voltage limits from the simulation diagram, just after the PID controller subsystem, was also noticeable (also shown in Figure 3).

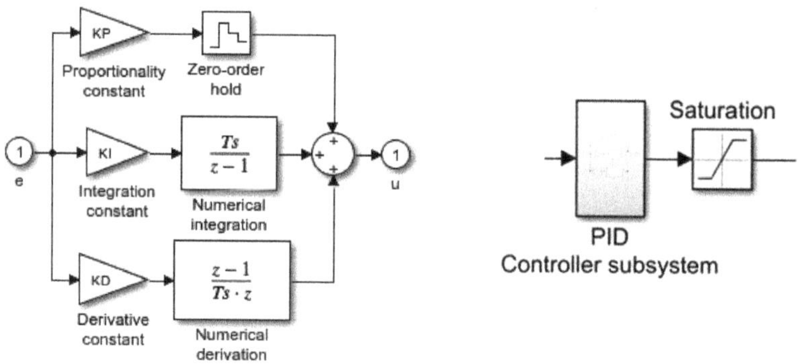

Figure 5. Modifications in the simulation diagram: (**a**) implementation of the digital PID subsystem; (**b**) saturation block placed just after the PID subsystem.

4.3. Positioning Regime

For single-axis rapid positioning, the motions of the moving elements of the system were usually along a trapezoidal velocity profile that included an acceleration phase, a constant-velocity phase, and a deceleration phase, as shown in Figure 6.

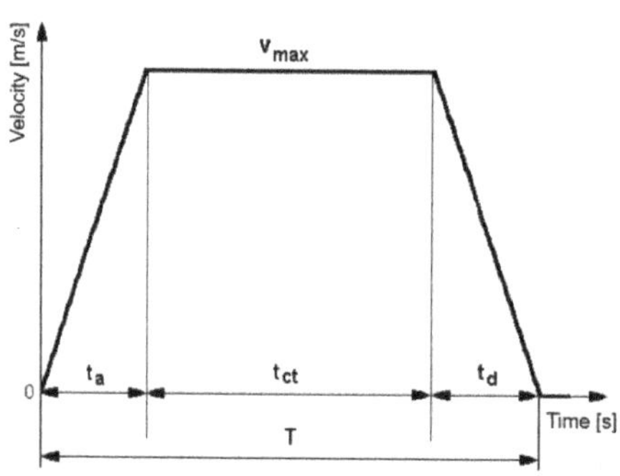

Figure 6. Trapezoidal velocity profile.

The notations employed in Figure 6 were defined as follows:

v_{max} is the maximum speed during the positioning cycle;

t_a is the duration of the acceleration phase;

t_{ct} is the duration of the constant speed travel phase;

t_d is the duration of the deceleration phase;

T is the total positioning cycle time.

The value of the parameters used to simulate the behavior of the system during the positioning regime are presented in Table 4.

Figure 7 shows the comparative evolution of the imposed and actual speeds. It can be seen that the actual velocity followed the imposed velocity relatively closely, and most importantly, the maximum imposed value was reached. The shape of the variation of the effective velocity also ensured that jerking was avoided.

Table 4. Parameters of the positioning regime.

Parameter	Values
v_{max}	0.0083 m/s (500 mm/min)
t_a	0.0167 s
t_{ct}	4.8 s
t_d	0.0167 s
T	4.8333 s
K_P	140
K_I	0
K_D	0.05
T_s	0.0028 s

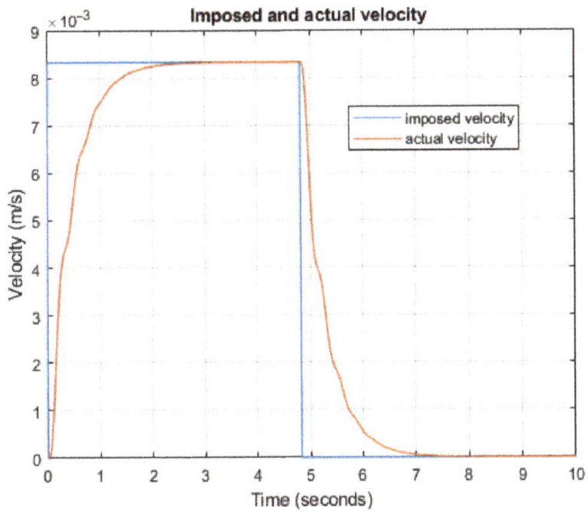

Figure 7. Comparison of the evolution of the imposed and actual velocities.

Figure 8 shows the comparative evolution of the imposed and actual positions. It can be seen that there was a difference in the time axis between the two positions, but it should be noted that this was only an instantaneous deviation and not a real positioning error. The fact that in the end, the two positions, imposed and actual, coincided shows that the positioning was carried out without error.

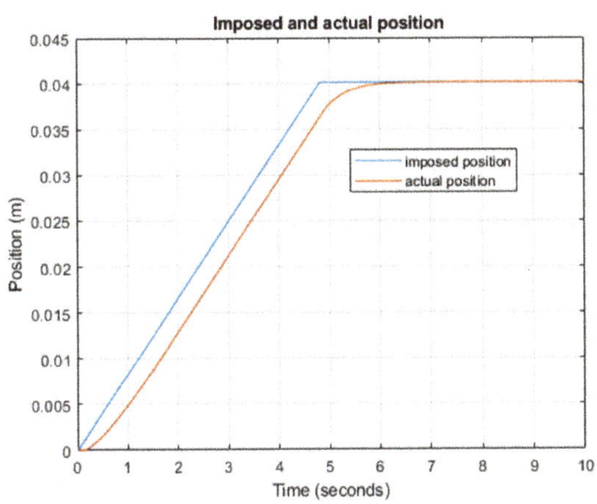

Figure 8. Comparison of the evolution of the imposed and actual positions.

Another feature of the motion control system that could be studied by simulation in this case was its behavior in the presence of disturbances. In the simulations presented so far, the value of the disturbance torque was assumed to be zero, with only the resisting torque due to the load drive (the rotating or translating masses driven by the axis drive servo motor) acting in the system.

Figure 9a–c shows the evolution of the effective velocity for three values of the disturbance torque, 1% (0.0014 Nm), 2% (0.0028 Nm), and 5% (0.007 Nm), respectively, of the nominal torque developed by the axis drive servo motor (0.14 Nm).

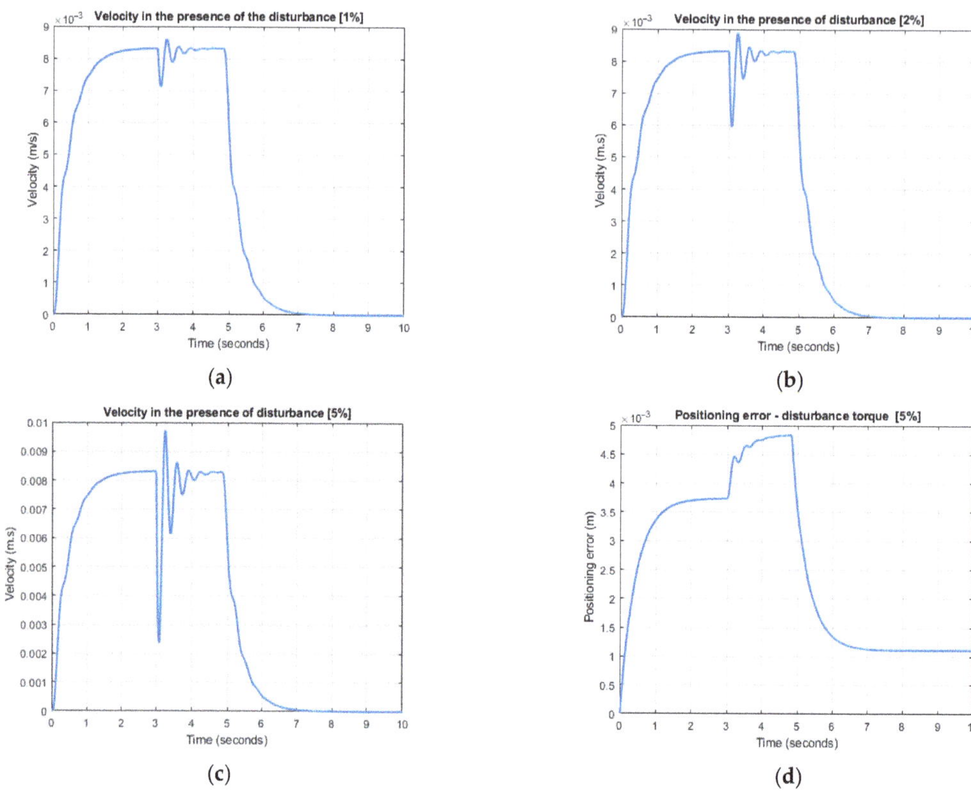

Figure 9. Positioning regime simulation results: (**a**) evolution of the actual velocity in the presence of the disturbance torque [1%]; (**b**) evolution of the actual velocity in the presence of the disturbance torque [2%]; (**c**) evolution of the actual velocity in the presence of the disturbance torque [5%]; (**d**) evolution of the positioning error in the presence of the disturbance torque [5%].

It can be seen that in all three cases, although the resistant torque disturbance had a significant effect on the way in which the effective velocity varied, the system was eventually able to cancel its effect.

To determine whether the disturbance also affected the positioning accuracy, the evolution of the actual position in the presence of a resistive torque disturbance equal to 5% of the rated motor torque was plotted in Figure 9d.

From Figure 9d, it can be seen that the time evolution of the position signal was apparently insignificantly affected by the occurrence of the disturbance, but the final position value was different from the imposed value, x_{imp} = 40.1 mm = 0.0401 m, so there was a positioning error in this case. The amount of positioning error in this case was also very large for such a system, but it must also be taken into account that the disturbance torque was very large.

In conclusion, it can be stated that in the positioning regime, the system demonstrated an adequate level of resilience to disturbances that did not exceed 2% of the nominal torque developed by the motor. However, when this threshold was exceeded, the system's performance was found to be less optimal.

4.4. Contouring Regime and Circular Interpolation

The basic principle of motion control on CNC machines, from the point of view of generating the trajectories along which the tool must move, is as follows: each trajectory, regardless of its complexity, is broken down into straight line segments and circular arcs.

Similarly, in the case of translational movements, regardless of the number of numerical control axes available on the CNC, the trajectories are usually generated by combining the movements of two translational axes, in the XY, XZ, and YZ planes. Although it is theoretically possible, the displacement obtained by combining movements on three axes simultaneously is very rarely used in practice.

The movement mode in which simultaneous movements on several axes take place and machining operations (cutting) are performed during these movements is called the contouring mode. As mentioned above, a contour in a plane (the XY plane for the case study) is generated by decomposing it into entities, such as line segments and circular arcs.

There are two categories of interpolation algorithms, depending on the elementary trajectory generated, line segment or arc, linear interpolation algorithms, and circular interpolation algorithms. Even within these two categories, there are several types of algorithms. In the following, the construction of a simulation diagram will be pursued to allow the study by simulation of the contouring regime in general, with particularization on circular interpolation.

On the basis of the studies presented in [29,30], considering the coordinates x(t) and y(t) of a point on the circular arc, the relations can be written as:

$$\begin{cases} x(t) = R[K_1 sin(f\cdot t+\varphi) - K_2 sin(f\cdot t)] \\ y(t) = R[K_1 sin(f\cdot t) + K_2 sin(f\cdot t+\varphi)] \end{cases}. \tag{9}$$

where:

R is the radius of the circle [m];

f is the sine function frequency [Hz]

φ is the sine function phase [rad];

K_1 and K_2 are constants to be specified later.

Further relations can be written as:

$$v = \omega\cdot R = \frac{2\pi}{T}\cdot R. \tag{10}$$

where:

v is the feed speed on the circular trajectory (constant) [m/s];

T is the time to generate a full circle (period) [s].

ω is the angular velocity [rad/s].

Also, it can be stated that:

$$T = n\cdot T_S. \tag{11}$$

where:

Ts is the sampling period [s];

N is the total number of samples.

Combining relations (10) and (11), the total number of samples can be calculated as:

$$n = \frac{2\pi \cdot R}{v \cdot T_s}. \tag{12}$$

The smallest value of the angular displacement on the axis (angular increment) α_{incr} can be calculated with the following relation:

$$\alpha_{incr} = \frac{2\pi}{T} \cdot T_s = \frac{v}{R} \cdot T_s. \tag{13}$$

Also, the smallest value of the linear displacement (linear displacement increment) l_{incr} can be calculated with the following relation:

$$l_{incr} = 2R \cdot \frac{\alpha_{incr}}{2\pi} = \frac{v}{\pi} T_s. \tag{14}$$

Since the sine function used and the interpolation algorithm, respectively, have a numerical character, the relations can be written further as:

$$f = \frac{2\pi}{NT_s}. \tag{15}$$

where:

N is the number of samples per period.

For N, the relation can also be written as:

$$N = round(n). \tag{16}$$

where *round* is a MATLAB function that rounds the bracketed value to the nearest integer.

Also, to preserve the numeric character of the sine function, the number of offset samples N_{off} can be defined as:

$$N_{off} = \frac{\varphi N}{2\pi}. \tag{17}$$

Since for this algorithm, the phase ρ is equal to $\pi/2$, one can rewrite the relation (17) as follows:

$$N_{off} = \frac{-\frac{\pi}{2} round(n)}{2\pi}. \tag{18}$$

The relation (9) can now be rewritten as follows:

$$\begin{cases} x(t) = R\left[K_1 sin\left(\frac{2\pi}{round(n)T_s} \cdot t - \frac{\pi}{2}\right) - K_2 sin\left(\frac{2\pi}{round(n)T_s} \cdot t\right)\right] \\ y(t) = R\left[K_1 sin\left(\frac{2\pi}{round(n)T_s} \cdot t\right) + K_2 sin\left(\frac{2\pi}{round(n)T_s} \cdot t - \frac{\pi}{2}\right)\right] \end{cases}. \tag{19}$$

For K_1 and K_2, the relations can be written as:

$$K_1 = cos\alpha_{incr} \text{and } K_2 = sin\alpha_{incr}. \tag{20}$$

The smallest value of the angular displacement on the axis (angular increment), α_{incr}, can be calculated as a function of the total number of pulses emitted by the incremental numerical encoder of the displacement control system at one complete rotation, N_{imp}:

$$\alpha_{incr} = \frac{360^0}{N_{imp}} = \frac{360^0}{2024} = 0.1758^0. \tag{21}$$

For the smallest value of the linear displacement (linear displacement increment) l_{incr}, the relation can be written as:

$$l_{incr} = \frac{\alpha_{incr} \cdot p_{sb}}{360^0} = \frac{0.1758^0 \cdot 6 \cdot 10^{-3}}{360^0} = 2.92 \cdot 10^{-6}\ m = 2.92\ \mu\text{m}. \quad (22)$$

Knowing the value of l_{incr} and using the relation (14), one can calculate the value of the sampling period Ts for a trajectory forward velocity v = 0.0033 m/s as follows:

$$T_s = \frac{l_{incr}\pi}{v} = \frac{2,92 \cdot 10^{-6} \cdot \pi}{0.0033} = 0.0028\ \text{s}. \quad (23)$$

Figure 10 shows the diagram used to simulate the behavior of the system under the circular interpolation regime.

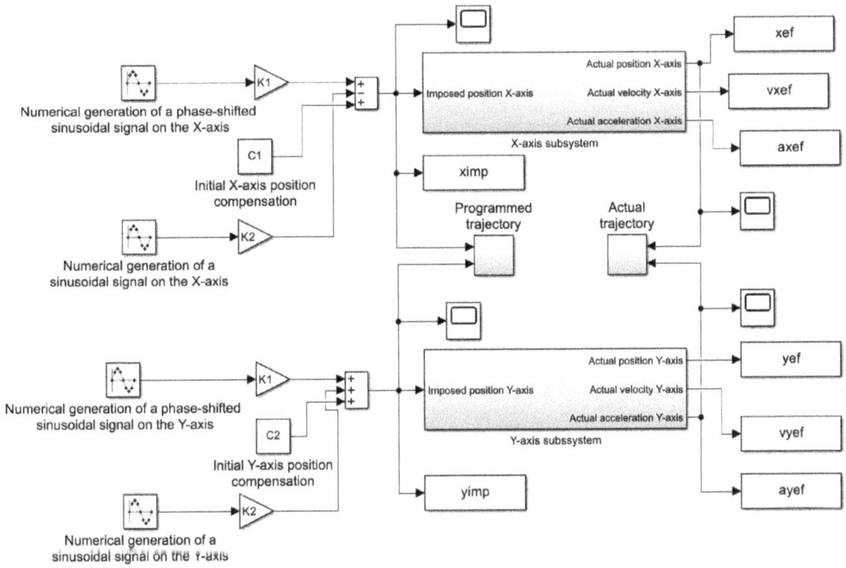

Figure 10. Diagram used to simulate system behavior under the circular interpolation regime.

The simulations aimed to study the behavior of the equipment when generating a full circle in the XY plane. In addition to the data calculated above, the following numerical values were used (Table 5).

Table 5. Parameters used for circular interpolation simulation.

Parameter	Meaning	Values (for X- and Y-Axes)
R	Circle radius	0.030 m (30 mm)
v	Feed velocity	0.0033 m/s (198 mm/min)
K_1	Constant	1
K_2	Constant	$3.068 \cdot 10^{-4}$
C_1	Initial position compensation constant	0.03 m (R)
C_2	Initial position compensation constant	0
T	Generation period of a full circle	56.5487 s
n	Total number of samples at full-circle generation	20,480
N	Total number of samples during the period of generating a full circle	20,480

Figure 11a,b shows the comparative evolution of the imposed and actual position signals on the X- and Y-axes.

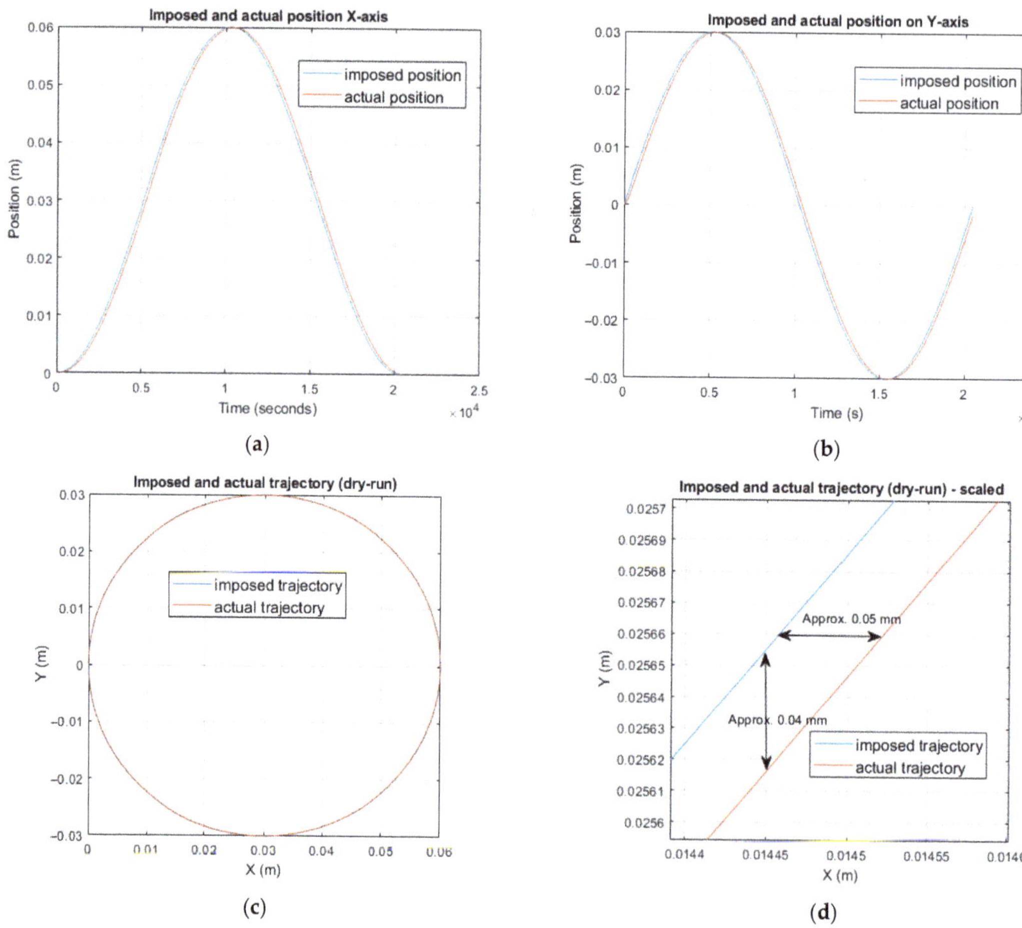

Figure 11. Circular interpolation: (**a**) imposed and actual positions on the X-axis; (**b**) imposed and actual positions on the Y-axis; (**c**) imposed and actual trajectories (dry-run); (**d**) imposed and actual trajectories (dry-run)—scaled.

Figure 11c shows the full-circle, imposed, and effective trajectories obtained by simulation under zero technological resistant torques (dry-run). It can be seen that at the scale used in the figure, the two trajectories were practically superimposed.

In Figure 11d, which shows a scaled detail of Figure 11c, it can be seen that there were, however, some contouring deviations in the X-axis and Y-axis, which, according to the figure, were less than 0.05 mm, which is an acceptable value for the role and purpose of the proposed system.

Figure 12a shows the full-circle, imposed, and effective trajectories obtained by simulation, under conditions of technological resistant torques equal to 2% of the value of the nominal torques developed by the axis actuating servo motors (0.0028 Nm), acting on each axis.

From Figure 12a, it is relatively easy to notice the occurrence of path deviations, even with the naked eye, especially on the sectors of the circle located in quadrants I and IV.

Figure 12b shows a scaled detail of Figure 12a, which highlights the occurrence of deviations between the imposed and actual paths that exceeded 1 mm.

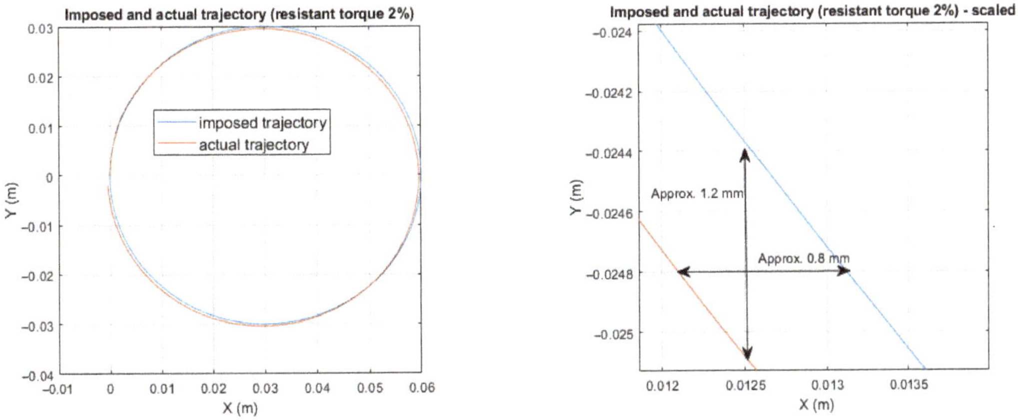

(a) (b)

Figure 12. Circular interpolation: (**a**) imposed and actual trajectories (resistant torque 2%); (**b**) imposed and actual trajectories (resistant torque 2%)—scaled.

Obviously, for a CNC system, deviations of more than 1 mm are large, but the following aspects should be taken into account

- The proposed CNC equipment was a development system with open architecture, dedicated to the study of modular control solutions;
- The size and rigidity of the equipment were relatively small, compared to industrial equipment, which, to some extent, explains the significant influence of the technological resistance moments on the contouring accuracy.

In the experimental program, measures to reduce the number of deviations between the imposed and the actual trajectory were aimed at reducing the number of these technological resistant moments using an appropriate choice of cutting regime parameters.

In practical terms, the aim was not to optimize the behavior of the CNC equipment but to rigorously determine the conditions under which it can be used as technological equipment for machining, its limits in terms of shape accuracy, and methods of reducing errors.

4.5. Physical Implementation of the System

The physical implementation of the open-control architecture CNC equipment (built by the authors) is illustrated in Figure 13a–c.

The notations in Figure 11a–c represent the following:

1. The Y-axis incremental encoder;
2. Y-axis DC servo motor (Dunkermotoren);
3. X-axis DC servo motor (Dunkermotoren);
4. X-axis incremental encoder;
5. Z-axis DC servo motor (Pololu);
6. Main spindle unit;
7. Support structure (frame);
8. Z-axis ball screw transmission unit;
9. Workpiece fixing system;
10. Y-axis ball screw transmission unit;

11. X-axis ball screw transmission unit;
12. Arduino control boards together with DC motor power supply boards.

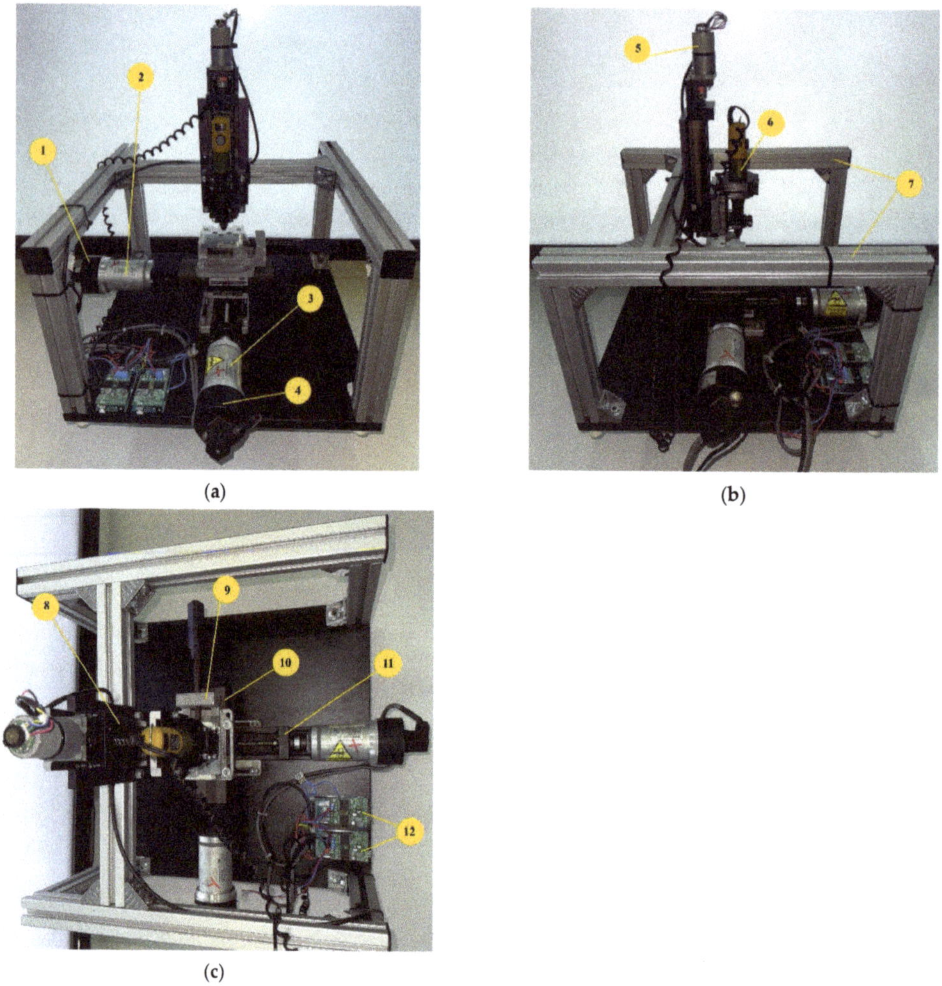

Figure 13. Circular interpolation: (**a**) main view; (**b**) side view; (**c**) view from above.

The system was realized in a modular form and consisted of the following modules:

- A supporting structure composed of extruded aluminum profiles;
- Kinematic feed kinematic chains on the X-, Y-, and Z-axes, structured as closed-loop motion control systems. The drive equipment was the DC servo motor, and the feedback device was the incremental encoder. For the X- and Y-axes, Dunkermotoren GR63x25 servo motors (Dunkermotoren GmbH, Bonndorf, Germany)were employed, while a Pololu servo motor (Pololu, Las Vegas, USA) was utilized for the Z-axis. THK LM Guide Actuator (THK Co., Ltd., Tokyo, Japan) KR ball screw–nut type THK LM Guide Actuator KR were used as transmission devices on all three axes.
- Main kinematic chain (main spindle drive unit) achieved through the integration of a Proxxon mini drilling machine, comprising a drive system powered by a single-phase AC motor with manual speed adjustment capability and a chuck-type tool clamping

system for the cutting tool. This system enabled the manual clamping and replacement of cutting tools, including cylindrical mill, center drill, twist drill, and tapping drill types. The working unit provided the primary cutting motion for the CNC system.

- Command and control module achieved through the implementation of two Arduino MEGA development boards, each equipped with an 8-bit AVR ATMega 2560 microcontroller (Microchip Technology, Chandler, AZ, USA). The Arduino boards were responsible for issuing commands to the drive modules of the three axes (as reference inputs for the automatic motion control systems) and to the main spindle module. The term "control" in this context refers to the implementation of PID-type automatic control algorithms on each motion axis. The utilization of these command-and-control systems guaranteed the system's open-architecture nature, facilitating the implementation of flexible control algorithms and the straightforward modification of control parameters. The command-and-control module incorporated Pololu Dual VNH5019 boards (drivers, Pololu, Las Vegas, USA), which facilitated the power supply for DC servo motors, with voltages ranging from 5.5–24 V and currents ranging from 12–24 A. Each of these two drivers was capable of powering up to two DC servo motors and was compatible with Arduino MEGA boards (Arduino S.R.L., Monza, Italy).

It is imperative to reiterate that the two Arduino MEGA boards facilitated the coordination and control of all three drive modules, as well as the main spindle module.

5. Developing a Digital Twin of the System

5.1. CAE Simulation

The process of creating a digital twin for CNC equipment entailed the sequence of steps illustrated in Figure 14.

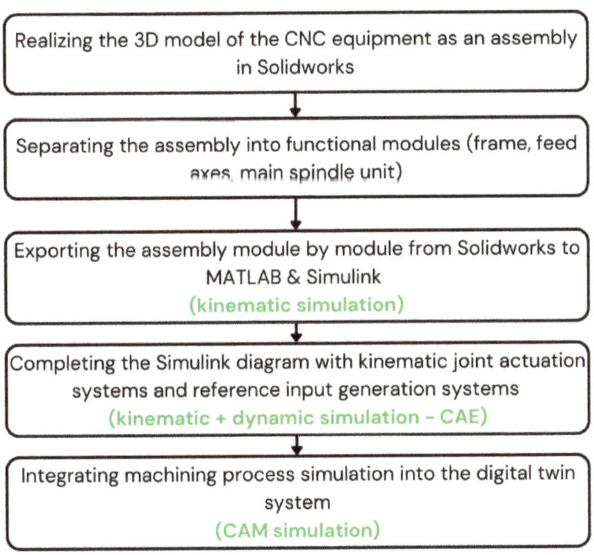

Figure 14. Steps required to realize a digital twin of the CNC equipment.

The steps shown in Figure 13 are detailed below.

Realizing the 3D model of the CNC equipment as an assembly in Solidworks

The first step required making a 3D model of the CNC equipment in a CAD program. The Solidworks software (v. 2019) package was chosen for this task.

The use of the Solidworks software package to realize the 3D model as an assembly, later broken down into functional modules, was necessary because it was directly integrated with MATLAB and Simulink, allowing the automatic generation of a kinematic model in the latter by exporting the 3D assembly from Solidworks.

Separating the assembly into functional modules

The separation into subassemblies of the CNC equipment assembly was necessary in order to correctly identify its functional modules in the export process from Solidworks to MATLAB/Simulink. The identification of movable components (feed axes, main spindle unit) and fixed components (frame, fixtures) was performed by defining the corresponding constraints in Solidworks (v. 2019). Figure 15 shows the 3D model of the CNC equipment together with the functional modules.

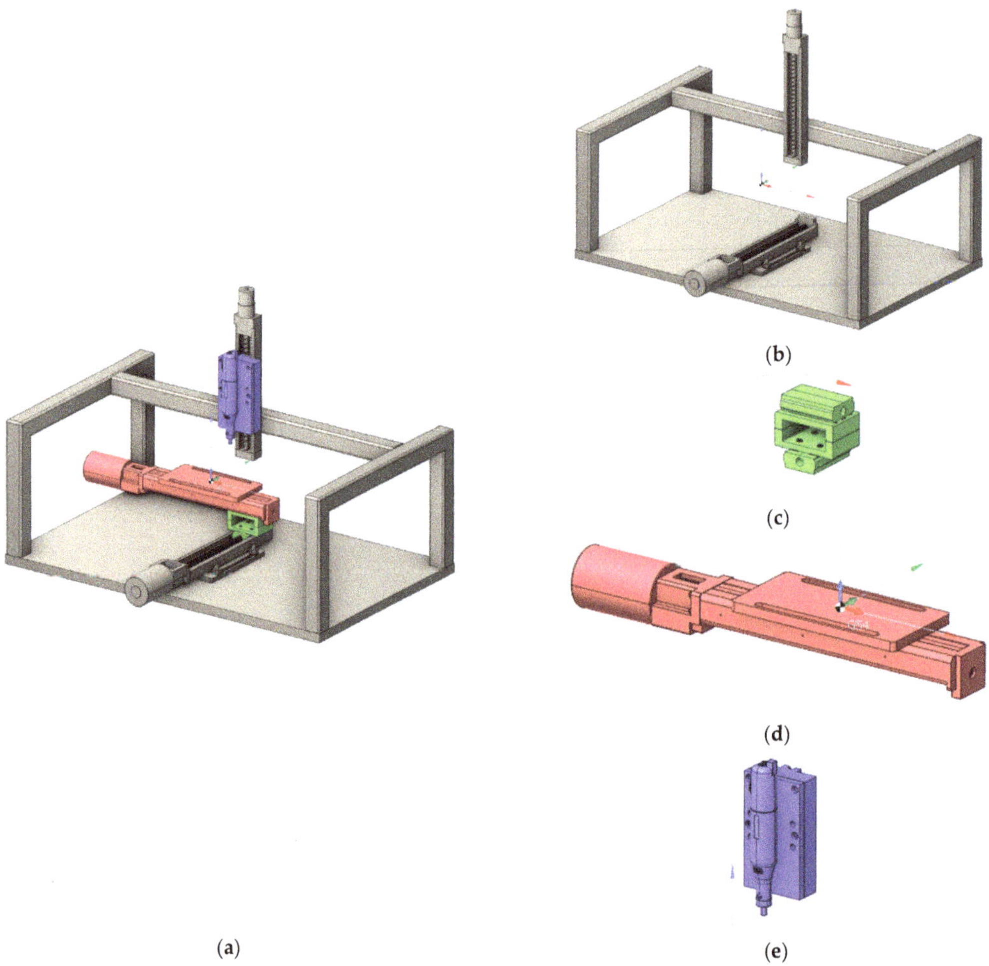

(a) (b) (c) (d) (e)

Figure 15. The 3D model of CNC equipment, divided into functional modules: (**a**) whole assembly; (**b**) frame; (**c**) X-axis; (**d**) Y-axis; (**e**) Z-axis.

Exporting the assembly module by module from Solidworks to MATLAB/Simulink

The Solidworks interface includes specific commands for managing the export of assemblies to MATLAB/Simulink.

To export the 3D model from Solidworks, we went to the main menu Tools, submenu Simscape Multibody Link, submenu Export, and command Simscape Multibody.

These commands generated a file with the .xml extension to be imported into the MATLAB workspace with the command:

$$\text{smimport('filename')} \tag{24}$$

The import generated a Simulink file (slx file) and a data file (m file).

The slx file included all the necessary information for the kinematic simulation of the modeled CNC equipment using Simulink and its modules Simscape Multibody or Simscape Multibody First Generation (Simmechanics), depending on the MATLAB version used.

Completing the Simulink diagram with kinematic joint actuation systems and reference input generation systems

At this stage the diagram could only simulate the motions executed by the CNC equipment, but only in an uncontrolled way. By supplementing the diagram with kinematic joint actuation systems and reference frame generation systems, the diagram in Figure 15 was obtained, which allowed the simulation of the system behavior both kinematically and dynamically.

Kinematic simulation, in this case, means the possibility to generate the kinematic reference quantities, either in a positioning or in a circular interpolation regime, and to study how they are processed in the system from input to output.

The dynamic simulation allowed the evaluation of how motion control systems influenced the behavior of the CNC equipment and how the technological resisting forces acting at the level of the main spindle influenced the operation of the X- and Y-axes feed systems.

Most of the component subsystems of the diagram in Figure 15 were already presented in the previous figures, as follows:

- The axis drive system (X, Y, Z) was presented in Figure 2;
- The PID controller was presented in Figure 5a;
- The input variable generation system (X, Y) was presented in Figure 2 or Figure 10, depending on the operating mode (fast positioning or contouring).

The diagram in Figure 16 made it possible to simulate the movements of the CNC equipment, but this simulation was complex: on the one hand, the laws of motion describing the displacements of the moving elements (in different working regimes, either rapid positioning or plane contouring) were introduced; on the other hand, the diagram made it possible to study the effects caused by the dynamics of the drive systems for each axis on the movements. The diagram also made it possible to study the effects of changing the parameters of the PID controllers on the behavior of the system for each axis. It also made it possible to study the effects of technological resisting forces and moments, which can be introduced either as quantities that vary according to a certain law of variation with time (at the top level of the diagram, using the technological resistance force signal generation subsystems) or as constant quantities that vary stepwise (using the disturbance torque-type blocks in the axis drive subsystems).

Figure 17 shows a screenshot of the system motion simulation in the Mechanics Explorer interface.

The last step from Figure , **integrating machining process simulation into the digital twin system, is explained** in the next section.

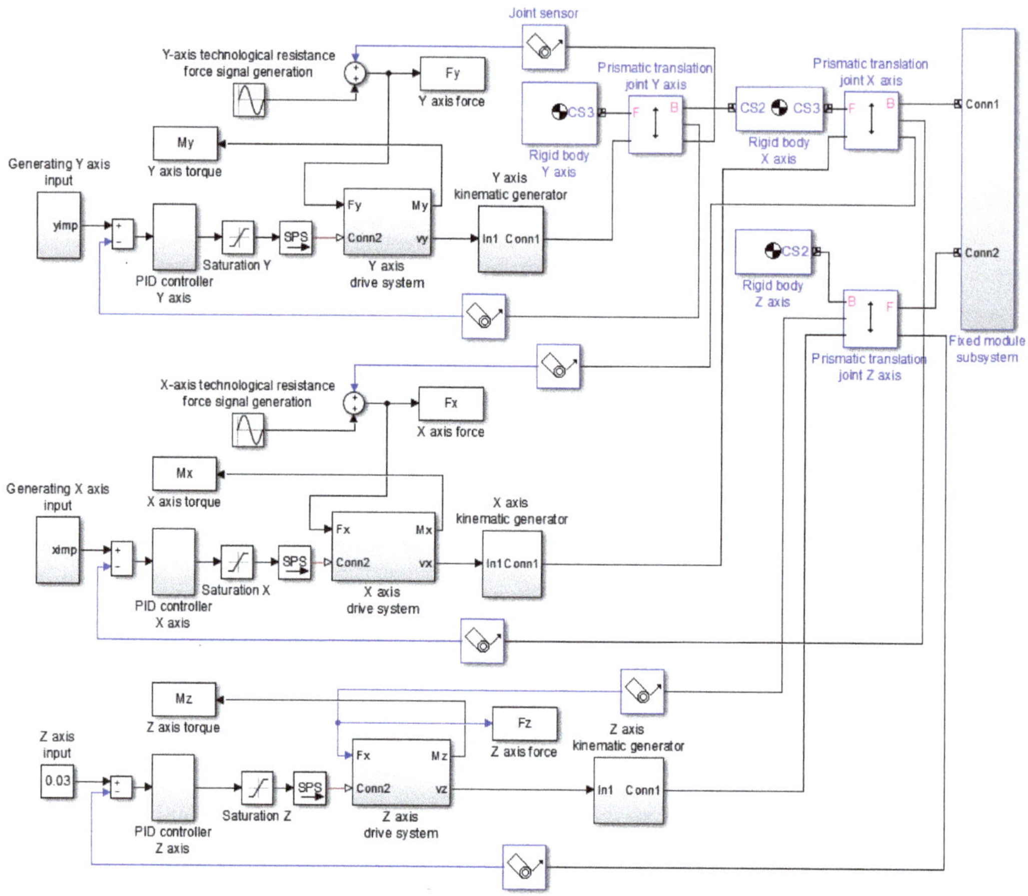

Figure 16. Digital twin simulation diagram for CNC equipment (CAE component).

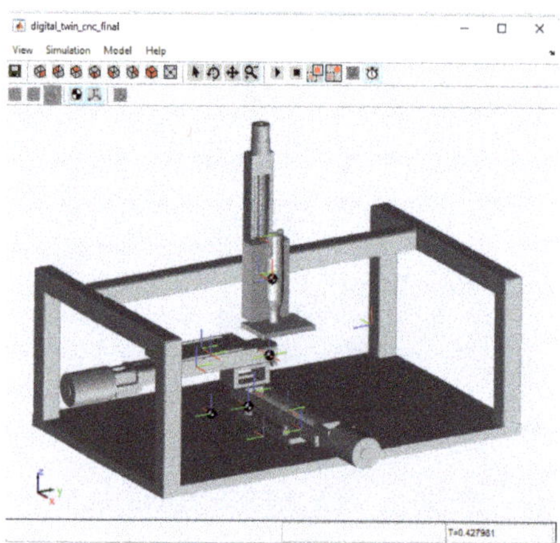

Figure 17. Digital twin simulation diagram for CNC equipment (CAE component).

5.2. CAM Simulation

In order for the integrated digital twin simulation system to also allow CAM simulation, a kinematic model of the CNC equipment was created using the MachineMaker module of the CAM ENCY software (v. 1) package.

Figure 18 shows a screenshot of MachineMaker showing how the components of the kinematic model of the CNC equipment were defined.

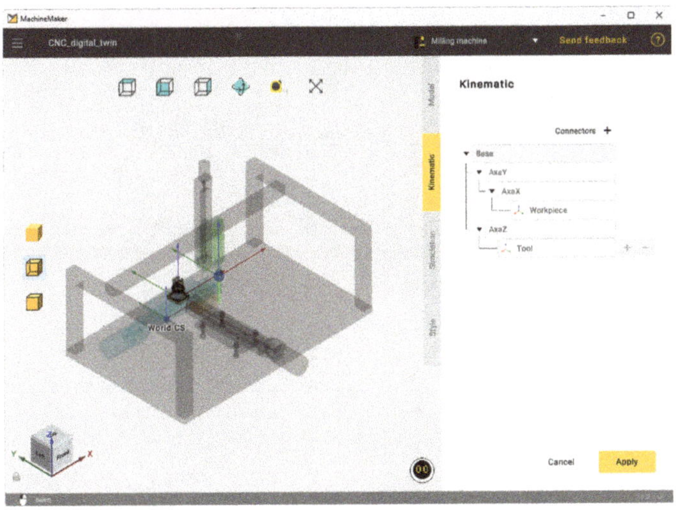

Figure 18. Defining kinematic model components in MachineMaker (v. 2).

Once the kinematic model was created, it was used to simulate machining in the CAM ENCY program. Figures 19 and 20 show screenshots from this program showing how the workpiece was clamped on the machine table and the simulation of a milling operation.

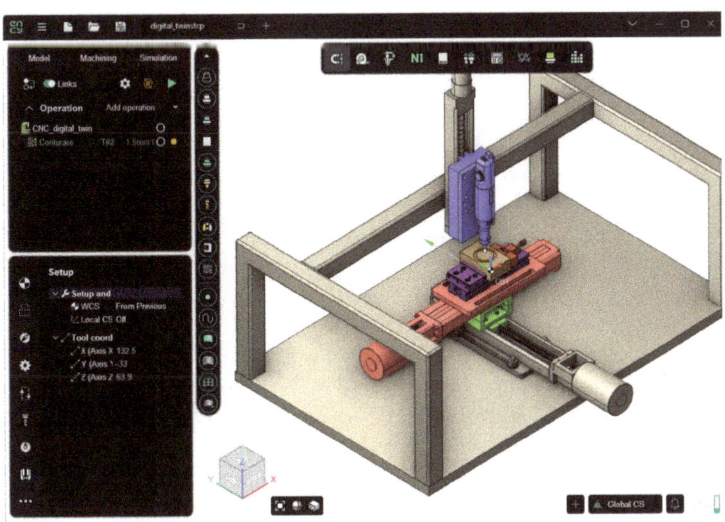

Figure 19. Preparation for the milling processing in the ENCY software (v. 1) package.

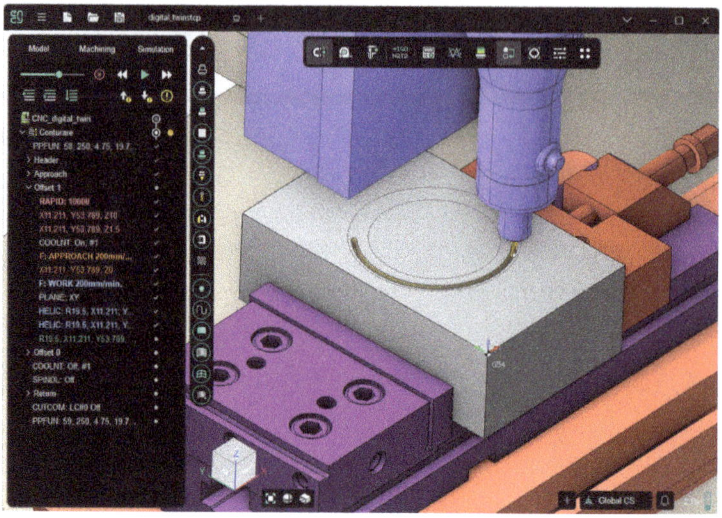

Figure 20. Simulation of milling machining in the ENCY software package.

It should be noted that the simulation process was a complete one, simulating the movements performed by the tool, those performed by the workpiece, and those performed by the moving elements of the machine. It also simulated how the tools removed material from the workpiece during the milling process.

5.3. Features of the Integrated CAE/CAM Digital Twin Simulation Environment

The previous sections described the steps taken to develop the integrated digital twin simulation system.

Table 6 describes the capabilities it provided for simulating the behavior of the CNC equipment.

Summarizing the information in Table 6, it can be stated that the realized digital twin system can perform the following functions.

Virtual commissioning tool function

The proposed digital twin system allowed the study of the motions of the moving elements of the CNC equipment in both CAE (Mechanics Explorer interface) and CAM components. In the first case (CAE), dynamic simulations in different regimes were complemented by visual confirmation of whether or not the moving parts were moving (if the loads exceeded the allowed limits). In the second case (CAM), it was verified that the correlation of the simultaneous motions required for machining complex-shaped parts did not lead to collisions between the moving parts of the equipment or between the moving parts and tool/tool holders, workpieces, or fixtures of the workpieces.

Real-time dynamic modeling function

The correlation between the CAE and CAM components of the digital twin made up for the shortcomings of each component individually. Thus, the CAE simulation, while allowing real-time study of the system dynamics, did not allow a realistic visualization by simulation of the milling machining process. On the other hand, the CAM component, which allowed the visualization of the machining process, did not provide information related to the influence of the dynamic stresses occurring in the equipment and the tuning parameters on the machining accuracy.

Simulation-based design aid function

The digital twin system can also be used for the design (or redesign) of CNC equipment, especially if it is realized in the open-architecture version. Any structural design solution (or in the case of re-design, any constructive change) can be tested and validated by CAM simulation, while the CAE component allows testing and validation of changes in actuation and motion control systems in the feed drive kinematic chains.

Table 6. Facilities offered by the integrated digital twin simulation environment.

Simulation Type	Objective	What Can Be Simulated/Studied	Benefits
CAE simulation	Study of the behavior of motion control systems in the structure of the CNC equipment	- Dynamic behavior of motion control systems in different operating regimes (fast positioning and contouring). - Influence of controller tuning parameters on positioning accuracy and contouring accuracy. - Determination of the required inputs (displacements, velocities, and accelerations). - Values of the quantities occurring in the system during its operation (voltages, currents, forces, moments, displacements, velocities, and accelerations). - Visualization of the movement of the moving parts of the equipment and how static and dynamic loads influence these movements.	Motion control system controllers can be tuned. Positioning accuracy can be improved. Contouring accuracy can be improved. Cutting regime can be designed according to the required machining accuracy. Dynamic overloading of the equipment can be avoided.
CAM simulation	Study of the movement of mobile elements of CNC equipment during the milling of workpieces	- How the blank is clamped to the table of the equipment (including the vise is simulated). - Movement of the cutting tool. - Movement of the workpiece. - Movement of the mobile parts of the equipment. - How the material is removed from the blank.	Collisions between tool and workpiece blanks can be identified and eliminated. Collisions between tool and fixtures (vice and/or flanges) can be identified and eliminated. Collisions between moving parts of the machine can be identified and eliminated. Overtravel on the axes of motion of the equipment can be identified. NC code for part machining can be generated and verified.

At present, the interaction between the digital twin system and the CNC equipment is limited by the absence of an advanced sensor system to accurately measure the physical quantities occurring in the real system and to transfer this information to the digital twin for simulation testing.

The CAE and CAM components of the digital twin system are useful both as independent modules (as summarized in Table 6), but they can also exchange data between them, as shown in the following examples:

- The NC numerical code generated by the CAM component (which describes the trajectories followed for machining a part) can be used to generate kinematic input quantities for the CAE model that are more complex than a simple circular in-plane trajectory;

- The values of voltages, currents, and moments indicated by the CAE model can be used to design and test in the CAM model cutting regimes that do not lead to overstressing the CNC equipment.

6. Experimental Tests

As mentioned in the previous sections, the command and control of the system were realized using two Arduino MEGA boards, equipped with 8-bit AVR ATMega 2560 (Microchip Technology, Chandler, AZ, USA) (Figure 21a) and AVR ATMega 2560 microcontrollers, together with Pololu Dual VNH5019 (Pololu, Las Vegas, USA) (Figure 21b) boards, which provided the DC servo motors with voltages ranging from 5.5–24 V and currents ranging from 12–24 A.

(**a**) (**b**)

Figure 21. The control boards: (**a**) Arduino MEGA development board; (**b**) Pololu Dual VNH5019 board.

The Pololu Dual VNH5019 VNH5019 Pololu Dual boards (acting as DC motor control drivers) were connected to both the Arduino development boards and the motors they control. As previously stated, the two Arduino MEGA boards, together with the two Pololu Dual VNH5019 boards, provided command and control for all three axis drive modules, as well as the main spindle module.

Incremental quadrature photoelectric digital transducers (encoders) were used to realize closed-loop control. The proposed resolution of the CNC system determined the number of pulses/rotation required. Thus, the Dunkermotoren encoder (which equips the motors) was used for the X- and Y-axes, with a resolution of 2048 pulses/rotation, and for the Z-axis, which, at this stage of the system development, was used only for positioning. There was the possibility of reading 64 pulses/rotation.

Some of the characteristics of the encoders used are shown below:

- Resolution: 2048 pulses/rotation (X, Y), 64 pulses per rotation (Z);
- Input voltage: 5–12 V, DC;
- Maximum rotational speed: 6000 rpm;
- Permissible radial load: 5 N;
- Permissible axial load: 3 N;
- Cable length: 50 cm;
- Output shaft diameter: 6 mm.

The MATLAB–Simulink software package, together with the additional module dedicated to Arduino hardware-based systems, Simulink Support Package for Arduino Hardware, was used to control the X- and Y-axes motions of the proposed CNC system.

To ensure the real-time control of the X- and Y-axes movements, the diagram presented in Figure 22 was employed. This ensured the real-time generation and execution of the requisite executable code for the operation of the AVR ATMega 2560 microcontrollers.

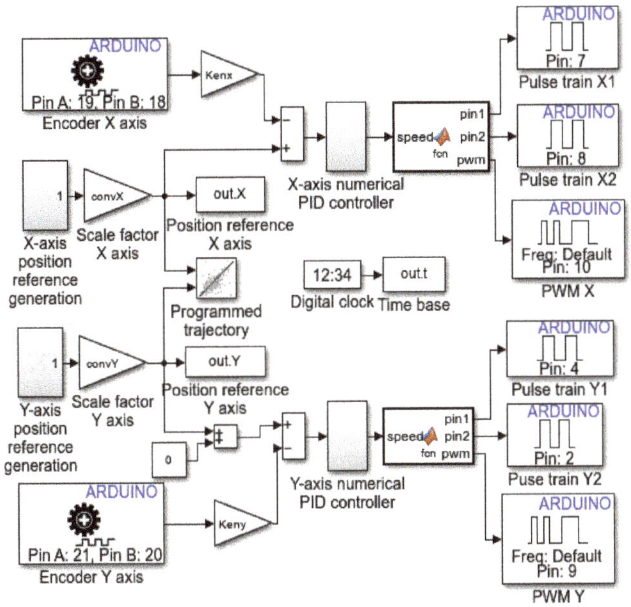

Figure 22. MATLAB–Simulink diagram for real-time control of X- and Y-axes motions.

The diagram used specific blocks from Simulink, as well as blocks from the Simulink Support Package for Arduino Hardware, which are explained in Table 7.

Table 7. Main blocks from the Simulink Support Package for Arduino Hardware used in the structure of the simulation diagram in Figure 21.

Block	Block Name	Function
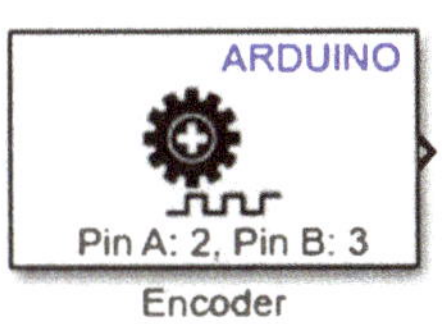	Encoder	The Encoder block outputs the number of pulses from a quadrature encoder on a rotary motor connected to an Arduino board. Each increase in the number of encoder pulses indicates that the motor is rotating clockwise. Each decrease in the number of encoder pulses indicates that the motor is rotating counterclockwise. The total number of pulses represents the incremental position of the rotating motor.
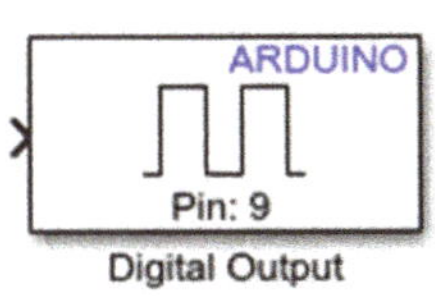	Digital Output	Sets the logic value of a digital pin on the Arduino board: - Transmitting value 1 to the block input sets the logic value of the HIGH digital pin to 5 V or 3.3 V, depending on the board voltage. - Transmitting value 0 to the block input sets the logic value of the LOW digital pin to 0 V.
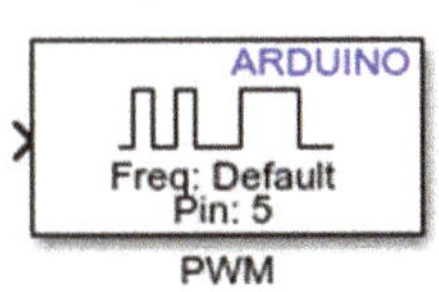	PWM	The PWM block generates variable duty-cycle rectangular pulses (width modulated) depending on the input value sent to the block on the Arduino board hardware terminal. This block allows a digital output to provide a range of different power levels, similar to that of an analog output.

The diagram in Figure 22 also used two "Numerical PID controller" subsystems of the type shown in Figure 3 and two "Reference position generation" subsystems of the type shown in Figure 9. The sampling period used for the real-time control system was T = 0.001 s.

Also used in the diagram in Figure 22 was the user function (introduced via a MATLAB Function block) "Speed", which had the following syntax:

```
function [pin1, pin2, pwm] = fcn(speed)
pin1 = 1;
pin2 = 0;
if speed < 0
pin1 = 0;
pin2 = 1;
end
pwm = abs(sped);
```
(25)

Design of the Experiments

Simulations were carried out under varying working conditions, including feed rates, moving masses, and cutting forces, to adapt the tuning of the PID controllers process to the real behavior of the system. For each configuration, multiple PID sets were tested, either directly from autotuning or refined manually, based on the Ziegler–Nichols logic.

Each simulation output was evaluated using three quantitative indicators:

- Settling time (T_s);
- Steady-state error (E_{ss});
- Maximum trajectory deviation (E_{max}).

Using a weighted objective function, the resulting performance data were aggregated:

$$J = \alpha_1 \cdot T_s + \alpha_2 \cdot E_{ss} + \alpha_3 \cdot E_{max} \tag{26}$$

where $\alpha_1 = 0.4$, $\alpha_2 = 0.4$, and $\alpha_3 = 0.2$, accentuating stability and the positional accuracy.

This concept allowed a balanced comparison between configurations, avoiding subjective or single-criterion selection. The configuration Kp = 140, Ki = 0, Kd = 0.05 achieved the lowest global score (J = 0.245). Excellent performance indicators were registered: settling time = 0.566 s, steady-state error = 0.025 mm, and maximum deviation = 0.183 mm. These values were consistent across both positioning and 2D circular interpolation regimes.

Although not based on a complete factorial design, the procedure reflected a practical applied statistical method. The process enabled not only the identification of the best-performing configuration in absolute terms but also an evaluation of trade-offs under different scenarios. Two final configurations—one PI and one full PID—were selected and implemented on the real CNC system. Subsequently, experimental tests, run under the same conditions as the simulations, were designed to test the predicted behavior.

Due to the limited stiffness of the system, soft materials were chosen for the experimental tests. The chosen material (NECURON 651, NECUMER GmbH, Bohmte, Germany) fell into this category and also had a good machining behavior. (The chips generated were discontinuous and did not affect the surface of the tool and/or the machined part.) This was important because the CNC equipment had no cooling or chip evacuation systems.

The tool clamping system also limited the maximum diameter of tools that could be mounted in the main spindle to 6 mm. It was decided to use a 1 mm diameter tool, both for reasons of low rigidity of the equipment and to limit the amount of chips generated, since, as mentioned above, there was no chip evacuation system.

To experimentally test the contouring accuracy of the proposed CNC system, a circular groove with an outside diameter of 31 mm and an inside diameter of 29 mm was machined, as shown in Figure 23a. A picture of a machined circular groove is shown in Figure 23b. The workpiece material was NECURON 651, a composite polymer and a mild material with a compressive strength of 25 N/mm^2.

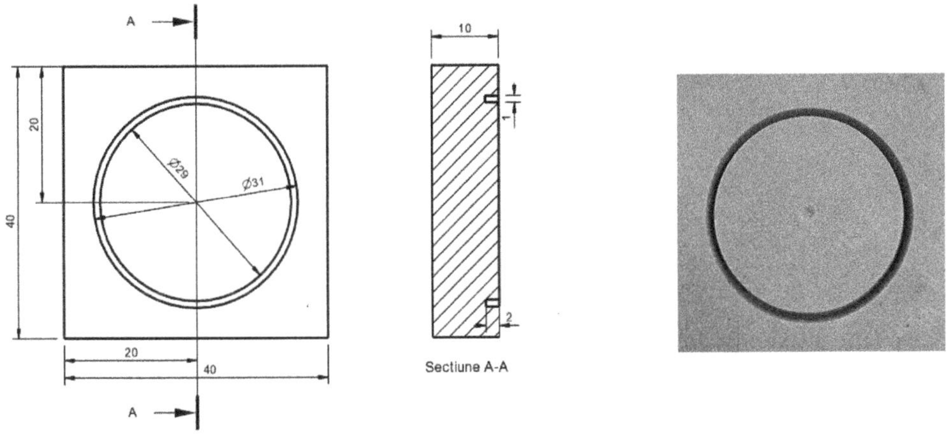

(a) (b)

Figure 23. Machining tests: (**a**) circular groove (2D model); (**b**) machined circular groove.

A 1 mm diameter cylindrical end mill was used for machining. The center of the cutter described a circle with a diameter of 30 mm, exactly as in the simulation of the contouring regime. Two sets of controller parameters were used, the measurement of the inner and outer diameters of the flute was performed with the GOM Atos Core 200 optical measuring device, and the GOM Inspect Suite 2020 software package was used for data processing and visualization.

Figures 24 and 25 show the results of measuring the inner and outer diameters of the circular groove for the two sets of controller parameters on the X- and Y-axes.

The first set of controller parameters was as determined from the simulation study, and a zero value of the derivative constant was used for the second set. This was tried in order to reduce the computational effort of the software system in real time and to study the influence of this on the processing accuracy.

It can be observed that the system had a dimensional accuracy that is typical of equipment in this category, which, of course, can still be improved, in particular by intervening on the mechanical structure to increase its rigidity. However, experimental tests showed that the system can work in contouring mode (simultaneous displacements on two axes, X and Y), and the hardware control system (Arduino boards) worked efficiently with the software control system (MATLAB–Simulink) implemented in real time.

An industrial CNC system designed for steel cutting in the machine-building industry typically exhibits deviations from the nominal dimensions of the circular paths in the order of hundredths of a millimeter, practically an order of magnitude smaller than the proposed system. On the other hand, however, the costs of such systems are much higher. (It can be estimated that they exceed the costs of the proposed system by up to 100 times.) Also, such high accuracies are often not necessary in industrial practice.

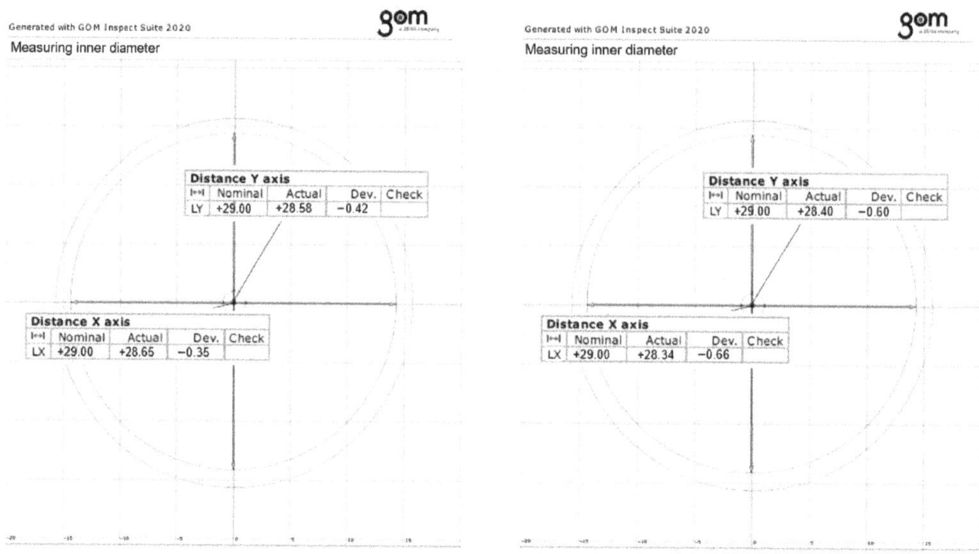

(a) (b)

Figure 24. Measuring inner diameter: (**a**) X-axis: K_{PX} = 140, K_{IX} = 0, K_{DX} = 0.05/Y-axis: K_{PY} = 140, K_{IY} = 0, K_{DY} = 0.05 (parameters set in the simulation study); (**b**) X-axis: K_{PX} = 140, K_{IX} = 0, K_{DX} = 0/Y-axis: K_{PY} = 140, K_{IY} = 0, K_{DY} = 0.

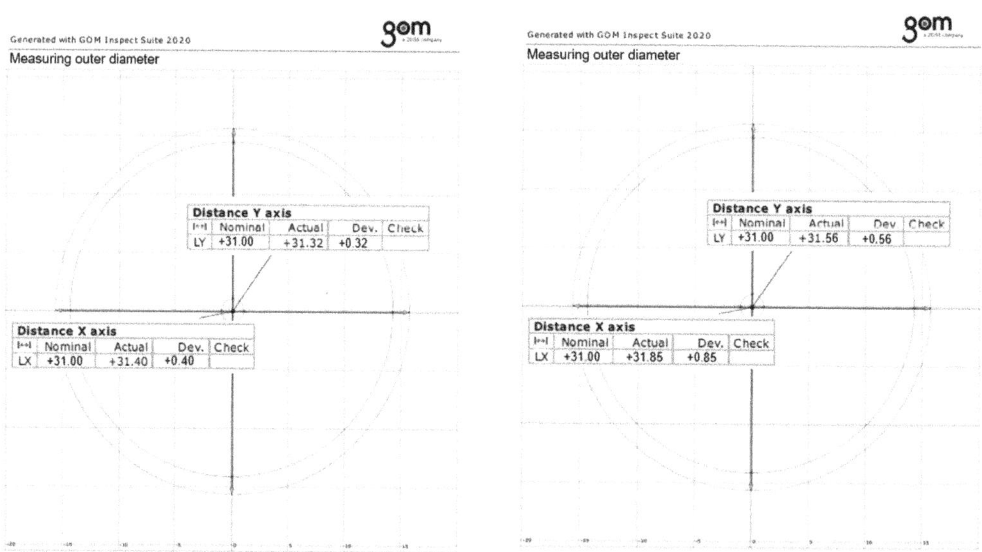

(a) (b)

Figure 25. Measuring outer diameter: (**a**) X-axis: K_{PX} = 140, K_{IX} = 0, K_{DX} = 0.05/Y-axis: K_{PY} = 140, K_{IY} = 0, K_{DY} = 0.05 (parameters set in the simulation study); (**b**) X-axis: K_{PX} = 140, K_{IX} = 0, K_{DX} = 0/Y-axis: K_{PY} = 140, K_{IY} = 0, K_{DY} = 0.

It can therefore be stated that the proposed system can be used in a wide range of applications for which lower accuracies (in the order of tenths of a millimeter) are acceptable and can be listed here as machining of injection molds for plastic packaging, machining of parts made of soft materials (mainly wood, but also other types of materials), engraving operations, and machining operations of parts for decorative purposes.

7. Conclusions

Within this paper, an experimental open-architecture CNC system was proposed and realized. An integrated digital twin CAE/CAM simulation environment was also developed for this system.

The CAE simulation part was implemented in MATLAB–Simulink–Simscape, where simulation diagrams were developed to study the behavior of the motion control systems in the structure of the CNC equipment under different working regimes (fast positioning and contouring in the XY plane). The CAM simulation part was implemented in the ENCY software package, where a kinematic model of the system was developed, which allowed the study of the motion of the moving elements of the CNC equipment during the milling machining of parts.

Experimental tests were also carried out to verify the efficiency of the multi-axis (X- and Y-axes) motion control systems, realized on the basis of Arduino MEGA development boards using the MATLAB–Simulink package as real-time control software. The experimental tests also validated the accuracy of the simulation studies in terms of tuning the numerical controllers.

Of course, the proposed system had limitations, among which the following can be listed:

- The rigidity of the proposed equipment was relatively low, mainly also due to the small size of the structural elements;
- The possibility of adapting control systems based on Arduino control boards to an equipment of increased size and rigidity (scalability) needs to be confirmed by further research;
- The experimental tests were performed only on soft materials, which somewhat limited practical applications.
- Future research directions to be addressed include:
- Development of digital twins that also include software modules for processing information from sensors and transducers mounted on the physical system;
- Implementation of more advanced control strategies than PID strategies on the developed CNC system (feed-forward controllers, fuzzy controllers, etc.) or even AI-based control strategies;
- Use of more advanced microcontroller control cards (16 or 32 bit) as command-and-control systems for the developed CNC equipment;
- Integration of sensor feedback for real-time correction;
- Considering a wider range of workpiece materials.

Author Contributions: Conceptualization, C.-D.P., F.M., R.-E.B. and S.-G.R.; methodology, R.-E.B., S.-G.R. and M.C.; software, C.-D.P. and M.C.; validation, C.-D.P., F.M., S.-G.R. and R.-E.B.; formal analysis, C.-D.P., F.M. and C.-E.G.; investigation, C.-D.P. and F.M.; resources, R.-E.B., S.-G.R., C.-E.G., A.B. and C.-M.B.; data curation, C.-D.P., F.M. and M.C.; writing—original draft preparation, C.-D.P. and R.-E.B.; writing—review and editing, C.-D.P. and R.-E.B.; visualization, C.-D.P., F.M., C.-E.G., A.B. and C. B; supervision, R.-E.B. and S.-G.R.; project administration, R.-E.B.; funding acquisition, R.-E.B. and S.-G.R. All authors have read and agreed to the published version of the manuscript.

Funding: This research was funded by Lucian Blaga University of Sibiu (Knowledge Transfer Center) and Hasso Plattner Foundation research grant LBUS-HPI-ERG-2023 and grant LBUS-HPI-ERG-2023-02.

Informed Consent Statement: Not applicable.

Data Availability Statement: The data presented in this study are openly available.

Conflicts of Interest: The authors declare no conflicts of interest. The funders had no role in the design of the study; in the collection, analyses, or interpretation of data; in the writing of the manuscript; or in the decision to publish the results.

Abbreviations

The following abbreviations are used in this manuscript:

CNC	Computer Numerical Control
PID	Proportional Integral Derivative
CAE	Computer-Aided Engineering
CAM	Computer-Aided Manufacturing

References

1. Papadopoulos, K.G.; Papastefanaki, E.N.; Margaris, N.I. Explicit analytical PID tuning rules for the design of type-III control loops. *IEEE Trans. Ind. Electron.* **2012**, *60*, 4650–4664. [CrossRef]
2. Borase, R.P.; Maghade, D.K.; Sondkar, S.Y.; Pawar, S.N. A review of PID control, tuning methods and applications. *Int. J. Dyn. Control* **2021**, *9*, 818–827. [CrossRef]
3. Díaz-Rodríguez, I.D.; Sangjin, H.; Bhattacharyya, S.P. *Analytical Design of PID Controllers*; Springer: Berlin/Heidelberg, Germany, 2019; ISBN 978-3-030-18227-4.
4. Gai, H.; Li, X.; Jiao, F.; Cheng, X.; Yang, X.; Zheng, G. Application of a New Model Reference Adaptive Control Based on PID Control in CNC Machine Tools. *Machines* **2021**, *9*, 274. [CrossRef]
5. Sun, X.; Liu, N.; Shen, R.; Wang, K.; Zhao, Z.; Sheng, X. Nonlinear PID Controller Parameters Optimization Using Improved Particle Swarm Optimization Algorithm for the CNC System. *Appl. Sci.* **2022**, *12*, 10269. [CrossRef]
6. Yu, Z.; Liu, N.; Wang, K.; Sun, X.; Sheng, X. Design of Fuzzy PID Controller Based on Sparse Fuzzy Rule Base for CNC Machine Tools. *Machines* **2023**, *11*, 81. [CrossRef]
7. Liu, L.; Xu, Z.; Qu, X. A Reconfigurable Architecture for Industrial Control Systems: Overview and Challenges. *Machines* **2024**, *12*, 793. [CrossRef]
8. Latif, K.; Adam, A.; Yusof, Y.; Kadir, A.Z.A. A review of G code, STEP, STEP-NC, and open architecture control technologies based embedded CNC systems. *Int. J. Adv. Manuf. Technol.* **2021**, *114*, 2549–2566. [CrossRef]
9. Zhou, X. An optimized data interaction structure for SERCOS-based open CNC system. *Int. J. Adv. Manuf. Technol.* **2023**, *129*, 4643–4661. [CrossRef]
10. Liu, Q.; Liu, H.; Xiao, J.; Tian, W.; Ma, Y.; Li, B. Open-architecture of CNC system and mirror milling technology for a 5-axis hybrid robot. *Robot. Comput.-Integr. Manuf.* **2023**, *81*, 102504. [CrossRef]
11. Dharmawardhana, M.; Ratnaweera, A.; Oancea, G. STEP-NC Compliant Intelligent CNC Milling Machine with an Open Architecture Controller. *Appl. Sci.* **2021**, *11*, 6223. [CrossRef]
12. Attar, H.; Abu-Jassar, A.T.; Amer, A.; Lyashenko, V.; Yevsieiev, V.; Khosravi, M.R. Control system development and implementation of a CNC laser engraver for environmental use with remote imaging. *Comput. Intell. Neurosci.* **2022**, *1*, 9140156. [CrossRef]
13. Khan, Z.H.; Rahman, T.; Arabi, S.; Mohammad, S.; Khan, M.M.; Dey, R.; Khan, M.M.; Mukherjee, D. Development of a Low-Cost CNC Machine Laser Engraver. In Proceedings of the 2021 IEEE 12th Annual Ubiquitous Computing, Electronics & Mobile Communication Conference (UEMCON), New York, NY, USA, 1–4 December 2021. [CrossRef]
14. Das, U.C.; Shaik, N.B.; Suanpang, P.; Nath, R.C.; Mantrala, K.M.; Benjapolakul, W.; Gupta, M.; Somthawinpongsai, C.; Nanthaamornphong, A. Development of automatic CNC machine with versatile applications in art, design, and engineering. *Array* **2024**, *24*, 100369. [CrossRef]
15. Cao, X.; Zhao, G.; Xiao, W. Digital Twin–oriented real-time cutting simulation for intelligent computer numerical control machining. *Proc. Inst. Mech. Eng. Part B J. Eng. Manuf.* **2022**, *236*, 5–15. [CrossRef]
16. Yao, K.-C.; Chen, D.-C.; Pan, C.-H.; Lin, C.-L. The Development Trends of Computer Numerical Control (CNC) Machine Tool Technology. *Mathematics* **2024**, *12*, 1923. [CrossRef]
17. Luo, W.; Hu, T.; Zhang, C.; Wei, Y. Digital twin for CNC machine tool: Modeling and using strategy. *J. Ambient Intell. Humaniz. Comput.* **2019**, *10*, 1129–1140. [CrossRef]
18. Luo, W.; Hu, T.; Ye, Y.; Zhang, C.; Wei, Y. A hybrid predictive maintenance approach for CNC machine tool driven by Digital Twin. *Robot. Comput.-Integr. Manuf.* **2020**, *65*, 101974. [CrossRef]
19. Tao, F.; Xiao, B.; Qi, Q.; Cheng, J.; Ji, P. Digital twin modeling. *J. Manuf. Syst.* **2022**, *64*, 372–389. [CrossRef]

20. Heo, E.; Yoo, N. Numerical Control Machine Optimization Technologies through Analysis of Machining History Data Using Digital Twin. *Appl. Sci.* **2021**, *11*, 3259. [CrossRef]
21. Pan, L.; Guo, X.; Luan, Y.; Wang, H. Design and realization of cutting simulation function of digital twin system of CNC machine tool. *Procedia Comput. Sci.* **2021**, *183*, 261–266. [CrossRef]
22. Wei, Y.; Hu, T.; Zhou, T.; Ye, Y.; Luo, W. Consistency retention method for CNC machine tool digital twin model. *J. Manuf. Syst.* **2021**, *58*, 313–322. [CrossRef]
23. Pantelidakis, M.; Mykoniatis, K. Extending the digital twin ecosystem: A real-time digital twin of a LinuxCNC-controlled subtractive manufacturing machine. *J. Manuf. Syst.* **2024**, *74*, 1057–1066. [CrossRef]
24. Vishnu, V.S.; Varghese, K.G.; Gurumoorthy, B.J.P.C. A data-driven digital twin of CNC machining processes for predicting surface roughness. *Procedia CIRP* **2021**, *104*, 1065–1070. [CrossRef]
25. Liu, K.; Song, L.; Han, W.; Cui, Y.; Wang, Y. Time-varying error prediction and compensation for movement axis of CNC machine tool based on digital twin. *IEEE Trans. Ind. Inform.* **2021**, *18*, 109–118. [CrossRef]
26. Xue, R.; Zhang, P.; Huang, Z.; Wang, J. Digital twin-driven fault diagnosis for CNC machine tool. *Int. J. Adv. Manuf. Technol.* **2024**, *131*, 5457–5470. [CrossRef]
27. Yang, X.; Ran, Y.; Zhang, G.; Wang, H.; Mu, Z.; Zhi, S. A digital twin-driven hybrid approach for the prediction of performance degradation in transmission unit of CNC machine tool. *Robot. Comput.-Integr. Manuf.* **2022**, *73*, 102230. [CrossRef]
28. Yu, H.; Yu, D.; Wang, C.; Hu, Y.; Li, Y. Edge intelligence-driven digital twin of CNC system: Architecture and deployment. *Robot. Comput.-Integr. Manuf.* **2023**, *79*, 102418. [CrossRef]
29. Masory, O.; Koren, Y. Reference-Word Circular Interpolators for CNC Systems. *J. Eng. Ind.* **1981**, *104*, 400–405. [CrossRef]
30. Suh, S.-H.; Kang, S.K.; Chung, D.-H.; Stroud, I. *Theory and Design of CNC Systems*; Springer: London, UK, 2008.

applied sciences

MDPI

Article

Proactive Approach to Production Control Utilizing Heterogeneous Shop-Level Production Data

Fedor Burčiar *, Monika Herchlová, Bohuslava Juhásová *, Martin Juhás and Pavel Važan

Institute of Applied Informatics, Automation and Mechatronics, Faculty of Material Science and Technology in Trnava, Slovak University of Technology in Bratislava, 91724 Trnava, Slovakia; monika.herchlova@stuba.sk (M.H.); martin_juhas@stuba.sk (M.J.); pavel.vazan@stuba.sk (P.V.)

* Correspondence: fedor.burciar@stuba.sk (F.B.); bohuslava.juhasova@stuba.sk (B.J.)

Abstract

This paper presents an approach for integrating data between a production system and its digital twin, focusing on achieving proactivity in production control. Recognizing the unique nature of each production system, this research highlights that a universal, plug-and-play solution is only partially feasible, primarily through general guidelines. The study successfully applied and automated proposed data acquisition methods, resulting in a functional, simulation-based digital twin that adheres to the latest ISO standards. The developed solution incorporates multiple data acquisition strategies, including files containing comma-separated values, a permanent connection to the production control system database, open platform communications unified architecture, and external command files for scenario alteration. The main motivation behind the presented implementation is its application on the shop-floors of small and medium enterprises, where it could provide useful tools for keeping up with the ever-rising competition in the manufacturing sector. This integrated approach allows for affordable and accurate system representation within the proactive simulation concept. The methodology was empirically validated across two distinct production systems: a lab-scale food and beverage line focusing on product tracking, and a sub-assembly line with automated guided vehicle optimization. Despite system variability, the core data acquisition methods demonstrated remarkable adaptability.

Keywords: data acquisition; digital twin; simulation; production data

1. Introduction

The existing body of research has established the undisputable advantage of Digital Twins (DTws) for facilitating organizational efficiency in manufacturing [1–3]. The necessity to advance further in terms of analytics, real-time monitoring, and optimization has been growing alongside the identified need for a more efficient approach to manufacturing in the context of simulations [4].

Simulation and modeling has become a key instrument for improving industrial processes [5–7] and has remained one of the pillars of most modern enterprises. Simulation and modeling lies at the heart of innovation and agility and consistently yields positive results for respective use cases.

The most widely used approach to simulating manufacturing production lines are discrete event simulations (DES); however, depending on the system features, modelers also opt for continuous [8–11] or agent-based simulation. The best solution to realistically capture the system flow is a hybrid multi-layer simulation of a given cyber–physical production line. Hybrid simulations offer the most adaptive and high-fidelity simulations that unify the control flow and physical processes and can serve as a base for DTw implementation.

DTw technology improves on simulation technology with its core being real-time data integration and subsequent data-driven analytics for decision support. DTws help track and manage asset data flow and enable for reconfiguration at a keystroke [12–15]. DTws serve to represent a physical object or a system, virtually, with real-time data and are particularly useful for ease-of-use by utilizing simulation. Simulations enable visualization of relationships among subsystems, inner components of systems, and their ability to interact with their environments.

The benefits of DTw technology depend on specific uses and implementation. It can simulate production processes, optimize urban planning, monitor operations and continuously assess system variables. Studies have proven that DTws' application in enterprise resource planning (ERP) and manufacturing execution systems (MES) aids in decision-making and reduces response latency levels [16]. The combination of DTws with simulation achieves real-time control and optimization [17] of workflows [18], enabling for a closed-loop manufacturing system if integrated with artificial intelligence (AI) or machine learning (ML) This facilitates sizeable improvements in various aspects of industry use cases, such as error mitigation with real-time predictive maintenance [19] and safety monitoring of safety-critical infrastructures [20].

The aim of this study is to apply the Industry 4.0 concepts of cyber–physical systems (CPS) with real-time monitoring, data-driven control, and adaptive planning to demonstrate innovative approaches toward DTw creation for smart manufacturing systems where data sources are not only more unified than DTw solutions have been until this day but also to do so in a manner that complies with industry standards.

Analysis of common challenges and drawbacks faced in established implementations or presented case studies can be summarized as follows [21–24]:

- Limited scope of data types—Research focuses on specific data types and does not fully address the complexities of handling heterogeneous data.
- Focus on specific vendors—While the OPC UA architecture is aimed at vendor neutrality, implementations are still tied to specific vendors of PLCs or OPC UA software development kits [25].
- Economic viability—Implementations often lack cost–benefit analysis of data acquisition and integration solutions.
- Security simplification—The cyber-security vulnerability of presented applications is often outside of the primary scope of research.

In an attempt to solve some of these drawbacks, we have opted into proposing an approach which can help alleviate decision-making pressure from management workers responsible for production planning. Our approach revolves around incorporating multiple data acquisition methods, which are readily available in systems commonly found in small and medium enterprises (SMEs) and are often left underutilized. This is mainly due to the lack of trained workers or shop-floor-level expertise in data governance. While our proposed methodology still requires a higher degree of knowledge about the utilized systems, the all-in-one-place reference point of a complex solution can still be beneficial to the general base of knowledge. Keeping the full solution within a local network addresses the security aspect of the data acquisition process itself; however, the threat of outside breaches is still a possibility. This was only partially addressed by transferring minimal amounts of data, while its context is hard-coded into individual entities.

2. Data Management as a Prerequisite for Success

Industries and smart manufacturing organizations generate large amounts of data. After their acquisition and proper use, data allow for efficiency in organizational workflows as an asset creating value, central to innovation, development, and growth.

As defined by the standard, observable manufacturing elements (OMEs) are items within the manufacturing environment associated with production, materials, equipment use and other articles of the industrial processes. This suggests that companies must invest and focus on data management and an appropriate data strategy given the urge to make profit and gain insights from the values. However, data within organizations lack centered panoplies of their streams which has been an ongoing issue resulting in heterogeneity of secondary data sources, which is detrimental to the design process. It is at this phase of the model development process where operation comes to a halt due to the lack of an integrated unified center of data concerning the modeled entity, which hinders the process of information exchange between device communication entities [26] and OMEs in previous DTw adoption case studies [27–29].

Data is either available, acquirable or unavailable, which from that point onwards is a tall order to manage. The limiting conditions of unavailable data have been addressed by approaches of manual model development and by analyzing production logs. However, modern data-driven methods utilize advanced analytical techniques of statistical analysis and machine learning (ML) as they are suitable for complexity and dynamics representation [30,31].

In the creation of DTws, reliable data of the observable manufacturing domain are fundamental to enabling accurate system representation by OMEs' states, and the data ought not only to be available but also synchronized with the DTw in the device communication domain. The data stream, i.e., the communication update period, must also be continuous and seamlessly integrated with the DTw system. Therefore, it is beneficial to assume a proactive stance toward data management. Although studies have previously been conducted to aid in the development of DTws [4,32], a more straightforward approach for requirements and methodologies for proper implementation has been finally supplied by the ISO/IEC 30173 standard [33].

In order to avoid the possible emergence of inconsistencies, as observed in textual sources of multiple case studies and our empirical knowledge, we have investigated the applicability of the standard for our solution. The common framework proposed by the ISO/IEC 30173 standard has from then onwards helped in providing the following:

1. A basis for understanding the concept and composition of a DTw through definitions of digital twin-related concepts;
2. An overview of the life cycle of a DTw in relation to the target entity it represents;
3. A basis for the development of standards, specifications and uses of DTws.

The standard draws together a comprehensive definition of DTw concepts with terminology that can be applied in multiple domains. All relevant parts of the ISO 23247 standard, "Automation systems and integration—Digital twin framework for manufacturing", were considered. The series defines a framework to support the creation of DTws of OMEs, including personnel, equipment, materials, manufacturing processes, facilities, environment, products, and supporting documents. The individual parts considered include

- ISO 23247-1 [34]: An overview and general principles for developing DTws in manufacturing, including concepts and definitions, as well as requirements for designing a DTw.
- ISO 23247-2 [26]: A reference architecture for implementing DTws in manufacturing. It includes a reference model with domains and entities. There are four domains,

each with a logical group of tasks and functions performed by functional entities. The four domains include OME, data collection and device control, DTw, and user domains. Functional entities may include data acquisition, simulation, synchronization, etc.

- ISO 23247-3 [35]: A description of the basic informational attributes for a typical OME. The digital representation of OMEs may include both static and dynamic information. It is recommended to use existing standards to represent OME, based on use case requirements.
- ISO 23247-4 [36]: A description of the technical requirements for information exchange between individual elements (entities) within the defined reference architecture.

According to the standard, to ensure that our solution complies with digital twin entity requisites, it was necessary to assess our data collection sub-entity. For this purpose, we had to consider the data management of existing systems up until the time of DTw creation. To support the study, if necessary, data were aggregated so that they could be mapped onto the entity for seamless information exchange. Accessing the data entailed choosing the acquisition method and the rate of transfer. This played an important role in expediting the attempt for a proactive approach to task scheduling and production process simulation [37]. The act of transferring production data between the production control system (PCS) and the simulation platform can be facilitated in multiple ways, utilizing communication protocols on different technological levels, depending on the data subset transfer method.

3. Methodology

The four approaches discussed in this paper are aimed at the transference of selected production data between the PCS and the simulator. Each approach is proposed with the aim of achieving proactivity and prevention in production control as opposed to reactive issue management.

In our case studies, the first system is a lab-scale food and beverage production line with both discrete and continuous elements. As the system is in laboratory conditions, it is amenable enough to serve educational purposes and also as a testbed for further academic research and development. The other system is a sub-assembly line in an assembly shop in the automotive industry.

For funneling the necessary production data into the simulator, the data transference had to be done in a custom-made manner. Using industry standards, such as files containing comma-separated values (CSV), an open platform communications unified architecture (OPC UA) as well as a direct connection to the PCS database were a part of our proposed solution. In conjunction with the aforementioned data transference options was a simulator-specific command file, utilized as a tool for simulating different scenarios in rapid succession.

Figure 1 illustrates the architectural schema for data exchange among the systems detailed in this study. Continuous data flow is maintained between the PrS, the MES located on the PCS, and the database system throughout all phases of the production process. A system supervisor or operator can, upon request, initiate specific data streams linked to the data management interfaces and the simulation platform. This action triggers the generation and redirection of required data files to the simulator. The simulation process commences immediately upon validating all required input data files.

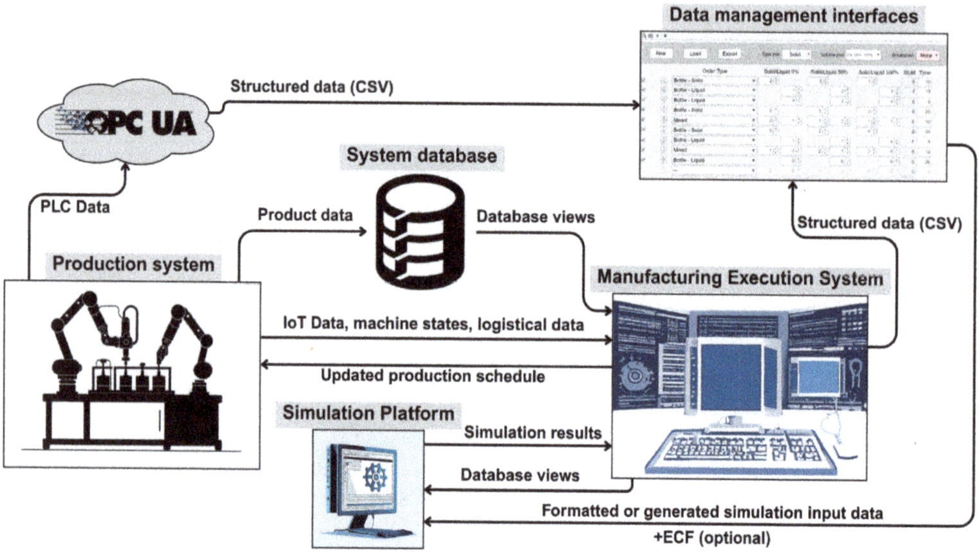

Figure 1. Data exchange schema for the proposed approach.

3.1. *CSV for Production Orders*

Production simulations necessitate the input of large amounts of data, which must adhere to predefined structure to ensure proper functionality. Some of the examples of this kind of data are production orders (PO), machine tasks or logistics schedule. Utilizing CSV files is a common approach to interconnect PCS and a simulator in general. Developing a DTw for a production system (PrS), regardless of whether concurrently with the PrS or for an existing one, defining a communication protocol is paramount. This protocol is used to establish explicit rules for generating and interpreting CSV file handling during the process. Unifying CSV file read and write rules for both the PCS and the simulator allows for better data interpretation and allocation within the respective entities.

When attempting to achieve seamless data file generation within the PCS and consequently integrating the generated data files with the simulator, multiple prerequisites for both entities need to be fulfilled.

3.1.1. Viewing and Modification of Incoming Production Orders

After performing the initial system analysis on the studied PrS, the first identified requirement was to integrate with a control platform, enabling a PrS operator to actively adjust the ongoing production. It could be done through, e.g., a Human–Machine Interface (HMI) panel or direct access to various interfaces of the manufacturing execution system (MES). In our studied use cases, the operator had access to data related to the production orders (PO), particularly to the incoming request execution priority. The operator was then able to adjust the priority of PO based on the current state of the PrS. The importance of fulfilling this requirement emerged from the fact that the control activity was performed at the start and at the end of the data integration and production process simulation.

To achieve full control over PO modification at the start of the data integration process, a custom multi-functional application tailored to the studied PrS, codenamed Order Management Interface (OMI), was proposed and implemented. The application shared a functionality similar to the overarching MES system while also allowing us to quickly generate multiple production scenarios for process analysis. To achieve this functionality, the OMI has been equipped with multiple control elements, which allow it to generate

CSV files and ECFs based on the state of these elements, e.g., setting a different priority for bottle filling from 0%–>50%–>100% to 50%–>0%–>100% or changing a filling priority from solid filling to liquid filling to be executed first, which changes the output file composition and creates a new scenario for the simulator. Similarly, the OMI operator can opt into simulating a partial breakdown using elements related to this functionality. This occurrence creates a potential scenario in the simulator, where the production system can only produce one type of products, while the broken-down subsystems related to the specified resource are undergoing maintenance. It served the purpose of an offline research alternative, while the MES still fulfilled the functionality of directly controlling the production process in the physical system.

The following figure (Figure 2) contains a simplified showcase of the data file structures created by the OMI at the start of the simulation process. Individual modules of the simulation model then parse the CSV files. Concurrently, the command file provides scenario-specific parameters. This process is described in a later chapter.

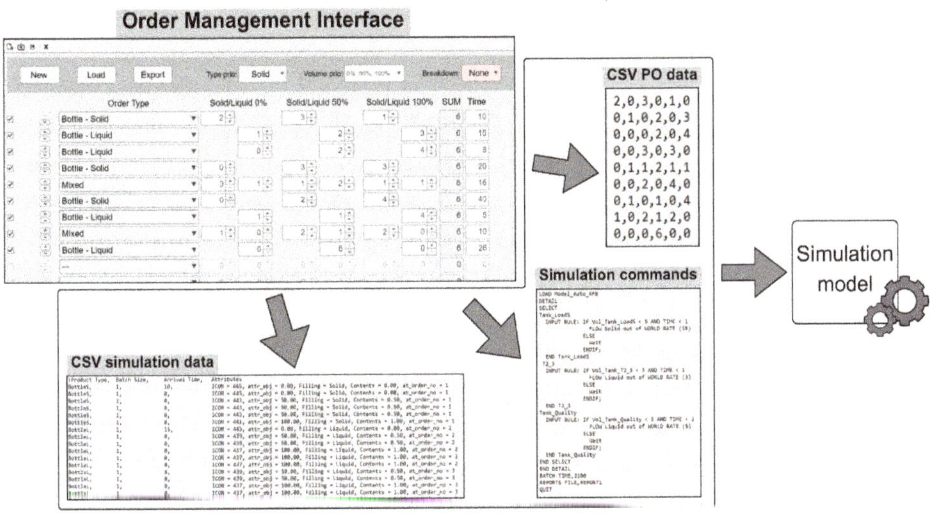

Figure 2. Data file creation through the OMI.

3.1.2. Providing a Coding Environment or a Platform to Run User-Defined Scripts

Another important prerequisite for seamless data integration is the existence of an integrated development environment (IDE) within the PCS, or at the very least an option to run custom user-defined scripts. This functionality also removes dependency on third-party interfaces, which means that localized databases are not exposed to external influence and only the items important to the production simulation process are being accessed. Considering the data integration process between the PCS and the simulation platform, the benefit of being able to write and modify the user scripts plays a significant role. This prerequisite originates from the fact that production process and PO data are stored directly within the database of the PCS system. The user is then able to access and extract the data using various system variables or attributes, which mitigates the time pressure being put on the data acquisition process. Data extracted by this method is used for two different data file types, which are then forwarded into the simulator; first the CSV PO file, then the command file.

3.1.3. Ability to Load Data from External Files

By utilizing specialized algorithms for altering the shop-floor production schedule, with stochastic factors present, it is possible to create different scenarios and replications without the loss of initial data. The main benefits to this approach identified thus far directly correlate to the possibility of running an offline simulation without the need for active connections to the PCS or any other data sources.

To achieve this functionality, it is important to implement a software module for a PCS which allows the operator to create various scenarios for the ongoing production process. Alternatively, building a custom-built interface or a series of interfaces with different functionalities is also an option, as these can be iterated and expanded upon, allowing for further production process customization. The only requirement for this approach is the ability to load the customized scenarios or production schedules back into the PCS, which implies that the involved systems must be fully integrated for data to be exchangeable between the subsystems. To prove the concept, as mentioned in previous chapters, we have developed a fully customizable GUI for the PCS at our disposal, which allowed us to proactively run multiple simulations with different scenarios using the DTw of our monitored PrS. The GUI was a MATLAB R2023b application, equipped with the functionality to load the specialized data files from the PCS, allowing the user, or a production operator, to make suitable adjustments to the shop-floor schedule before initiating the simulation. The endpoint of the data-altering process is followed by generating the data files containing the modified schedule for the simulator. In addition, the simulation process is initiated directly from the GUI using the command files.

As presented in the conceptual model in Figure 3, after a successful simulation run, the data transfer back into the MES was narrowed down to a single data file containing the PO data. It was then a much simpler process, as our MES was equipped with a functionality which allowed it to identify the creation of new files with the specified parameters. Upon identifying a successful reading of the incoming data file, the extraction algorithm was initiated. The extraction algorithm was a custom script implemented to purposefully apply the simulated changes to the upcoming production schedule. A scenario where the simulation process can be deemed as unsuccessful (the "no" branch on the MES and Production system level in Figure 3) is a possible occurrence. This state is related to the prolonged or delayed acquisition of simulated data from the simulator (the simulator or the PCS was temporarily disconnected from the local network) or the inability to read the simulated data inside the MES (an error occurred during the simulation, preventing the generation of an output file in accordance with the pre-specified protocol). In this scenario, a relevant error message is displayed, presenting details about the cause of the error to the operator. Figure 4 displays an example of how the raw data collected from the MES is interpreted within the simulator after reading the CSV files. The interpretation adheres to the rules set up for the first case study presented further in this paper.

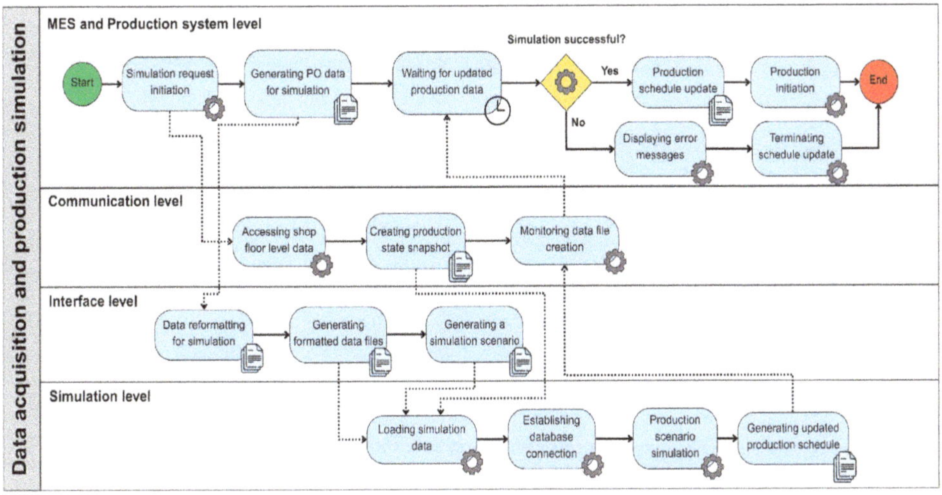

Figure 3. Conceptual model of the data acquisition process including simulation.

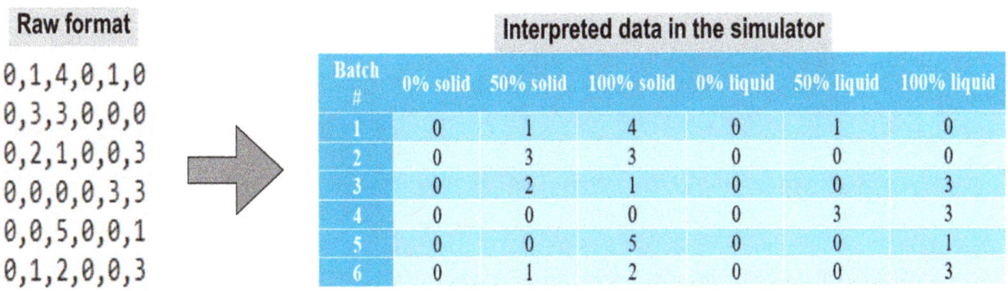

Figure 4. Interpretation of raw data in the simulator (current production plan).

3.2. CSV for Supervisory Data

To date, many SME category enterprises still operate on manual data transfer [38], most commonly using spreadsheets as a tool of choice for collecting and storing production data. Although the process lacks automation, it is still possible to point out some clear benefits to this approach, such as detailed data analysis or data reusability for future validation processes. However, to integrate the PCS with the simulator for a proactive approach to production control requires a certain level of automation. A similar approach can also be applied to data without a predefined structure, e.g., data located in PLC data blocks, where the format and contents vary depending on the workstation [39]. An ability to access the remote data located within the PLC data block is not considered a common functionality of PCSs, as this data can be easily exploited. In this instance, however, it is possible to implement custom scripts with the purpose of accessing this type of data, which can then be used to further calibrate the simulation models. Alternatively, it is possible to use an OPC server to access and gather specific production data, while having the ability to store the data in spreadsheets, if necessary. By utilizing this approach to gather and transfer data, it is possible to gain a real-time digital image of the current production state. It is also manageable to get insights into its very near future by snapshotting the real-time data and sending it directly into the simulator with the in-app tools of the simulator.

3.3. Direct Database Connection

As per all the data acquisition methods described in this paper, utilizing a local database connected to the PCS alongside a simulation-based DTw is an approach used for accessing, acquiring and displaying supervisory data to the operator. It can be considered an alternative approach to data transfer, where intermediary data storage is not available or required [40]. One of the benefits of a simulation-based DTw is the option to display the data located in databases directly within the simulation environment, providing useful information for decision-making in production control. Our simulator of choice had a functionality that allowed DTws to be directly connected to the database and offered a range of elements to display or manipulate the data by custom algorithmic coding practices.

The volume of data being accessed and transferred between the PCS and the simulator is highly dependent on the type and volume of data being stored in the PCS itself. It is also important to note that not all of the data stored in the PCS is required or applicable for use in the simulation using the DTw. Therefore, data selection and preliminary data formatting algorithms should be implemented before the full integration of a DTw with the PCS database. A good practice is to create highly specific views or tables inside the database, which are then directly read by the DTw. This reduces the risk of potentially exposing the full scale of vulnerable data to a possible malicious interception, as the simulation process can be performed remotely. Furthermore, interpretation of said data within the DTw, its alteration and re-formatting for the purpose of sending the data back to the PCS, is a part of the requirement for implementing the backwards communication of the DTw with the PCS, thus having the communication process fully automated.

Unlike other methods described in this paper, which utilize data file transfer to ensure their targeted functionality, when initializing a simulation process within the DTw utilizing the PCS database, an active connection to the PCS is required throughout the full duration of the simulation. This is due to the DTw being directly dependent on the current data located within the database for the purpose of displaying the product data to the operator or the system supervisor. Based on this type of connection, utilizing a database does not require any additional data files as the functional elements within the simulator handle the production data autonomously. A common scenario for utilizing this type of approach is when attempting to transfer production batch data, logistical unit data or recent historical data from individual production entities into the simulator, e.g., machines, storage units, transport vehicles, etc. This data is then used to reconfigure the simulation model where necessary, to adjust it to the current state of production.

3.4. Utilizing the Transport Layer Communication

A key focus is accessing raw production data from shop-level Programmable Logic Controllers (PLCs) and the communication protocols facilitating this data transfer to a control PC. For the purpose of this paper, we have considered communication protocols, which would allow us to gather the following data from the PrS in order to achieve our simulation goals:

- Timestamped events: Recording the precise moment of occurrence for process-defining events within the production system, e.g., product entering a machine, start/end of an operation, etc. This is crucial for analyzing the sequence of occurring events, potential identification of production disruptions, and quality control [41].
- Product movement tracking: Monitoring the flow of products as they progress through the production line. This data can involve tracking individual items, batches, or pallets, and is essential for inventory management, identifying potential bottlenecks throughout the PrS and, lastly, process optimization.

- Cycle time measurement: Determining the duration required for individual workstations or processes to complete their tasks. This specific type of data is highly valued during the process of creating a DTw for already existing and developed systems, as not all the data is available to the end user or documented in the PrS technical documentation [42].

This production data can be accessed in its raw state directly from the shop-floor PLCs. It provides a detailed and unfiltered view of the production process, minimizing data loss and maximizing accuracy.

To facilitate the production data transfer from the PrS to the control PC and further to the simulation platform, we have considered the use of the following three communication protocols. The main criteria for choosing a suitable data transfer protocol were reliability and efficiency.

- User Datagram Protocol (UDP)—a connectionless protocol that prioritizes speed over reliability. When using this method, data is transferred in packets without establishing a dedicated connection or ensuring its delivery to the designated receiver. While offering low latency, the potential for packet loss devalues the choice of this protocol for the resulting solution. Due to the requirements of a DTw fully integrated with the production system, any potential loss of transferred network packets is a significant drawback, as it can compromise the simulation accuracy and cause failed simulation runs, therefore possibly negatively affect subsequent decision-making.
- Transmission Control Protocol (TCP)—a connection-oriented protocol that prioritizes reliability and data integrity. In order to ensure that the packets are delivered in the correct order and retransmit any lost packets, it establishes a connection between the sender and receiver prior to the data transmission process. As data integrity is a crucial part of a successful simulation when using a DTw, the reliability of this communication protocol makes it more suitable for acquiring production data from the PrS, mainly due to the overhead actions associated with the connection establishment and error checking capabilities.
- Open Platform Communications Unified Architecture—the OPC UA protocol is a platform-independent, service-oriented architecture for industrial communication [43,44]. It provides a comprehensive framework for data exchange, encompassing not only raw data but also metadata, security, and information modeling. Among the three communication protocols considered, OPC UA operates with the highest volume of data. This does not necessarily pose a problem, since modern network infrastructures can easily handle the required volume without noticeably affecting the speed of the data acquisition process.

Figure 5 represents a topological layout of our established PLC network with a connection to the simulation platform through an OPC UA interface.

The data acquired through the OPC UA server is formatted into a JSON format which is then further adjusted for use in the simulation through our custom data integration interfaces. In the case depicted in Figure 6, the main attributes utilized within the simulator are the names of elements (element identification) and volume levels for individual tanks (initial value at the start of simulation). The remaining attributes extracted from the production system serve as control attributes for data integrity and trigger alarms, in case of data inconsistency occurrence. An example of data acquired from two PLCs in the system and their respective interpretation into an external command file is illustrated in the following figure.

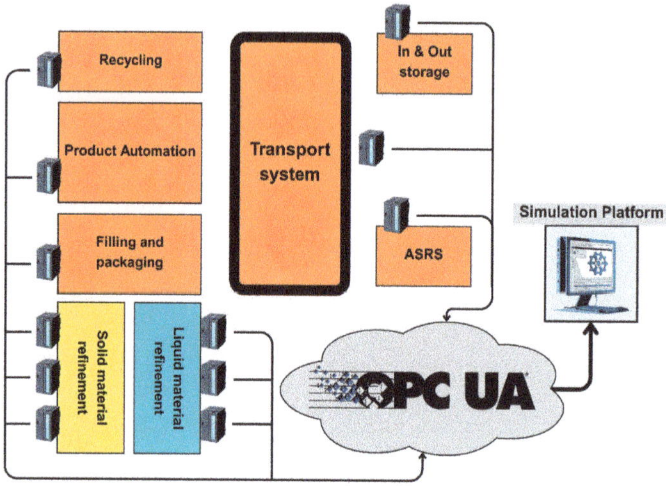

Figure 5. PLC network topology.

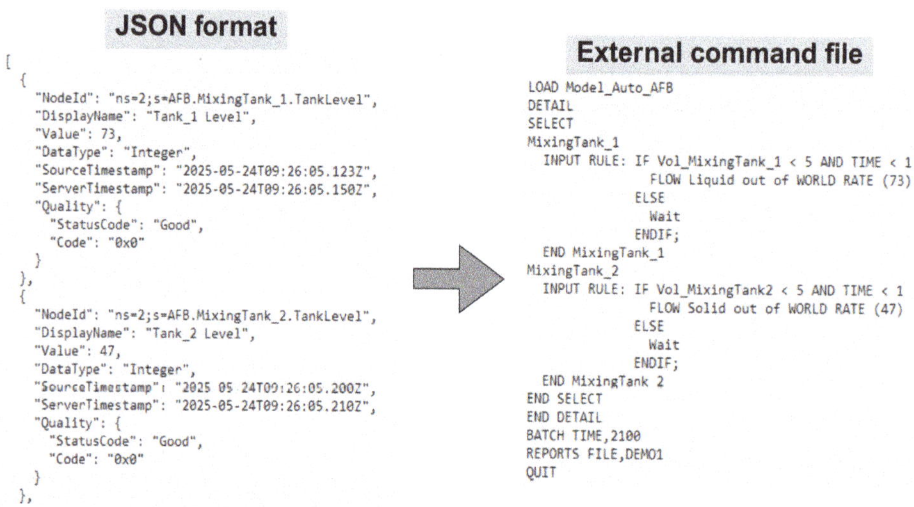

Figure 6. JSON format interpretation for simulation purposes.

A similar study by Coito, et al. (2022) [45] has shown that this protocol can be utilized to capture and transfer data related to machine availability and work states while keeping the data binary in nature and adjusting the interpretation where deemed necessary.

3.5. External Command Files

An alternative approach to production problem solving and production control using simulation tools and practices is by utilizing external command files (ECF). This approach exploits the functionalities of the chosen simulation platform to perform systematic modifications on existing models for the duration of a single simulation run, thus leaving the base model unchanged outside of the ongoing experiment.

The main prerequisite for utilizing this approach is to establish a default PrS state, which can be recreated within the simulator and used as a starting point able to be further altered through ECFs. Furthermore, it is possible to create a fully functional simulation model using ECFs exclusively, although this approach is not recommended by the developers of our simulator. This highly specific use case lacks proper visual representation of the model, i.e., it becomes unsuitable for the purpose of showcasing the ongoing processes [46].

By taking advantage of the ability to run the simulation through ECFs, it is possible to run multiple simulations in quick succession, while also being able to generate data or log files for further system analysis. For simulators which allow this type of simulation control, it is possible to create multiple ECFs specifically tailored to the requirements of the problem at hand. Furthermore, having sufficient familiarity with the simulated physical system and its digital model allows for the creation of ECFs which alter the functionality of selected workstations, functional entities, or even the entirety of the simulated process. Considering all the options and applications listed above, it is important to note that the exact size, structure, or usage of ECFs directly correlates to the specific needs of the problem at hand. This also holds true for other simulation platforms, though the syntax may differ for every individual software being used for this task.

In order to fully utilize the strengths of using ECFs for production process simulation, we opted into implementing various custom scripts for the system platform of the production control system (PCS), which would then automatically create these files. However, this puts a requirement on PCS choice, or any software directly connected to the PCS interface, to allow for the creation of such files. Furthermore, it is necessary to understand the syntax required for defining these CFs, and the syntax within the PCS scripting interface, as well as finding a way to combine these to get the desired output.

This approach comes with several benefits over standard methods used for data transfer between systems (e.g., through direct database connection). The most notable difference is in the way the PCS data is handled. By creating an ECF inside the PCS, the production data is accessed directly through internal variables and attributes, which store the sensor data from the PrS. This means that the raw data stays within the MES and only the data required by the simulator is being stored and transferred as a part of the ECFs. Due to the structure of these files, the risk of important data loss or potential harmful interceptions from outside sources is reduced.

4. Pilot Solution

After exploring available data acquisition options related to the proactively oriented DTws, we have concluded that a complex simulation task cannot be confined to a single source of production data. The ability to exploit multiple data acquisition methods simultaneously is necessary for carrying out accurate simulations. The most important result of our analysis was comprehending the appropriate applications of data acquisition methods for specific types of production data. For example, while production orders in the form of CSV data are only needed during the initiation process of a simulation task, a permanent connection to the PCS database has proven to be mandatory to identify and potentially react to any ongoing changes inside the PrS during the simulation process. Furthermore, it seems that capturing the real-time images of the production system through raw data located in the PLCs is imperative for a correct setup of the initial DTw state. Lastly, having an option to alter the existing model through ECFs for the specific scenario, whenever necessary and without directly modifying the base model, has proven to drastically affect how frequently we were able to simulate different production outcomes. However, when combining all the data acquisition methods described in this paper, even though the majority are sufficiently functional in an offline mode, the necessity of a permanent connection to the PrS database makes the implementation of our pilot solution in an environment which allows this type of connection a fundamental requirement. This presents additional hardware- as well as software-related challenges. The main challenge being the necessity for data acquisition process supervision by a skilled operator with real-time access to all related systems and the ability to perform adjustments in a timely manner. The biggest challenge related to shop-floor hardware was its age and wear. A similar occurrence can be identified in current-age SMEs, where hardware updates are not performed regularly due to the high costs of an ever-evolving industrial market and associated technology [47]. On the other hand, an overarching technological challenge related to software is the safety and reliability of used methods when exposed to external influence. A temporary solution to this problem was to keep the proposed approach in the laboratory environment for the duration of this research, although, it is important to note, that safety of vulnerable data is of the utmost importance when implementing any new technology into an ongoing production process [48]. To help alleviate time constraints, various alarms and control points have been set up throughout the entire data acquisition process monitoring failed or corrupt data file creation. Table 1 displays the proposed data transfer method allocation for every data type required for successful production process simulation.

Table 1. Application of proposed methods for the first use case.

Transfer Method	Production Orders	Supervisory Data	PLC-Level Data	Detailed Process Data for Simulation
CSV	(+)	(*)		(*)
Database	(*)	(+)		
UDP/TCP/OPC		(*)	(+)	(*)
ECF				(+)

(+)—chosen method for the proof of concept. (*)—method only partially viable for the specified type of data.

Our proposed methodology was empirically validated across two distinct production systems, each presenting unique operational characteristics. The initial system under investigation focused on the granular tracking and movement of individual products throughout its operational flow. In contrast, the second system's primary objective was the optimization and dynamic adjustment of workload distribution among automated guided vehicles (AGVs) and human operators.

The inherent variability in data provision within the MES across these two systems necessitated minor adaptations in data acquisition protocols. Despite these system-specific adjustments, the fundamental methodology for data acquisition remained consistent. This methodological uniformity across diverse operational contexts suggests a high degree of generalizability for our analytical framework.

4.1. Use Case #1: Agro-Food and Beverages Training Model

For the first model, the extraction of topological data was performed using two different approaches. Initial system parameters, such as the connections between production machinery, their spatial coordinates or the sequence of performed operations, were directly integrated into the simulation environment without requiring extensive pre-processing. However, the more intricate aspects of the fluid system, including tank levels and flow rates, along with precise positional data related to individual products within the PrS, necessitated a more detailed extraction process. This involved generating CSV files containing comprehensive product-specific data and utilizing ECFs to capture the real-time operational states of the machinery within the PrS. The final implementation of the simulation model, including an indication of individual data type utilization, is represented in Figure 7.

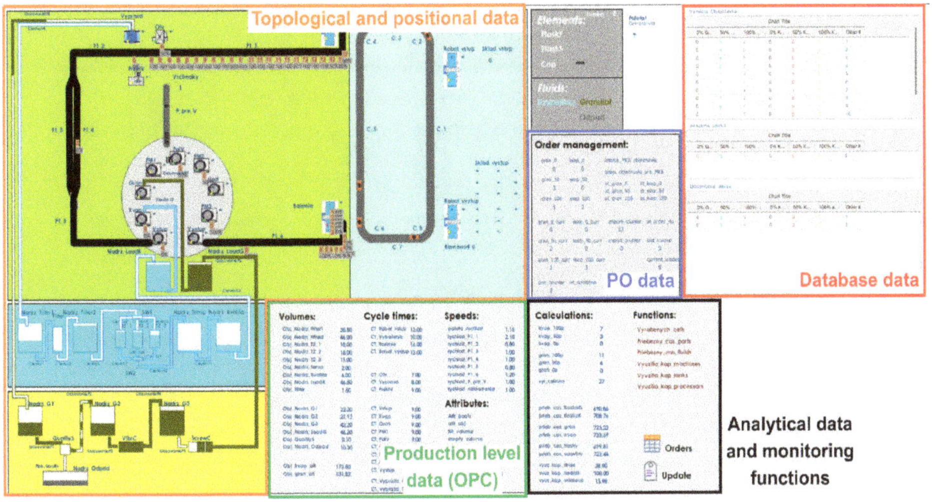

Figure 7. Simulation model with data utilization for the AFB Production Line.

The section of our simulation model dedicated to PO data integrates raw input from the MES. This data is then separated into singular values, which are subsequently allocated to individual products as they enter the PrS, or, in our case, the simulation model. Production level data, which provides a real-time snapshot of the PrS at the precise moment a simulation request is initiated, represents a critical component of each specific production scenario analyzed. This data was seamlessly integrated into the simulation model via the transport communication layer of the shop-floor PLCs.

The database interface within our model facilitates a real-time comparison between the current state of the simulated scenario and the anticipated outcomes. This interface presents comprehensive information to the operator, including details regarding the types and volumes of both planned and completed products.

Finally, the analytical and monitoring functionalities embedded within our model enable expeditious generation of reports and data files. These files can be subjected to further in-depth analysis after the simulation process is concluded. It is important to note that this analytical data is generated internally by the simulator itself and is not extracted from the physical PrS, serving instead as a direct reflection of the ongoing simulation's performance.

Exploration of possibilities for data utilization and identification of the scope of data, which can be seamlessly integrated directly into the simulation model without the necessity of data pre-processing algorithms, could be considered as the main takeaway from this implementation. Within the context of a production system which consists of discrete (individual products) and continuous (filling material) elements, it is important to distinguish between the static and dynamic parts of the production process. The processing machines, transport stations or robotic manipulators could be considered static, as the likelihood of changes in the shop-floor topology during an ongoing production is minimal. On the other hand, movement of individual products or the flow of resources used during the production process is not pre-defined, as these are dependent on the production plan at the moment of the production snapshot in the PCS. Static elements fall into the data category, which is often modeled in the simulator, and serves as a base structure for the movement of aforementioned dynamic elements. Furthermore, as mentioned by Bambura (2020) [49] in their research, the importance of storing and accessing relevant process data related to products or machine states in some form of database is important for near-real-time simulation of a production system. Furthermore, a study by Wang et al. (2023) [50] provided an in-depth analysis and discussed the importance of action adjustments and work reassignment for system components in relation to predictive maintenance.

As the data required for simulation is being collected locally, it is possible to establish an architecture in which the delay between the simulation request and the process of carrying out the production simulation is seemingly non-existent. This type of architecture supports the idea of the proactive approach to simulation, where identification of potentially unexpected production outcomes could occur.

4.2. Use Case #2: Sub-Assembly Line of a Just-in-Time Production Process

Building upon the methodological framework described in this paper, a similar approach was then used with a functionally and topologically dissimilar PrS. The variability between not only the tested systems but also any PrS in general excludes the possibility of universal applicability for our proposed approach. While not being able to create a plug-and-play solution, the data acquisition process itself demonstrated remarkable adaptability, requiring only minor adjustments when transitioning to the new PrS. Due to the differences between the two investigated systems, a new DTw as well as new data processing interfaces had to be created.

The simulation model for this case study, even though simpler in terms of representation, fulfilled a unique requirement for abstraction and design of background calculations. Notably, this system was a sub-assembly line within a larger production environment, focused on a Just-In-Time (JIT) production strategy. The time-sensitive nature of JIT processes amplified the impact of any adjustments or interventions, regardless of their scope. Due to this fact, exploration and simulation using different scenarios, facilitated through adjustments to various attributes using ECFs, were limited to a notable extent; however, the enabled utility for rapid experimentation and validation still positively contributed to the overall agility of the DTw. Data utilization in our second system is showcased in Figure 8.

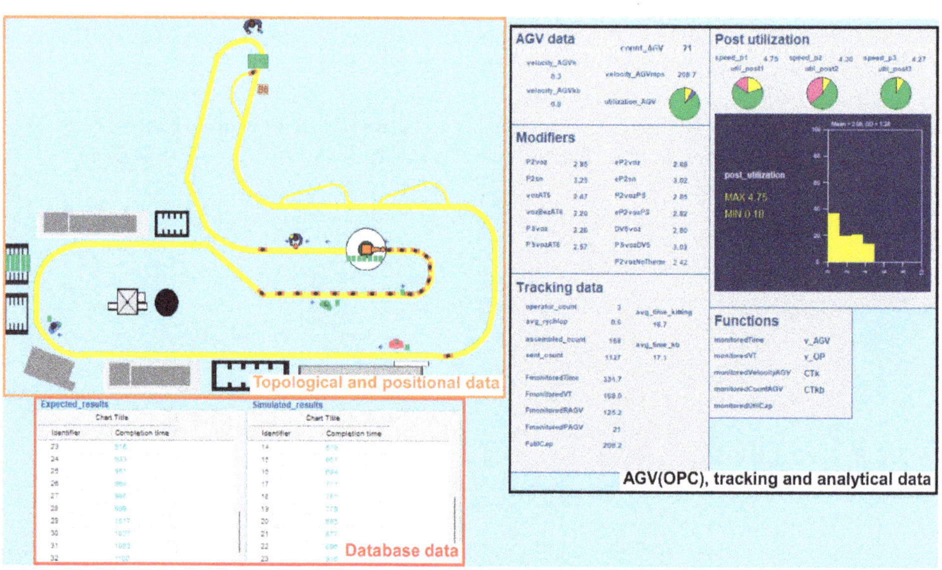

Figure 8. Simulation model with data utilization for the Sub-Assembly Line.

A key simplifying factor when creating a model for the second production system was the absence of fluid dynamics. Unlike the first PrS, this implementation allowed for a higher degree of abstraction in certain sections of the DTw, which further lowered the computational requirements for background processes without compromising its accuracy, while achieving the primary goal of the DTw: proactivity in production simulation.

A significant advancement in the data acquisition strategy for this system involved the direct utilization of an OPC UA server for real-time tracking of automated guided vehicles (AGVs). This approach exploited the inherent functionality of the AGVs' internal control systems and/or their associated fleet managers, which were designed to continuously publish their positional data [51]. This continuous stream of positional data (e.g., coordinates, velocity, status) was then leveraged by the manufacturing execution system (MES). The MES acted as an OPC UA client, subscribing to and collecting this dynamic data from the AGV's OPC UA server. This enabled the MES to generate snapshots of the current system status, which were then transmitted to the simulator [52], providing the DTw with up-to-date data on AGVs.

A similar approach was used by Zheng et al. (2019) [53], where they employed an OPC UA technology to gather data related to mechanical arms located within their respective manufacturing system.

The two presented iterations of our proposed methods adhere to the majority of recommendations and standards discussed in the ISO 23247 standard. Figure 9 showcases areas which our solution managed to incorporate (marked green). The remaining functional entities (FEs) are subject to future implementation.

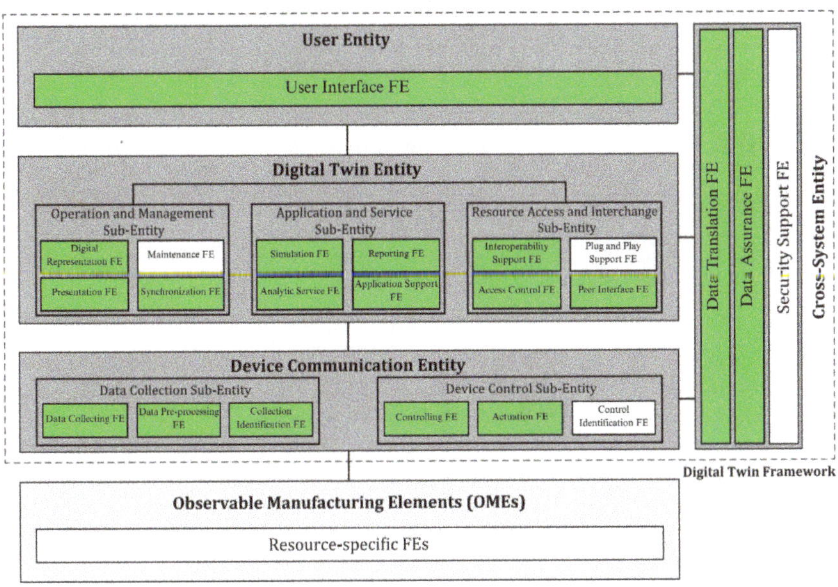

Figure 9. Achieved level of adherence to the ISO standards, adapted from [13].

5. Conclusions

This paper presented a unique method for facilitating data integration between a production system and its DTw. When compared to the existing case studies and research reports, we proposed a unique data type allocation for individual acquisition methods, while also incorporating a broader range of systems. While the proposed approach was specifically tailored to the presented use cases, several aspects offer reusable general guidelines for other systems, as demonstrated through testing on two distinct systems. Guidelines that can be deduced from our research can mainly be viewed as reference points for data allocation when attempting to build a DTw for an existing PrS. Our research indicates that a truly universal method applicable to all production systems is only partially achievable, primarily in the form of general guidelines, which can be derived from our procedures. This limitation stems from the inherent distinctive character of each production system, varying from topological configurations to data storage methodologies, therefore demanding custom-made solutions.

We have successfully applied and automated the proposed data acquisition methods, which lead to a fully functional simulation-based DTw with a focus on achieving proactivity in production control and adherence to the latest ISO standards.

The implementation, however, faced a few challenges and limitations, mainly related to old and outdated hardware throughout the production systems. Many of these challenges had to be overcome through the implementation of various software adaptations, which resulted in making the proposed methods a bit more system-specific than generally applicable. This was, however, a sub-optimal solution to the underlying problems, but, due to the high financial requirements of updating the old hardware, it was evaluated as a more suitable solution for the purpose of carrying out the necessary research.

Future research will focus on the optimization and thorough performance testing of these methods and algorithms.

Author Contributions: Conceptualization, F.B. and M.J.; methodology, F.B.; software, F.B. and M.H.; validation, F.B., M.H. and B.J.; formal analysis, M.J.; investigation, M.H. and B.J.; resources, F.B.; data curation, F.B.; writing—original draft preparation, F.B.; writing—review and editing, F.B. and M.H.; visualization, F.B. and M.H.; supervision, M.J.; project administration, P.V.; funding acquisition, P.V. All authors have read and agreed to the published version of the manuscript.

Funding: This research was funded by VEGA agency, grant number 1/0176/22 "Proactive control of hybrid production systems using simulation-based digital twin". This research was funded by Mladý výskumník, "Návrh inovatívneho prístupu k dátovej integrácii v procese riadenia výroby v kontexte Industry 4.0".

Institutional Review Board Statement: Not applicable.

Informed Consent Statement: Not applicable.

Data Availability Statement: The original contributions presented in this study are included in the article. Further inquiries can be directed to the corresponding author.

Conflicts of Interest: The authors declare that this study received funding from VEGA agency. The funder was not involved in the study design, collection, analysis, interpretation of data, the writing of this article or the decision to submit it for publication.

References

1. Pang, T.Y.; Pelaez Restrepo, J.D.; Cheng, C.-T.; Yasin, A.; Lim, H.; Miletic, M. Developing a Digital Twin and Digital Thread Framework for an 'Industry 4.0' Shipyard. *Appl. Sci.* **2021**, *11*, 1097. [CrossRef]
2. Aziz, S.; Jung, D.W.; Zaman, U.K.; Aqeel, A.B. Digital Twins in Smart Manufacturing. In *Handbook of Manufacturing Systems and Design*; CRC Press: Boca Raton, FL, USA, 2023; pp. 53–68. [CrossRef]
3. Korepin, V.; Mohamed, T.; Zhaksylyk, A.; Jie, L. Implementation of digital twins as a tool for increasing the efficiency of business operations. *Econ. Innov. New Technol.* **2024**, *34*, 689–704. [CrossRef]
4. Fantozzi, I.C.; Santolamazza, A.; Loy, G.; Schiraldi, M.M. Digital Twins: Strategic Guide to Utilize Digital Twins to Improve Operational Efficiency in Industry 4.0. *Future Internet* **2025**, *17*, 41. [CrossRef]
5. Moreira, B.M.D.N.; Gouveia, R.M.; Silva, F.J.G.; Campilho, R.D.S.G. A Novel Concept of Production and Assembly Processes Integration. *Procedia Manuf.* **2017**, *11*, 1385–1395. [CrossRef]

6. Ferrer, B.R.; Mohammed, W.M.; Martinez Lastra, J.L.; Villalonga, A.; Beruvides, G.; Castano, F.; Haber, R.E. Towards the Adoption of Cyber-Physical Systems of Systems Paradigm in Smart Manufacturing Environments. In Proceedings of the 2018 IEEE 16th International Conference on Industrial Informatics (INDIN), Porto, Portugal, 18–20 July 2018; pp. 792–799. [CrossRef]
7. Briceño, C.P.; Ponce, P.; Fayek, A.R.; Anthony, B.; Bradley, R.; Molina, A. Discrete event simulation for photovoltaic integration in sustainable manufacturing—A review and future directions. *Renew. Sustain. Energy Rev.* **2025**, *216*, 115676. [CrossRef]
8. Bandyopadhyay, S.; Bhattacharya, R. *Discrete and Continuous Simulation: Theory and Practice*, 1st ed.; CRC Press: Boca Raton, FL, USA, 2014.
9. Wang, Y. Virtual Simulation Modeling for Production Process of Continuous-discrete Hybrid Manufacturing System. *J. Syst. Simul.* **2008**, *20*, 2445–2453. Available online: https://en.cnki.com.cn/Article_en/CJFDTotal-XTFZ200809051.htm (accessed on 22 May 2025).
10. Rabelo, L.; Sarmiento, A.T.; Helal, M.; Jones, A. Supply chain and hybrid simulation in the hierarchical enterprise. *Int. J. Comput. Integr. Manuf.* **2015**, *28*, 488–500. [CrossRef]
11. Bursi, F.; Ferrara, A.; Grassi, A.; Ronzoni, C. Simulating Continuous Time Production Flows in Food Industry by Means of Discrete Event Simulation. *Int. J. Food Eng.* **2015**, *11*, 139–150. [CrossRef]
12. Moussa, M.; Abbas, M.; ElMaraghy, H. Industry 4.0 in Automotive Manufacturing: A Digital Twin Approach. *Procedia CIRP* **2025**, *134*, 825–830. [CrossRef]
13. Ricondo, I.; Porto, A.; Ugarte, M. A digital twin framework for the simulation and optimization of production systems. *Procedia CIRP* **2021**, *104*, 762–767. [CrossRef]
14. Attaran, M.; Celik, B.G. Digital Twin: Benefits, use cases, challenges, and opportunities. *Decis. Anal. J.* **2023**, *6*, 100165. [CrossRef]
15. Ginkel, M.; Yazdi, P.G.; Rupert, T.; Reijers, J.; Thiede, S. Digital Twins to Support Smart Manufacturing System Design and Ramp up. *Procedia CIRP* **2025**, *134*, 933–938. [CrossRef]
16. Mahankali, R. Digital twins and enterprise architecture: A framework for real-time manufacturing decision support. *Int. J. Comput. Eng. Technol.* **2025**, *16*, 578–587. [CrossRef]
17. Chen, Y.-P.; Karkaria, V.; Tsai, Y.-K.; Rolark, F.; Quispe, D.; Gao, R.X.; Cao, J.; Chen, W. Real-Time Decision-Making for Digital Twin in Additive Manufacturing with Model Predictive Control using Time-Series Deep Neural Networks. *arXiv* **2025**. [CrossRef]
18. Besigomwe, K. Closed-Loop Manufacturing with AI-Enabled Digital Twin Systems. *Cogniz. J.* **2025**, *5*, 18–38. [CrossRef]
19. Okpala, C.C.; Udu, C.E.; Nwankwo, C.O. Digital twin applications for predicting and controlling vibrations in manufacturing systems. *World J. Adv. Res. Rev.* **2025**, *25*, 764–772. [CrossRef]
20. Al-Ammari, W.A.; Sleiti, A.K.; Rahman, M.A.; Rezaei-Gomari, S.; Hassan, I.; Hassan, R. Digital twin for leak detection and fault diagnostics in gas pipelines: A systematic review, model development, and case study. *Alex. Eng. J.* **2025**, *123*, 91–111. [CrossRef]
21. Molina, A.; Vargas, D.; Rodas, A. Implementation of A Data-Acquisition System and Its Cloud-Based Registration Using the Unified Architecture of Open Platform Communications. *Eng. Proc.* **2023**, *47*, 20. [CrossRef]
22. Habib, K.; Saad, M.H.M.; Hussain, A.; Sarker, M.R.; Alaghbari, K.A. An Aggregated Data Integration Approach to the Web and Cloud Platforms through a Modular REST-Based OPC UA Middleware. *Sensors* **2022**, *22*, 1952. [CrossRef]
23. Klaeger, T.; Gottschall, S.; Oehm, L. Data Science on Industrial Data—Today's Challenges in Brown Field Applications. *Challenges* **2021**, *12*, 2. [CrossRef]
24. Ahoa, E.; Kassahun, A.; Verdouw, C.; Tekinerdogan, B. Challenges and Solution Directions for the Integration of Smart Information Systems in the Agri-Food Sector. *Sensors* **2025**, *25*, 2362. [CrossRef]
25. Mahdi, M.M.; Bajestani, M.S.; Noh, S.D.; Kim, D.B. Digital twin-based architecture for wire arc additive manufacturing using OPC UA. *Robot. Comput.-Integr. Manuf.* **2025**, *94*, 102944. [CrossRef]
26. *ISO 23247-2:2021*; Automation Systems and Integration—Digital Twin Framework for Manufacturing—Part 2: Reference Architecture. International Organization for Standardization: Geneva, Switzerland, 2021.
27. Lindow, K. Horizontal Value Creation through IoT and Digital Twin Data. In Proceedings of the Mechanical Engineering Congress, Japan, Okinawa, Japan, 11–14 September 2022; Volume 2022. [CrossRef]
28. West, S.; Stoll, O.; Meierhofer, J.; Züst, S. Digital Twin Providing New Opportunities for Value Co-Creation through Supporting Decision-Making. *Appl. Sci.* **2021**, *11*, 3750. [CrossRef]
29. Rantala, T.; Ukko, J.; Nasiri, M.; Saunila, M. Shifting focus of value creation through industrial digital twins—From internal application to ecosystem-level utilization. *Technovation* **2023**, *125*, 102795. [CrossRef]
30. Jamali, R.; Lazarova-Molnar, S. A Comprehensive Framework for Data-Driven Agent-Based Modeling. In Proceedings of the 2024 Winter Simulation Conference (WSC), Orlando, FL, USA, 15–18 December 2024; pp. 620–631. [CrossRef]
31. Bärring, M.; Johansson, B.; Flores-García, E.; Bruch, J.; Wahlström, M. Challenges of Data Acquisition for Simulation Models of Production Systems in Need of Standards. In Proceedings of the 2018 Winter Simulation Conference (WSC), Gothenburg, Sweden, 9–12 December 2018; pp. 691–702. [CrossRef]
32. Nath, S.V.; Van Schalkwyk, P.; Isaacs, D. *Building Industrial Digital Twins: Design, Develop, and Deploy Digital Twin Solutions for Real-World Industries Using Azure Digital Twins*; Packt Publishing Ltd.: Birmingham, UK, 2021.

33. *ISO/IEC 30173*; Digital twin—Concepts and Terminology. IEC Secretariat: Geneva, Switzerland, 2023.
34. *ISO 23247-1*; Automation Systems and Integration—Digital Twin Framework for Manufacturing, Geneva, Part 1: Overview and General Principles. IEC Secretariat: Geneva, Switzerland, 2021.
35. *ISO 23247-3*; Automation Systems and Integration—Digital Twin Framework for Manufacturing, Part 3: Digital Representation of Manufacturing Elements. IEC Secretariat: Geneva, Switzerland, 2021.
36. *ISO 23247-4*; Automation Systems and Integration—Digital Twin Framework for Manufacturing, Part 4: Information Exchange. IEC Secretariat: Geneva, Switzerland, 2023.
37. Machalski, A.; Szulc, P.; Błoński, D.; Nycz, A.; Nemś, M.; Skrzypacz, J.; Janik, P.; Satława, Z. The Concept of a Digital Twin for the Wały Śląskie Hydroelectric Power Plant: A Case Study in Poland. *Energies* **2025**, *18*, 2021. [CrossRef]
38. Alkhodair, M.; Alkhudhayr, H. Harnessing Industry 4.0 for SMEs: Advancing Smart Manufacturing and Logistics for Sustainable Supply Chains. *Sustainability* **2025**, *17*, 813. [CrossRef]
39. Radlbauer, E.; Moser, T.; Wagner, M. Designing a System Architecture for Dynamic Data Collection as a Foundation for Knowledge Modeling in Industry. *Appl. Sci.* **2025**, *15*, 5081. [CrossRef]
40. Caiza, G.; Sanz, R. Digital Twin to Control and Monitor an Industrial Cyber-Physical Environment Supported by Augmented Reality. *Appl. Sci.* **2023**, *13*, 7503. [CrossRef]
41. Lucchese, M.; Salerno, G.; Pugliese, A. A Digital Twin-Based Approach for Detecting Cyber–Physical Attacks in ICS Using Knowledge Discovery. *Appl. Sci.* **2024**, *14*, 8665. [CrossRef]
42. Grznár, P.; Papánek, L.; Marčan, M.; Krajčovič, M.; Antoniuk, I.; Mozol, Š.; Mozolová, L. Enhancing Production Efficiency Through Digital Twin Simulation Scheduling. *Appl. Sci.* **2025**, *15*, 3637. [CrossRef]
43. Mathworks. "OPC UA". Available online: https://www.mathworks.com/discovery/opc-ua.html (accessed on 2 June 2025).
44. OPC Foundation. "Unified Architecture". Available online: https://opcfoundation.org/about/opctechnologies/opc-ua/ (accessed on 2 June 2025).
45. Coito, T.; Faria, P.; Martins, M.S.E.; Firme, B.; Vieira, S.M.; Figueiredo, J.; Sousa, J.M.C. Digital Twin of a Flexible Manufacturing System for Solutions Preparation. *Automation* **2022**, *3*, 153–175. [CrossRef]
46. Lanner Group, Ltd. *Learning WITNESS Book Two Manufacturing Performance Edition*; Lulu Press, Inc.: Morrisville, NC, USA, 2013; ISBN 978.
47. Shi, Y.; Wen, Z.; Zhang, Z. Evolutionary Game Analysis of Empowering SMEs' Digital Transformation by Core Manufacturing Enterprises Under Government Subsidies. *Systems* **2025**, *13*, 225. [CrossRef]
48. Pochmara, J.; Świetlicka, A. Cybersecurity of Industrial Systems—A 2023 Report. *Electronics* **2024**, *13*, 1191. [CrossRef]
49. Bambura, R.; Šolc, M.; Dado, M.; Kotek, L. Implementation of Digital Twin for Engine Block Manufacturing Processes. *Appl. Sci.* **2020**, *10*, 6578. [CrossRef]
50. Wang, S.; Zhao, X.; Wu, C.; Wang, X. Joint optimization of multi-stage component reassignment and preventive maintenance for balanced systems considering imperfect maintenance. *Reliab. Eng. Syst. Saf.* **2023**, *237*, 109367. [CrossRef]
51. Yao, F.; Alkan, B.; Ahmad, B.; Harrison, R. Improving Just-in-Time Delivery Performance of IoT-Enabled Flexible Manufacturing Systems with AGV Based Material Transportation. *Sensors* **2020**, *20*, 6333. [CrossRef]
52. OPC Foundation. *Put OPC UA Pub/Sub & Companion Specifications to Work with Any HMI, SCADA, MES or Historian*. 2024. Available online: https://opcconnect.opcfoundation.org/2024/09/put-opc-ua-pub-sub-companion-specifications-to-work-with-any-hmi-scada-mes-or-historian/ (accessed on 2 June 2025).
53. Zheng, Y.; Yang, S.; Cheng, H. An application framework of digital twin and its case study. *J. Ambient. Intell. Humaniz. Comput.* **2019**, *10*, 1141–1153. [CrossRef]

applied sciences

MDPI

Article

Conceptually Simple Method for Optimizing Model Computations in MATLAB Simulink

Štefan Ondočko [1], Jozef Svetlík [1,*], Rudolf Jánoš [1], Ján Semjon [1], Marek Sukop [1], Tomáš Stejskal [2] and Peter Marcinko [1]

[1] Department of Manufacturing Machinery and Robotics, Faculty of Mechanical Engineering, The Technical University of Košice, Letná 9, 04001 Košice, Slovakia; stefan.ondocko@tuke.sk (Š.O.); rudolf.janos@tuke.sk (R.J.); jan.semjon@tuke.sk (J.S.); marek.sukop@tuke.sk (M.S.); peter.marcinko@tuke.sk (P.M.)

[2] Department of Engineering for Design of Machines and Transport Equipment, Faculty of Mechanical Engineering, The Technical University of Košice, Letná 9, 04001 Košice, Slovakia; tomas.stejskal@tuke.sk

* Correspondence: jozef.svetlik@tuke.sk; Tel.: +421-55-602-2195

Abstract

This article describes a procedure for enhancing computational accuracy in MATLAB's Simulink and Simscape environments, as illustrated through specific example cases. It builds on earlier published by the authors' team, which demonstrated the practical application of the Simscape Multibody tool—originally designed for dynamic and kinematic analyses—for making static computations in truss systems. Simscape Multibody serves as an effective platform for realistic and simplified simulations of mechanical components, incorporating various mechanical properties. Consequently, it is valuable in simulating mechatronic systems, where the integration of mechanics, electronics, control systems, and information technologies is essential. Multiple models were tested and analyzed across different scenarios to facilitate a comparative assessment of the results. The significance of this work lies in its achievement of highly accurate computational results without relying purely on theoretical calculations, with superior values in terms of accuracy. The primary objective was to provide a clear and practical description of a simple procedure for improving computational accuracy, based on scaling.

Keywords: static analyses; mechanics; beam; bars; trusses; MATLAB; Simulink; scaling

1. Introduction

This article builds upon our previous publication [1], which addressed the analysis of a truss system using MATLAB–Simscape Multibody [2,3] version 24.1.0.2628055 (R2024a) Update 4. It presents additional examples to further clarify and elaborate on this computational approach. The paper also provides a more detailed explanation of how an additional scaling [4] procedure can be utilized to enhance computational accuracy within the Simulink environment. To ensure the reliability of the simulation results, the system was also analyzed within the framework of Hooke's law (linear elastic behavior). Due to the inherent complexity of such computations, particularly when dealing with more sophisticated structural systems, a range of software tools are typically employed. When the primary objective is to determine deformations and internal forces in components that may be part of a broader mechatronic system [5,6], Simscape Multibody proves to be a useful toolbox. It serves as a successor to the formerly used SimMechanics [7], which has been widely applied in mechanical simulations, as demonstrated in [8,9]. The approach described herein enables a notable increase in computational accuracy without significantly extending the simulation time. Using this approach, computational accuracy can be improved to the level of approximately 10^{-6}. Our testing has also demonstrated that this approach can, to a certain extent, be applied more broadly across the Simulink platform. Most toolboxes already have scaling by Nominal Values [10] implemented.

While the Multibody toolbox does not constitute a fully fledged finite element analysis (FEA) tool—particularly with regard to large deformations, nonlinear material strength behavior, and overall computational capacity—it nonetheless offers considerable value to users operating within the Simulink environment, especially when integrating additional toolboxes. It allows users to avoid constructing excessively complex mechanical models. The Multibody library includes a set of components known as flexible bodies [3], which represent elements capable of elastic deformation under combined loading conditions, specifically tension, compression, bending, and torsion [11]. The stiffness and inertia of these bodies derive from their geometric and material properties. The computational principle applied to bars follows the Euler–Bernoulli beam theory [12] (pp. 174–220), which is suitable for modeling truss systems. The computational setup also includes the visualization of results based on user-defined input parameters. For comparative purposes, each example is accompanied by results obtained through analytical procedures. In this context, the term "bars" refers to structural members with uniform cross-sections, fixed at both ends via joints. The connection point of at least two such bars is defined as a node. It should also be noted that under compressive loading, buckling may occur [13] (pp. 483–507), resulting in a loss of stability and deflection of the bar through buckling from its original axis. This article does not address the assessment of buckling stability [14] in compression-loaded bars. For detailed analyses and theoretical background related to the stability of structures, the reader is referred to [15] (pp. 449–511).

Ultimately, the article deals in more detail with the explanation of a simple procedure for improving computational accuracy, using simplified scaling depending on a single input variable. Scaling by nominal values allows you to set the value of a variable for the solver in such a range that the dynamics of the calculation of all variables in a given system are matched. This increases the robustness of the simulation. The benefit is faster convergence of the solution and maintaining a minimum iteration step size. In Simscape Multibody™, there is no scaling system [10] in the sense that we know it for most Simscape toolboxes using nominal values. This can be solved by changing the proportion of the variable, but this may result in a change in the output values. If we do not want to change the number of parameters and adapt the model for numerical purposes, we only shifted the value of the input quantity. Due to this simple adjustment, it is of course not possible to maintain the calculation accuracy in the same class (10^{-6}) with a dynamically changing input variable (in our case $\mathbf{F}_{\text{load m}}$). But it is especially suitable for steady states. In our case, for example, for static calculations. The most effective is to use automatic solver selection. The software will use an algorithm to select the solver (method) and the maximum step size. We emphasize that this is not a description of a new method, but a better use of the potential of existing tools. For the input of the scaling procedure, we can simply write input$\{\mathbf{F}_{\text{load m}}/con\}$ and for the outputs$\{(\mathbf{F}, \mathbf{N}, \delta) \times con\}$. This procedure, which is used in the examples, can bring a benefit in terms of accuracy of about two orders of magnitude.

2. Illustrative Planar Examples

This section presents selected examples of truss systems, which are used to demonstrate methods for achieving maximum computational accuracy in Simscape Multibody. In each example, the models differ slightly in the type of bar employed and in the way in which the node is implemented. These variations are deliberate, as we will later examine how the toolbox handles the visualization of structural loading. The models are partially inspired by publicly available sources [16], which have been adapted for various configurations.

The first tested structure is the truss system shown in Figure 1. It consists of five nodes and seven bars, and is once externally statically indeterminate (see, e.g., [17], pp. 107–108).

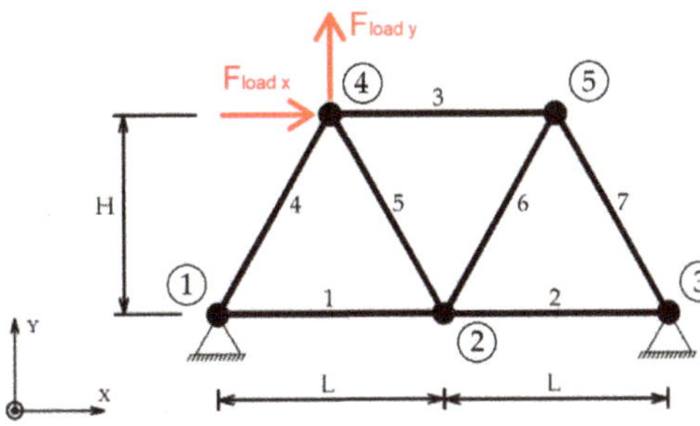

Figure 1. Example 1. A schematic of the truss system subjected to the force vector $\mathbf{F}_{load}$ at Node_4.

2.1. Example 1: Three Modeling Variants

A schematic of the truss system used in Example 1 is shown in Figure 1. The structure was experimentally modeled using three different approaches (models):

- Model 1.1 is composed of Euler–Bernoulli beams and joints.
- Model 1.2 serves as an experimental variant. It also employs Euler–Bernoulli beams, but it uses solid joints. This configuration does not conform to the assumptions of the theoretical model, as the nodes do not incorporate movable joints.
- Model 1.3 consists of rigid bars with one degree of freedom (1 DOF) and joints. This model most closely resembles the conventional standards used in static structural analysis software.

The truss system is subjected to a force vector applied at Node_4. All bars have identical cross-sections.

2.1.1. Model 1.1: Truss System Composed of Euler–Bernoulli Beams and Joints

Model 1.1, constructed in Simscape Multibody, is depicted in Figure 2. The same block structure is used for Model 1.2 and Model 1.3. All functional models are available at [18].

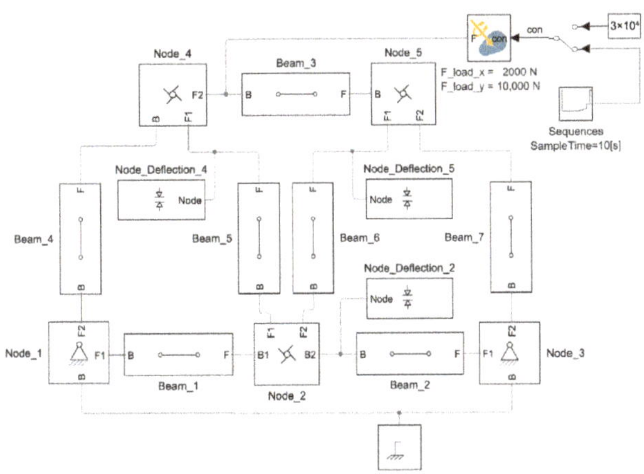

Figure 2. Models 1.1, 1.2, and 1.3. The truss system subjected to the force vector $\mathbf{F}_{load}$ at Node_4, modeled in MATLAB–Simscape Multibody as defined in Figure 1.

Figure 3 illustrates the subsystem of a truss bar, based on the Bernoulli beam model. Only axial force is considered. All truss bars in simulation Model 1.1 and Model 1.2 are identical in structure and constructed as follows:

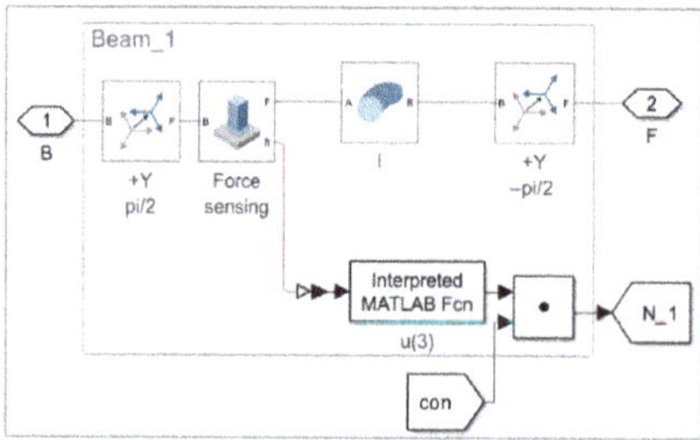

Figure 3. Models 1.1 and 1.2. An example of a truss bar subsystem, labeled Beam_1.

Blocks of the "Rigid Transform" type are used to rotate the truss bar into the desired plane about the y–axis of the World Frame. The value of the force vector $\mathbf{N}_1$ is obtained directly from the "Weld Joint" block using *Total force sensing*. The "Interpreted MATLAB Function" block—specifically, its output $u(3)$—extracts the z-component of the force vector $\mathbf{N}_1$ acting within the bar. Although the manufacturer recommends using "User-Defined Functions" [19] (p. 344) or their online counterparts [20], in order to improve computational speed, we have chosen the "Interpreted MATLAB Function" block, due to its greater flexibility in parameter syntax. The block that physically represents the truss bar, labeled *Beam*, is of the type "Flexible Cylindrical Beam". Within this block, key physical parameters are specified: the bar length l, the diameter d, the Young's modulus of elasticity E, the shear modulus G, the material density (*Density*), and the coefficients of proportional Rayleigh damping [21], denoted as b_m and b_k.

Figure 4 illustrates the subsystems by node in Model 1.1 and Model 1.3. While structurally similar, each node exhibits differences in configuration and in the specification of the initial position state (in radians). All nodes are constructed using joints.

The $\mathbf{F}_{\text{load m}}$ force essentially serves as a means of verifying the computed $\mathbf{F}_{\text{load}}$ force. It is derived from the internal forces within the system, specifically based on the equilibrium conditions at Node_4.

$$\sum F_x = 0;\ F_{\text{load }x} + N_3 - \frac{N_4}{2} + \frac{N_5}{2} = 0 \tag{1}$$

$$\sum F_y = 0;\ F_{\text{load }y} - \frac{\sqrt{3}\cdot N_4}{2} - \frac{\sqrt{3}\cdot N_5}{2} = 0 \tag{2}$$

The $\mathbf{F}_{\text{load}}$ force represents the external load applied to the system. In cases where multiple external forces act on the system, the one with the highest magnitude is selected for validation purposes.

2.1.2. Model 1.2: Truss System Composed of Euler–Bernoulli Beams and Solid Joints

The truss system defined as Model 1.2, constructed in Simscape Multibody, features the same block configuration as shown in Figure 2. The only distinction lies in the implementation of the nodes.

Figure 5 illustrates the subsystems by node in Model 1.2. Each node differs in its configuration and initial position state (in radians). The nodes are composed of solid joints.

As in the previous case, the $\mathbf{F}_{\text{load m}}$ force is derived from the equilibrium conditions at Node_4, as expressed in Equations (1) and (2). However, this interpretation is not entirely accurate in this context, since the nodes do not incorporate movable joints. Nevertheless, it is sufficient for the intended purpose.

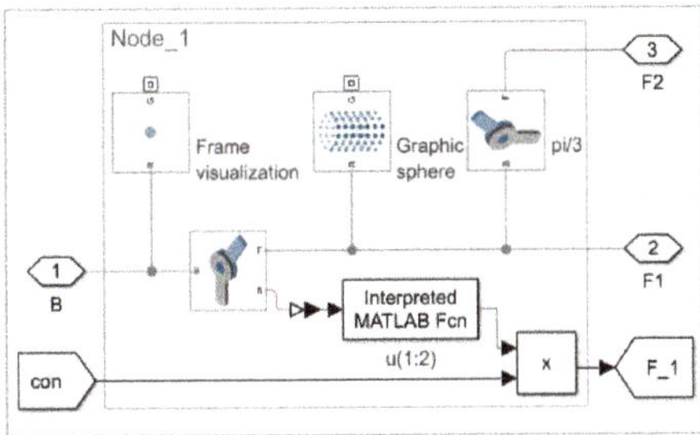

(a)

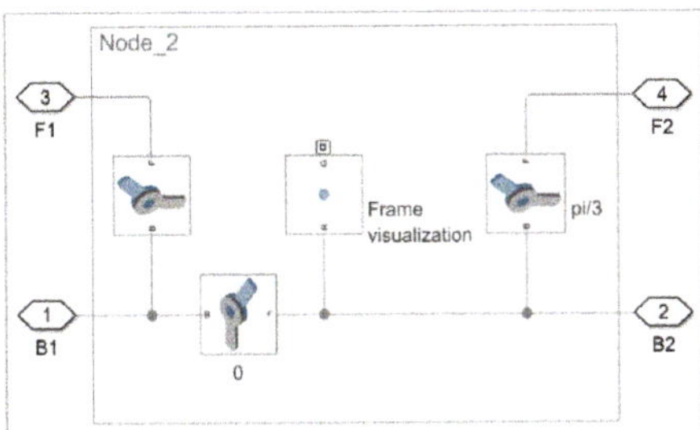

(b)

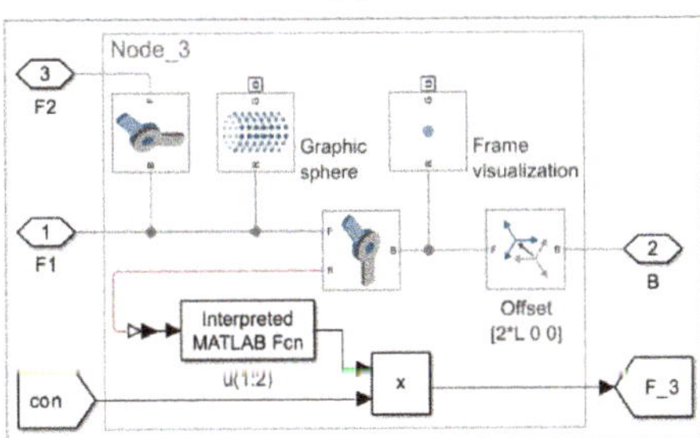

(c)

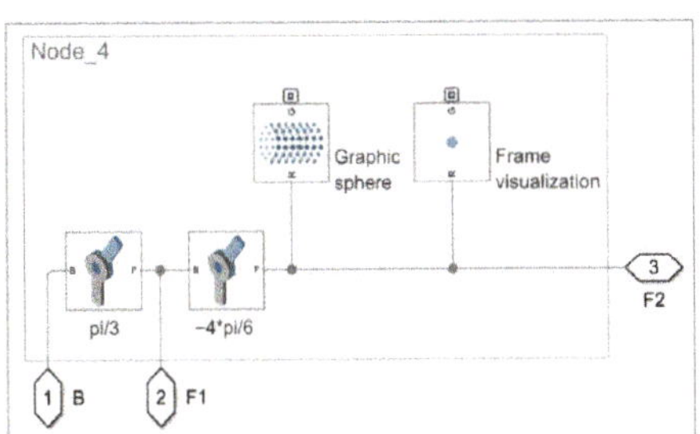

(d)

Figure 4. *Cont.*

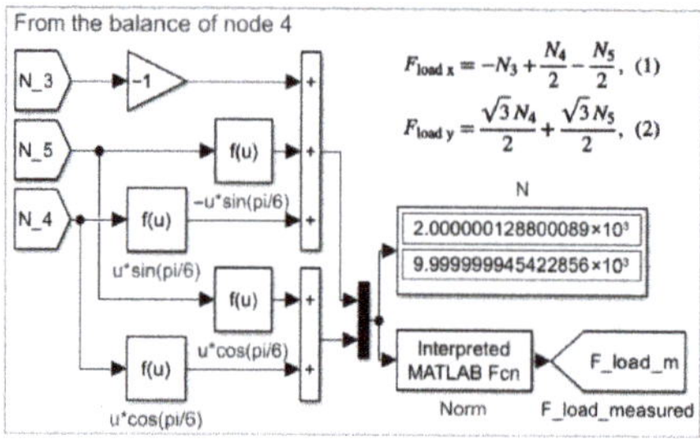

(e)

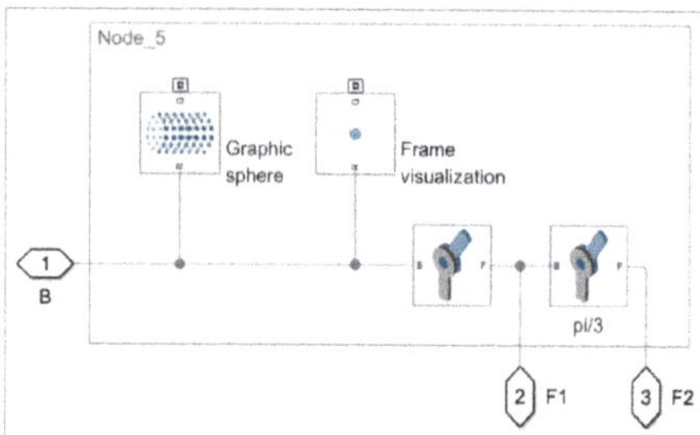

(f)

Figure 4. Model 1.1 and Model 1.3. Subsystems by node: (**a**) Node_1; (**b**) Node_2; (**c**) Node_3; (**d**) Node_4; (**e**) measurement of the $\mathbf{F}_{\text{load m}}$ force at Node_4; (**f**) Node_5.

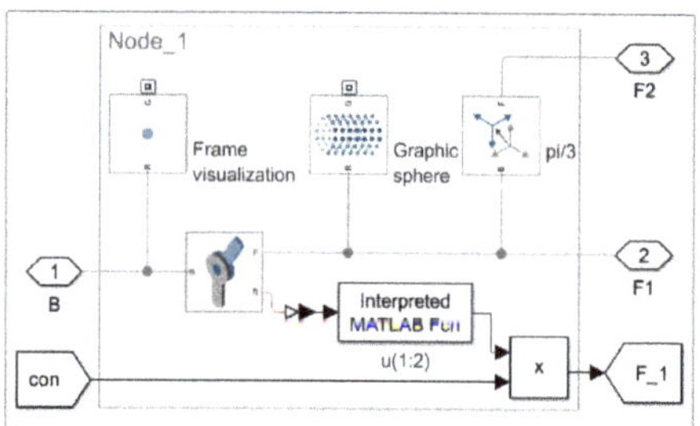

(a)

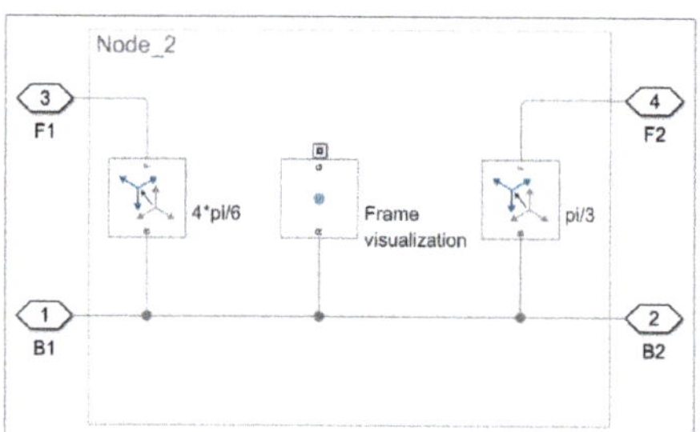

(b)

Figure 5. *Cont.*

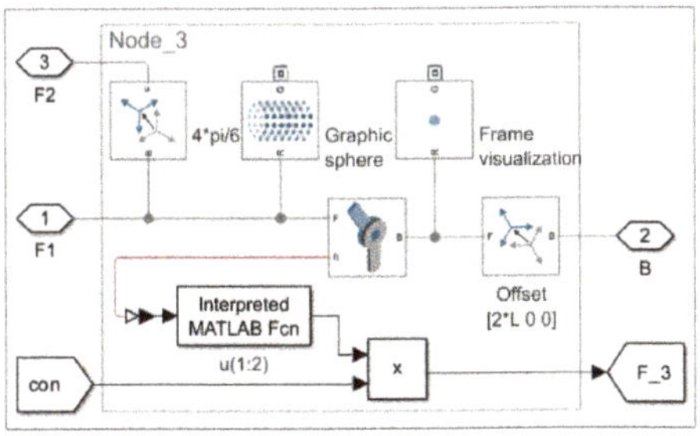

(c)

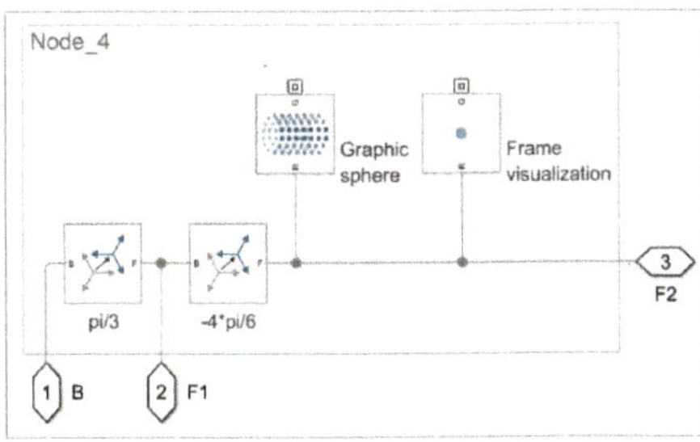

(d)

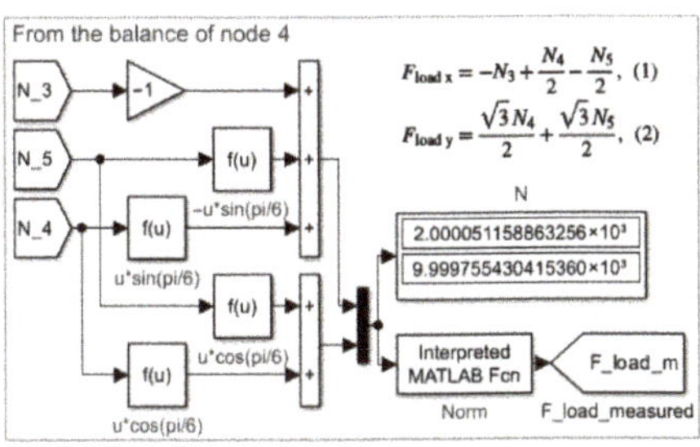

(e)

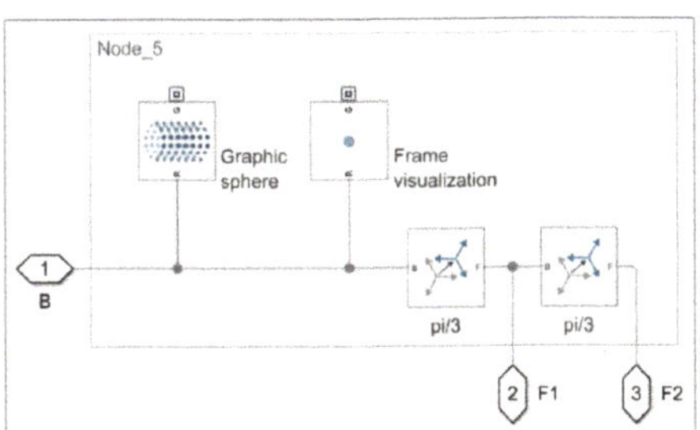

(f)

Figure 5. Model 1.2. Subsystems by node: (**a**) Node_1; (**b**) Node_2; (**c**) Node_3; (**d**) Node_4; (**e**) measurement of the $\mathbf{F}_{\text{load m}}$ force at Node_4; (**f**) Node_5.

2.1.3. Model 1.3: Truss System Composed of Solid Bars with One Degree of Freedom (1 DOF) and Joints

Model 1.3 of the truss system, constructed in Simscape Multibody, employs the same block configuration as shown in Figure 2. The nodes are constructed identically to those in

Model 1.1, as illustrated in Figure 4. The only difference lies in the structural implementation of the bars.

Figure 6 illustrates the subsystem of a truss bar, which consists of two rigid bodies connected by a prismatic joint with one degree of freedom (1 DOF). The joint defines a stiffness parameter k with a value of ES/l, damping b, and an initial position set to zero. All the truss bars in the simulation model—Model 1.3—share an identical structure and are constructed as follows:

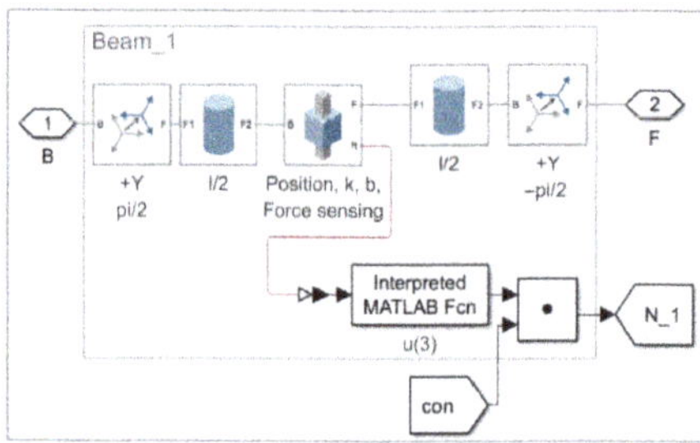

Figure 6. Model 1.3. An illustration of the bar subsystem named Beam_1. The subsystem comprises two solid bodies connected by a prismatic joint with one degree of freedom (1 DOF).

The measurement of nodal displacements δ follows the same principle across all models (Model 1.1, Model 1.2, and Model 1.3). The only difference lies in the (x, y) positioning of the nodes; therefore, only one representative example is shown. Figure 7 presents the subsystem named Node_Deflection_4.

"Transform Sensor"—*Translation* [x, y, z] are employed to measure the displacement relative to the World Frame. The remaining blocks are used to isolate the vector components x, y corresponding to displacement in the plane relative to the original position.

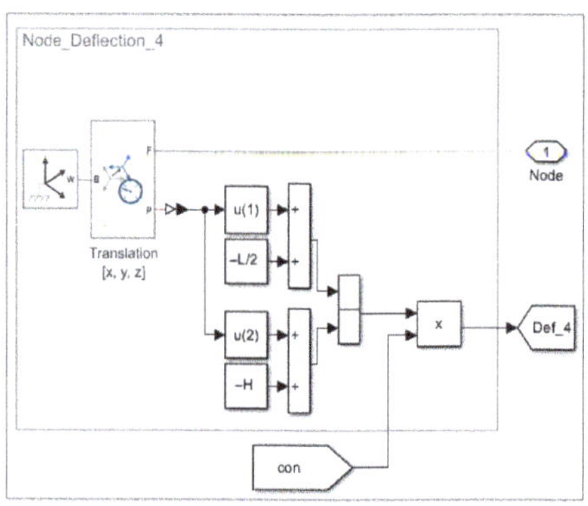

Figure 7. Model 1.1, Model 1.2, and Model 1.3. A detailed view of the deflection measurement subsystem labeled Node_Deflection_4 at coordinates $x = -L/2$, $y = -H$.

Figure 8 presents the subsystem responsible for applying the external load $\mathbf{F}_{\text{load}}$, including its proportional correction via a multiplication constant *con*. This configuration is identical across all models (Model 1.1, Model 1.2, and Model 1.3).

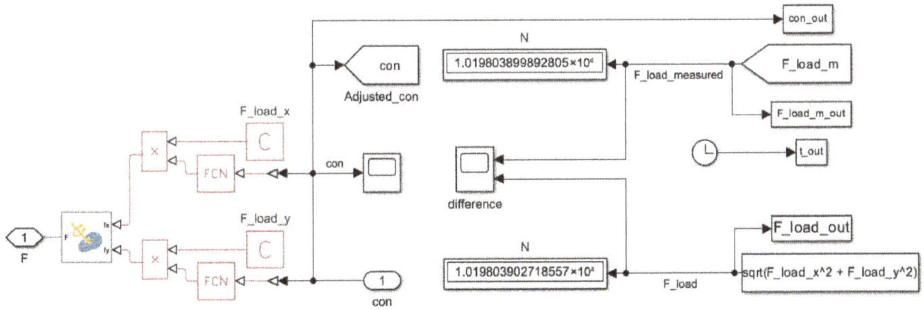

Figure 8. Model 1.1, Model 1.2, and Model 1.3. The subsystem under application of the external load F_{load}, including its proportional adjustment via the multiplication constant *con*.

2.2. Example 2: Three Modeling Variants

Figure 9 presents a schematic of the second truss system. The structure is also experimentally modeled using three different approaches:

- Model 2.1 consists of Euler–Bernoulli beams and joints.
- Model 2.2 is primarily experimental. It is composed of Euler–Bernoulli beams and solid joints. This model does not satisfy the assumptions of the theoretical model, as the nodes do not contain movable joints.
- Model 2.3 comprises solid bars with one degree of freedom (1 DOF) and joints.

The truss system is subjected to an external load applied at Node_11. All bars have an identical cross-section.

2.2.1. Model 2.1: Truss System Composed of Euler–Bernoulli Beams and Joints

Model 2.1, implemented using Simscape Multibody, is presented in Figure 10. The fully assembled functional models are available in [18]. The same block configuration is applied in both Model 2.2 and Model 2.3.

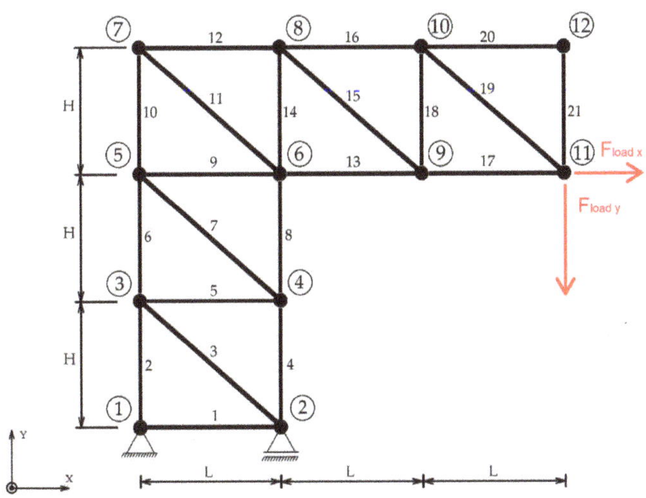

Figure 9. Example 2. A schematic of the truss system subjected to a force vector F_{load} at Node_11.

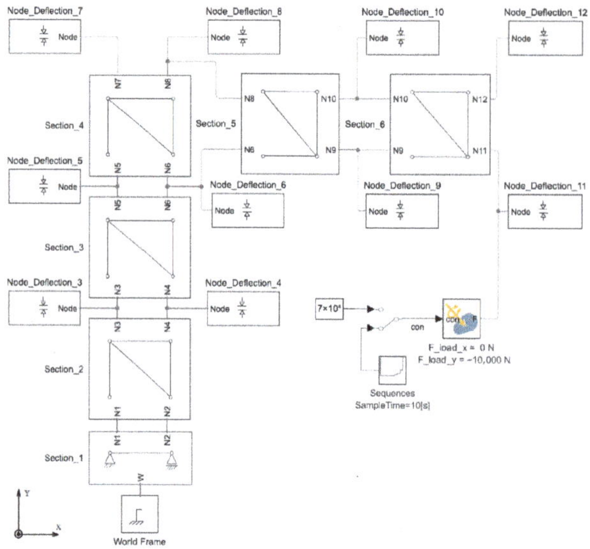

Figure 10. Model 2.1, Model 2.2, and Model 2.3. The structure of the truss system subjected to the force vector $\mathbf{F}_{\text{load}}$ at Node_11, implemented in MATLAB–Simscape Multibody, based on the schematic shown in Figure 9.

The truss bar subsystem, based on the Euler–Bernoulli beam model, is analogous to the configuration used in Model 1.1 and Model 1.2 (see Figure 3), and follows the same modeling principles. All the truss bars have an identical structural composition, and only axial force is taken into account.

Figure 11 presents the primary sections, i.e., subsystems of the structure, including those repeated throughout the model. All the components associated with the nodes are consistently labeled as Node_1, Node_2, and so forth.

The control measurement of the load force $\mathbf{F}_{\text{load m}}$ is depicted in Figure 12.

The load force $\mathbf{F}_{\text{load m}}$ also serves here as a control verification of the computed load force $\mathbf{F}_{\text{load}}$. The value of $\mathbf{F}_{\text{load m}}$ is derived from the internal forces within the system, specifically from the equilibrium condition at Node_11.

$$\sum F_x = 0;\ F_{\text{load }x} - N_{17} - \frac{\sqrt{2}\cdot N_{19}}{2} = 0 \tag{3}$$

$$\sum F_y = 0;\ F_{\text{load }y} - N_{21} + \frac{\sqrt{2}\cdot N_{19}}{2} = 0 \tag{4}$$

The $\mathbf{F}_{\text{load}}$ force represents the external load applied to the system. In cases where multiple external forces act on the structure, the one with the maximum magnitude is selected for control measurement.

2.2.2. Model 2.2. Truss System Composed of Euler–Bernoulli Beams and Solid Joints

The block configuration of Model 2.2, constructed in Simscape Multibody, is identical to the configuration shown in Figure 10. The only difference lies in the implementation of the nodes.

The subsystem of the truss bar, based on the Euler–Bernoulli beam model, is analogous to that used in Model 1.1 and Model 1.2 (see Figure 3). The same modeling principles apply. All truss bars in the simulation are structurally identical, and only axial force is considered.

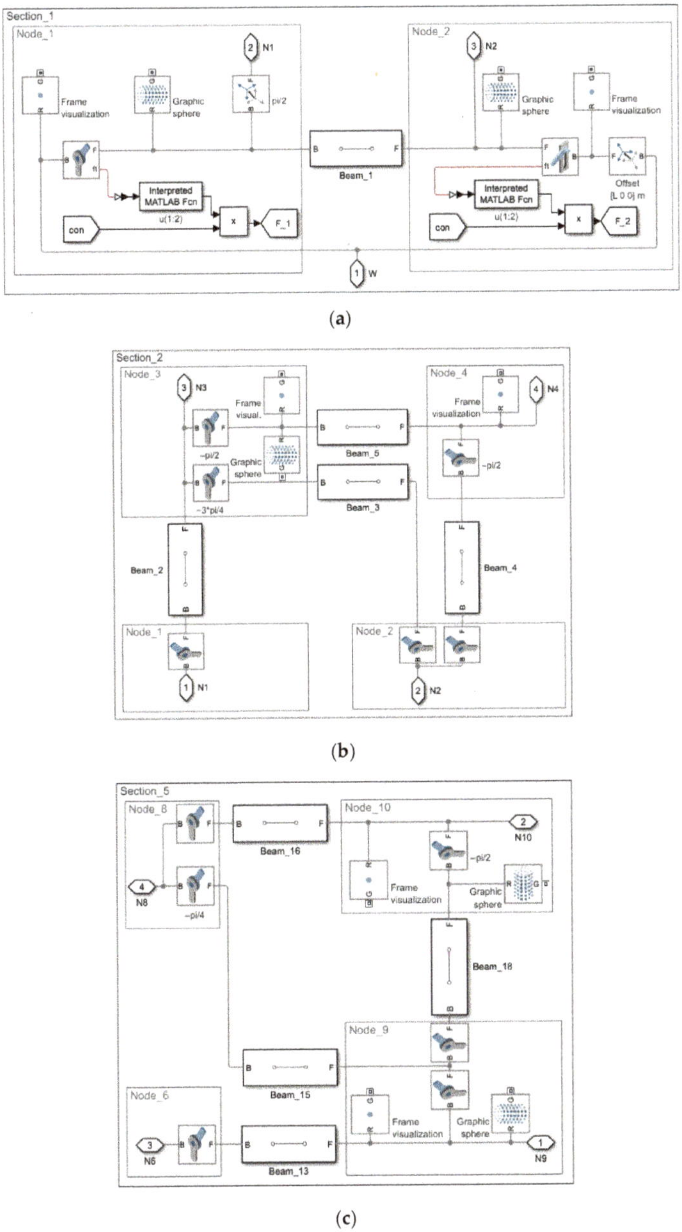

Figure 11. Model 2.1 and Model 2.3. Subsystem types by section: (**a**) Section_1; (**b**) Sections_2 to _4 are conceptually identical; (**c**) Sections_5 to _6 are conceptually identical.

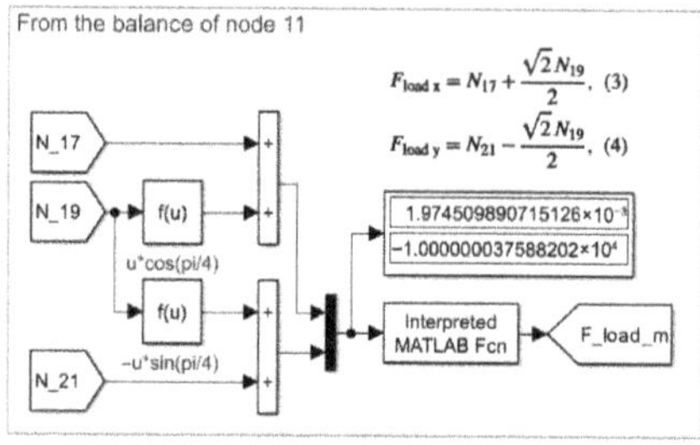

Figure 12. Model 2.1, Model 2.2, and Model 2.3. Measurement of the $\mathbf{F}_{\text{load m}}$ force in Section_6.

The subsequent illustration, Figure 13, presents the primary structural sections—the subsystems that define the construction and those that are repeated. All the elements associated with the nodes are labeled consistently as Node_1, Node_2, etc. Section_1, shown in Figure 11a, remains the same across Model 2.1, Model 2.2, and Model 2.3.

The control measurement of the load force $\mathbf{F}_{\text{load m}}$ is identical to that in Model 2.1, as shown in Figure 12. Equations (3) and (4) apply. However, this interpretation is not entirely accurate in this particular case, as the nodes do not contain movable joints. Nevertheless, it is sufficient for the intended purpose.

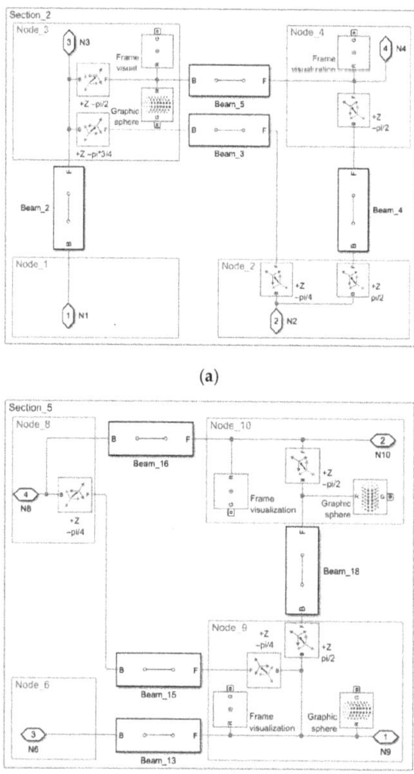

Figure 13. Model 2.2. The conceptual structure of the subsystems by section: (**a**) Sections_2 to _4 share an identical configuration; (**b**) Sections_5 to _6 are conceptually identical.

2.2.3. Model 2.3. Truss System Composed of Solid Bars with 1 DOF and Joints

The configuration of Model 2.3, assembled in Simscape Multibody, has the same block structure as that shown in Figure 10. The only difference lies in the representation of the bars, which is analogous to that used in Model 1.3; see Figure 6. The bar consists of two rigid bodies connected by a prismatic joint with one degree of freedom (1 DOF). Within the joint, the stiffness k is defined as ES/l, along with the damping b and the initial position set to zero. All the truss bars in the simulation model (Model 2.3) have an identical structure.

The measurement of node displacements is consistent across all three models (Model 2.1, Model 2.2, and Model 2.3); the nodes differ only in their (x, y) locations. The concept is identical to that for the structure shown in Figure 7.

The subsystem used to define the F_{load} force is identical to that shown in Figure 8, and is shared by all three models (Model 2.1, Model 2.2, and Model 2.3).

3. Computational Optimization Procedure

The optimization consists of identifying the appropriate input level—specifically, the value of the ratio $\mathbf{F}_{load}/con$ for a given model—at which the computational accuracy is maximized. The gain in precision can reach two to three orders of magnitude. The resulting values will be compared against the theoretical solutions obtained for Models 1.1, 1.2, and 1.3, and against the more complex structures represented by Models 2.1, 2.2, and 2.3. Additionally, a more complex test configuration derived from Model 1.1 will be evaluated to further assess its potential for improving the accuracy larger structural systems, considering both computational precision and processing time. The optimization is performed by applying a stepwise change (scaling) in the proportionality constant *con*. Alternatively, a dynamic adjustment of *con* based on the difference between $\mathbf{F}_{load}$ and $\mathbf{F}_{load\ m}$ may be implemented. In such cases, the value of *con* is determined proportionally to $\mathbf{F}_{load}$. Suitable computational solvers for this type of simulation within the toolbox include ode15s, ode23s, ode23t, and ode23tb (see [22] for further details). Among these, ode15s was selected for its proven reliability and computational efficiency. Another option for stiff systems is also solvers ode23s, ode23t, and ode23tb. The solvers ode45, ode23, and ode113 are not recommended for stiff systems (longer computational times).

The following Procedure 1 describes a simple algorithm for selecting the control measurement of the load force $\mathbf{F}_{\_load\_m}$. The entire set of measurements indexed by i is divided into time sequences, denoted as *Sequence*, with a total number k. During each sequence, the data $\mathbf{F}_{\_load\_m\_separ\_k}$ is selected outside the transient state *Lag*, caused by the natural vibration delay of the system. That is, to process the mean value $\mathbf{Mean\_F}_{\_load\_m}(k)$ from the given measurement sequence, data from the steady-state condition is used. This continues until a step change in the $\mathbf{F}_{\_load\_m}$ force occurs, at which point the end of the sequence yields undesired results. These are separated in advance using *Lead*. The proportionality of the step change in the load force $\mathbf{F}_{\_load\_m}$ is realized using the multiplication constant *con*. Additionally, the maximum $\mathbf{Max\_F}_{\_load\_m}(k)$ and minimum $\mathbf{Min\_F}_{\_load\_m}(k)$ are computed for each sequence.

The results are subsequently visualized in the plot $\mathbf{F}_{\_load\_m}$ = function(*con*), as shown in Figures 14–16. The closer the value of $\mathbf{F}_{\_load\_m}$ is to the applied force $\mathbf{F}_{\_load}$, the more accurate the computation is, provided that the variation in $\mathbf{F}_{\_load\_m}$ remains minimal. If the variation in $\mathbf{F}_{\_load\_m}$ exceeds the maximum and minimum values observed in the preceding sequence, the resulting error may increase (depending on the sample). In such cases, it is not meaningful to proceed with the computation.

Procedure 1: Visualization of the computational accuracy of $\mathbf{F}_{load\ m}$ as a function of *con*

Input: multiplication constant *con*, vector of load force $\mathbf{F}_{\_load}$, control measurement of the load force $\mathbf{F}_{\_load\_m}$, simulation sequence duration *SampleTime*, measurement sequence (*Sequence*), end of simulation *StopTime*, vector of sequences **Seq_vector**, *Lag*, *Lead*, sequence step *k*, *ith* sample of the $\mathbf{F}_{\_load\_m}$ variable.

Output: Sorted values $\mathbf{F}_{\_load\_m\_separ\_k}$, maxima of the measured values for the given sequence **Max_F**$_{\_load\_m}(k)$, minima of the measured values for the given sequence **Min_F**$_{\_load\_m}(k)$, mean of the measured values for the given sequence **Mean_F**$_{\_load\_m}(k)$, time for all sequences **TIME**$\{k\}$.

```
For k = 1 to size(Seq_vector) do
    If k < size(Seq_vector) then
        For i = 1 to size(F_load_m) do
            If ((Seq_vector(k)-Lag) ≤ F_load_m(i)) & (F_load_m(i) ≤ (Seq_vector(k+1)-Lead))
                F_load_m_separ (i) ← F_load_m (i)
            End
        End
    F_load_m_separ_k ← F_load_m_separ
    clear F_load_m_separ—delete previous values
    F_load_m_separ_1 = F_load_m_separ_k(1,:)—force vector extracted from matrix
    F_load_m_separ_k for the sequence
    F_load_m_separ_2 = F_load_m_separ_k(2,:)—time vector extracted from matrix
    F_load_m_separ_k for the sequence
    A{k} = F_load_m_separ_1(F_load_m_separ_1 ~=0)—exclusion of invalid values
    Max_F_load_m(k) = max(A{k})—maxima of F_load_m_separ_k
    Min_F_load_m(k) = min(A{k})—minima of F_load_m_separ_k
    Mean_F_load_m(k)—mean values computed from F_load_m_separ_1
    time = F_load_m_separ_2 (F_load_m_separ_2 ~= 0)—exclusion of invalid time values
    for the sequence
    TIME{k}—all time sequences
    Else
    End
End
```

The output plots below illustrate the computational accuracy of the control measurement of the load force $\mathbf{F}_{load\ m}$ as a function of the proportionality constant *con* for (a) Model 1.1, (b) Model 1.2, and (c) Model 1.3.

The output plots below illustrate the computational accuracy of the control measurement of the load force $\mathbf{F}_{load\ m}$ as a function of the proportionality constant *con* for (a) Model 2.1, (b) Model 2.2, and (c) Model 2.3.

The output plots below illustrate the computational accuracy of the control measurement of the load force $\mathbf{F}_{load\ m}$ as a function of the proportionality constant *con*. A more complex test variant, created based on Model 1.1 from Example 2, was used. A total of 39 bars were implemented.

The results indicate that for different multiples of the load force vector $\mathbf{F}_{load}$, the computational accuracy also varies. A similar situation occurs when other parameters are changed, such as the bar diameter, Young's modulus, or similar material properties. Our objective was to investigate all factors influencing this computation and to explore how such optimization could be standardized for future analyses. In the script "Dependence_Floadm_on_con.m" [18], we initially planned to calculate the standard deviation [23] in addition to the mean and extreme values; however, this approach proved to be impractical for the following reasons:

1. The computation did not consistently conform 100% to the theoretical probability of the Gaussian distribution [24]. For example, instead of the expected 99.73% probability set, the actual result was 98.98%.
2. During the computation, the simulation can be interrupted at any moment. As a result, no samples are excluded, and calculating the standard deviation does not provide meaningful insight in this context. Consequently, even samples that fall outside multiples of the standard deviation interval remain included in the data set. Therefore, we opted to use the actual maxima and minima derived from the given set of values.

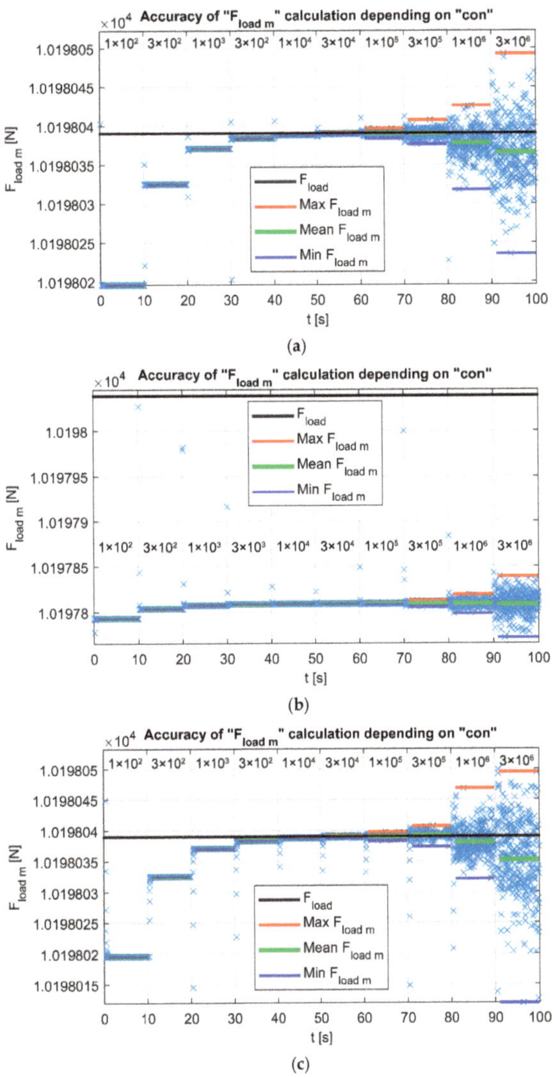

Figure 14. The computational accuracy of the control measurement of the load force $\mathbf{F}_{\text{load m}}$. The closer the value of $\mathbf{F}_{\text{load\_m}}$ is to the force $\mathbf{F}_{\text{\_load}}$, the more accurate the computation is, provided that the swing of $\mathbf{F}_{\text{load\_m}}$ values remains minimal: (**a**) Model 1.1; (**b**) Model 1.2; (**c**) Model 1.3.

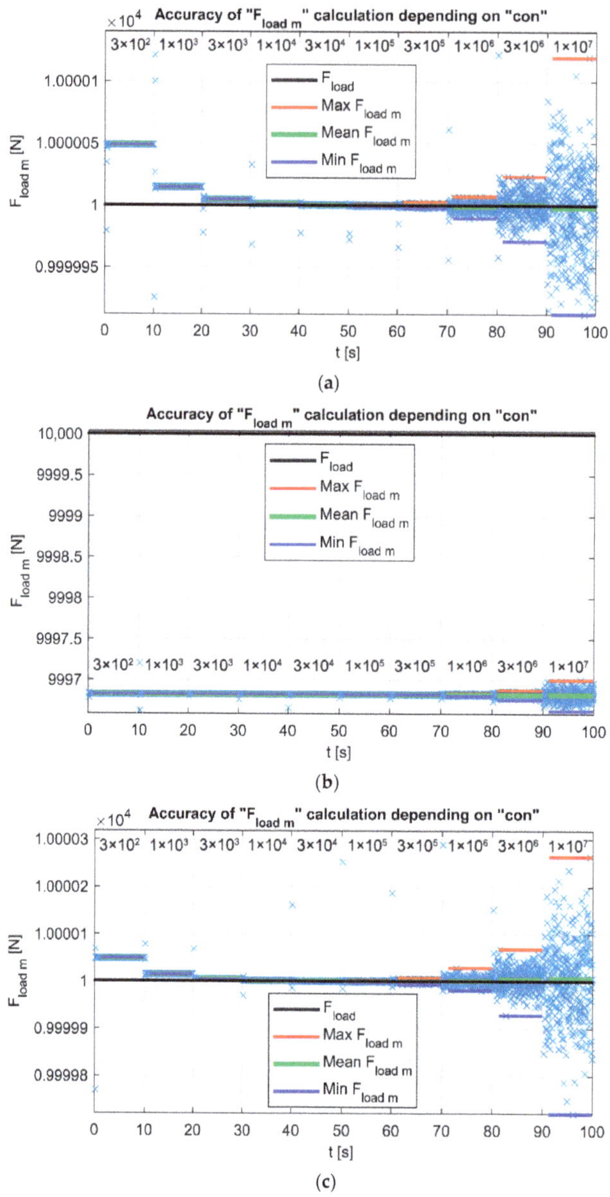

Figure 15. The computational accuracy of the control measurement of the load force $\mathbf{F}_{\mathrm{load\ m}}$. The closer the value of $\mathbf{F}_{\_\mathrm{load\_m}}$ is to the force $\mathbf{F}_{\_\mathrm{load}}$, the more accurate the computation is, provided that the swing of $\mathbf{F}_{\_\mathrm{load\_m}}$ values remains minimal: (**a**) Model 2.1; (**b**) Model 2.2; (**c**) Model 2.3.

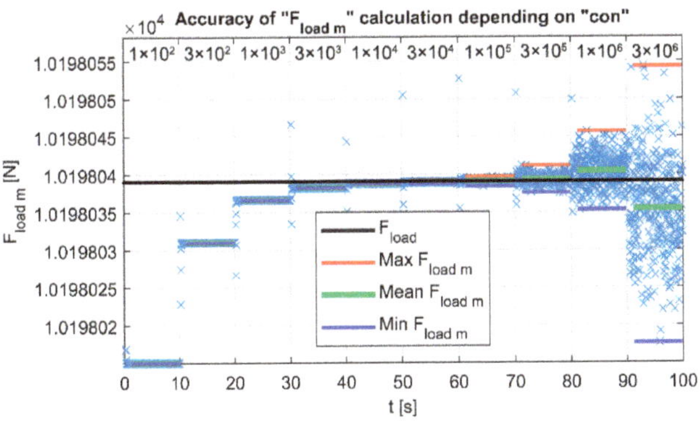

(a)

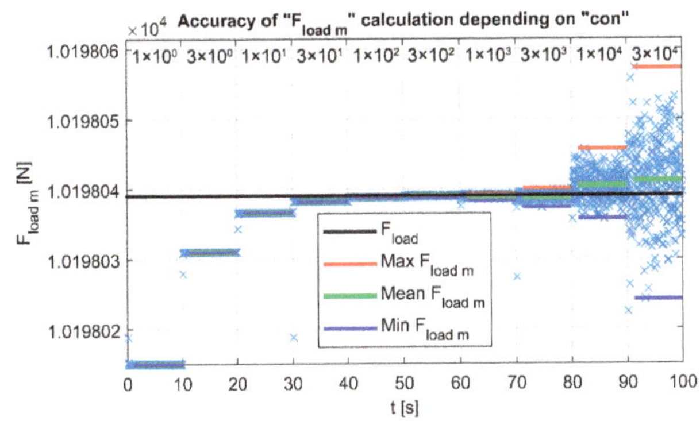

(b)

Figure 16. The computational accuracy of the control measurement of the load force $\mathbf{F}_{\text{load m}}$ for Example 3. This represents a test computation of a more complex structure consisting of 39 bars, created based on Model 1.1 from Example 2. The closer the value of $\mathbf{F}_{\_\text{load\_m}}$ is to the force $\mathbf{F}_{\_\text{load}}$, the more accurate the computation is, provided that the swing of $\mathbf{F}_{\_\text{load\_m}}$ values remains minimal: (**a**) bar diameter $d = 0.01$ m; (**b**) bar diameter $d = 0.1$ m.

4. Results

Analytical solutions were obtained using publicly available scripts from [25] for various truss systems analyzed through Matrix Structural Analysis in the MATLAB environment. Essentially, these involve two-dimensional problems derived from the principles of the Finite Element Method (FEM) [26] or, for example, [27] (pp. 141–143). Specific examples were selected or created for testing in Simscape Multibody, with the aim of achieving the highest possible computational accuracy. Additionally, we were interested in exploring how load visualization can be implemented in this toolbox. Load visualization is a representation of a system with bars deformation due to load (Figures 17 and 18). Within each example, the individual models differ slightly, either in the configuration of the bars or in the design of the nodes.

The outputs from Models 1.1, 1.2, and 1.3 in Example 1, as well as Models 2.1, 2.2, and 2.3 in Example 2, obtained through simulations in MATLAB–Multibody and from theoretical calculations, are summarized in the following tables: Tables 1 and 2. These outputs also include visualizations of the deformed state. Additionally, Table 3 presents a case involving an extended version of Model 1.1 with a larger number of bars. For this extended model, analytical solution formulas were not included; the results are provided only in numerical form.

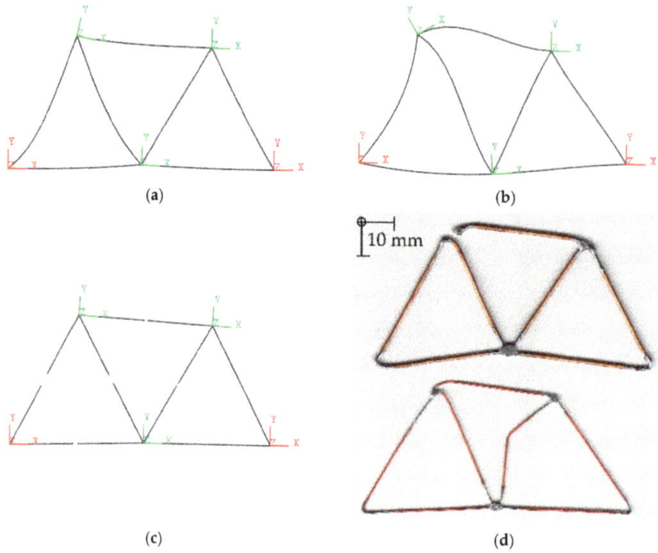

Figure 17. Deformed state visualization across the models: (**a**) Model 1.1; (**b**) Model 1.2; (**c**) Model 1.3 (Young's modulus E reduced by three orders of magnitude); (**d**) two physical models with cross-sectional areas of S_1 = 1.1309 mm^2 and S_2 = 0.3848 mm^2.

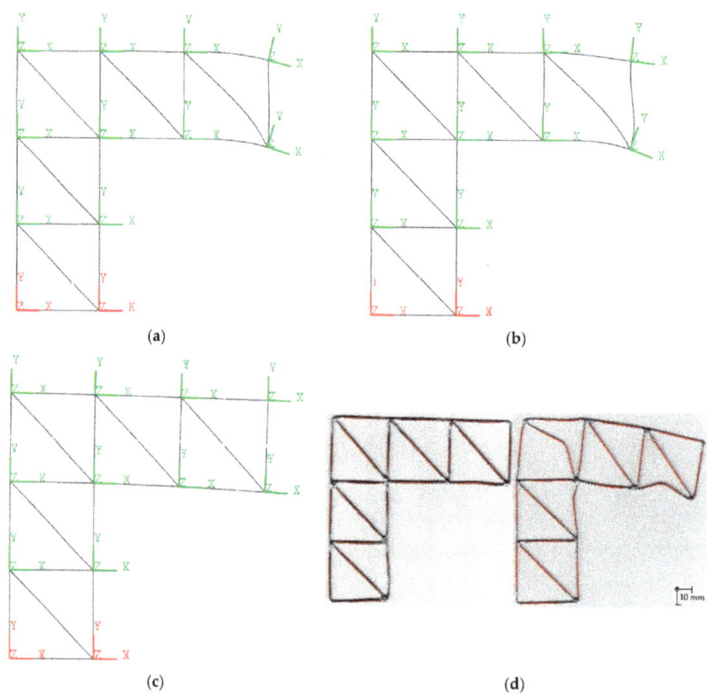

Figure 18. Visualization of the deformed state in the structural models: (**a**) Model 2.1; (**b**) Model 2.2; (**c**) Model 2.3 (Young's modulus, E, reduced by three orders of magnitude); (**d**) Two physical structures with different cross-sectional areas of the bars, S_1 = 1.5393 mm^2 and S_2 = 0.7853 mm^2.

Table 1. A comparison of the theoretical results for forces N_1–N_7 acting in the bars, forces $\mathbf{F}_1$, $\mathbf{F}_3$ (not reactions) at the supports of Node_1 and Node_3, and the deflections δ_2, δ_4, and δ_5, with the results from the individual models Model 1.1, Model 1.2, and Model 1.3 constructed in MATLAB–Multibody. The simulation time for the optimal *con* ranges from 2 to 5 s. Hardware configuration: processor i9–10900, 2.8 GHz, 32 GB RAM, MATLAB 2024a, 64-bit. The same solver (ode15s) was used across all models, with identical settings for absolute and relative tolerances (10^{-3}) and a maximum iteration step of 0.01 within a variable step.

Quantity [Unit]	Theoretical Calculation	MATLAB Model 1.1	MATLAB Model 1.2	MATLAB Model 1.3
N_1 [N]	$-0.94337567 \cdot 10^3$	$-0.94337566 \cdot 10^3$	$-0.94322189 \cdot 10^3$	$-0.94337567 \cdot 10^3$
N_2 [N]	$0.94337567 \cdot 10^3$	$0.94337562 \cdot 10^3$	$0.94322214 \cdot 10^3$	$0.94337563 \cdot 10^3$
N_3 [N]	$1.88675134 \cdot 10^3$	$1.88675121 \cdot 10^3$	$1.88649462 \cdot 10^3$	$1.88675121 \cdot 10^3$
N_4 [N]	$9.66025403 \cdot 10^3$	$9.66025400 \cdot 10^3$	$9.65990727 \cdot 10^3$	$9.66025394 \cdot 10^3$
N_5 [N]	$1.88675134 \cdot 10^3$	$1.88675131 \cdot 10^3$	$1.88681570 \cdot 10^3$	$1.88675120 \cdot 10^3$
N_6 [N]	$-1.88675134 \cdot 10^3$	$-1.88675129 \cdot 10^3$	$-1.88620302 \cdot 10^3$	$-1.88675108 \cdot 10^3$
N_7 [N]	$1.88675134 \cdot 10^3$	$1.88675124 \cdot 10^3$	$1.88657522 \cdot 10^3$	$1.88675128 \cdot 10^3$
F_{1x} [N]	$3.88675134 \cdot 10^3$	$3.88675125 \cdot 10^3$	$3.88666800 \cdot 10^3$	$3.88675120 \cdot 10^3$
F_{1y} [N]	$8.36602540 \cdot 10^3$	$8.36602540 \cdot 10^3$	$8.36602543 \cdot 10^3$	$8.36602535 \cdot 10^3$
F_{3x} [N]	$-1.88675134 \cdot 10^3$	$-1.88675122 \cdot 10^3$	$-1.88666800 \cdot 10^3$	$-1.88675126 \cdot 10^3$
F_{3y} [N]	$1.63397459 \cdot 10^3$	$1.63397452 \cdot 10^3$	$1.63397457 \cdot 10^3$	$1.63397461 \cdot 10^3$
$F_{\text{Load x}}$ [1] [N]	$2.00000000 \cdot 10^3$	$2.00000012 \cdot 10^3$	$2.00005115 \cdot 10^3$	$2.00000015 \cdot 10^3$
$F_{\text{Load y}}$ [1] [N]	$10.00000000 \cdot 10^3$	$9.99999994 \cdot 10^3$	$9.99975543 \cdot 10^3$	$9.99999980 \cdot 10^3$
δ_{2x} [m]	$-0.57197295 \cdot 10^{-4}$	$-0.57197293 \cdot 10^{-4}$	$-0.57187981 \cdot 10^{-4}$	$-0.57197300 \cdot 10^{-4}$
δ_{2y} [m]	$4.70248778 \cdot 10^{-4}$	$4.70248791 \cdot 10^{-4}$	$4.70200032 \cdot 10^{-4}$	$4.70248768 \cdot 10^{-4}$
δ_{4x} [m]	$0.35464965 \cdot 10^{-4}$	$0.35464940 \cdot 10^{-4}$	$0.35486893 \cdot 10^{-4}$	$0.35464980 \cdot 10^{-4}$
δ_{4y} [m]	$6.55838856 \cdot 10^{-4}$	$6.55838876 \cdot 10^{-4}$	$6.55801902 \cdot 10^{-4}$	$6.55838846 \cdot 10^{-4}$
δ_{5x} [m]	$1.49859556 \cdot 10^{-4}$	$1.49859564 \cdot 10^{-4}$	$1.49865932 \cdot 10^{-4}$	$1.49859564 \cdot 10^{-4}$
δ_{5y} [m]	$2.18612952 \cdot 10^{-4}$	$2.18612977 \cdot 10^{-4}$	$2.18604317 \cdot 10^{-4}$	$2.18612957 \cdot 10^{-4}$

[1] Verification of the calculated magnitude of the load force $\mathbf{F}_{\text{Load}}$ based on the equilibrium of Node_4, according to Equations (1) and (2), used solely for the optimization of the multiplication constant *con*.

Table 2. A comparison of the theoretical results for the forces N_1–N_{21} acting in the bars, forces $\mathbf{F}_1$ and $\mathbf{F}_2$ (not reactions) at the supports of Node_1 and Node_2, and displacements δ_2–δ_{12}. These results are compared with those obtained from Models 2.1, 2.2, and 2.3 simulated in MATLAB–Multibody. The simulation time for the optimal *con* value ranged from 10 to 22 s. The calculations were performed on a processor i9 10900, 2.8 GHz, with 32 GB RAM, using MATLAB 2024a, 64-bit. The same solver (ode15s) and identical settings for absolute tolerance, relative tolerance (10^{-3}), and the maximum iteration step (0.01) within the variable–step configuration were used throughout.

Quantity [Unit]	Theoretical Calculation	MATLAB Model 2.1	MATLAB Model 2.2	MATLAB Model 2.3
N_1 [N]	0	-0.00036038	0.06985929	-0.00092453
N_2 [N]	$2 \cdot 10^4$	$2.00000007 \cdot 10^4$	$1.99997963 \cdot 10^4$	$2.00000012 \cdot 10^4$
N_3 [N]	0	-0.00030489	-0.70545982	-0.00057530
N_4 [N]	$-3 \cdot 10^4$	$-3.00000005 \cdot 10^4$	$-2.99992002 \cdot 10^4$	$-3.00000008 \cdot 10^4$
N_5 [N]	0	-0.00069505	0.41597051	-0.00181315
N_6 [N]	$2 \cdot 10^4$	$2.00000004 \cdot 10^4$	$1.99999088 \cdot 10^4$	$2.00000008 \cdot 10^4$
N_7 [N]	0	-0.00084519	-3.72374373	-0.00178803
N_8 [N]	$-3 \cdot 10^4$	$-2.99999998 \cdot 10^4$	$-2.99971831 \cdot 10^4$	$-2.99999994 \cdot 10^4$
N_9 [N]	0	-0.00123336	-1.59349130	-0.00281782
N_{10} [N]	$2 \cdot 10^4$	$1.99999997 \cdot 10^4$	$1.99947667 \cdot 10^4$	$1.99999993 \cdot 10^4$
N_{11} [N]	$-2.82842712 \cdot 10^4$	$-2.82842702 \cdot 10^4$	$-2.82719856 \cdot 10^4$	$-2.82842694 \cdot 10^4$
N_{12} [N]	$2 \cdot 10^4$	$1.99999997 \cdot 10^4$	$1.99945650 \cdot 10^4$	$1.99999998 \cdot 10^4$
N_{13} [N]	$-2 \cdot 10^4$	$-1.99999982 \cdot 10^4$	$-1.99954651 \cdot 10^4$	$-1.99999958 \cdot 10^4$
N_{14} [N]	$-1 \cdot 10^4$	$-1.00000023 \cdot 10^4$	$-1.00000467 \cdot 10^4$	$-1.00000050 \cdot 10^4$
N_{15} [N]	$1.41421356 \cdot 10^4$	$1.41421358 \cdot 10^4$	$1.41343004 \cdot 10^4$	$1.41421362 \cdot 10^4$
N_{16} [N]	$1 \cdot 10^4$	$9.99999991 \cdot 10^3$	$1.00014635 \cdot 10^4$	$9.99999993 \cdot 10^3$
N_{17} [N]	$-1 \cdot 10^4$	$-9.99999820 \cdot 10^3$	$-9.99855709 \cdot 10^3$	$-9.99999543 \cdot 10^3$

Table 2. *Cont.*

Quantity [Unit]	Theoretical Calculation	MATLAB Model 2.1	MATLAB Model 2.2	MATLAB Model 2.3
N_{18} [N]	$-1 \cdot 10^4$	$-1.00000000 \cdot 10^4$	$-9.99689885 \cdot 10^3$	$-1.00000002 \cdot 10^4$
N_{19} [N]	$1.41421356 \cdot 10^4$	$1.41421358 \cdot 10^4$	$1.41391288 \cdot 10^4$	$1.41421366 \cdot 10^4$
N_{20} [N]	0	0.00019897	1.13419678	0.00002046
N_{21} [N]	0	−0.00019154	1.04655832	0.00011316
F_{1x} [N]	0	−0.00001027	−0.00000029	−0.00008829
F_{1y} [N]	$2 \cdot 10^4$	$2.00000007 \cdot 10^4$	$2.00001212 \cdot 10^4$	$2.00000012 \cdot 10^4$
F_{2x} [N]	0	0	0	0
F_{2y} [N]	$-3 \cdot 10^4$	$-3.00000007 \cdot 10^4$	$-3.00001212 \cdot 10^4$	$-3.00000012 \cdot 10^4$
$F_{Load\,x}$ [1] [N]	0	0.00197450	−0.68322370	0.00527883
$F_{Load\,y}$ [1] [N]	$-1 \cdot 10^4$	$-1.00000003 \cdot 10^4$	$-9.99682731 \cdot 10^3$	$-1.00000005 \cdot 10^4$
δ_{2x} [m]	0	0	0	0
δ_{2y} [m]	0	0	0	0
δ_{3x} [m]	$1.21260909 \cdot 10^{-3}$	$1.21260907 \cdot 10^{-3}$	$1.21268896 \cdot 10^{-3}$	$1.21260919 \cdot 10^{-3}$
δ_{3y} [m]	$1.21260909 \cdot 10^{-3}$	$1.21260913 \cdot 10^{-3}$	$1.21259634 \cdot 10^{-3}$	$1.21260913 \cdot 10^{-3}$
δ_{4x} [m]	$1.21260909 \cdot 10^{-3}$	$1.21260901 \cdot 10^{-3}$	$1.21269939 \cdot 10^{-3}$	$1.21260893 \cdot 10^{-3}$
δ_{4y} [m]	$-1.81891363 \cdot 10^{-3}$	$-1.81891360 \cdot 10^{-3}$	$-1.81886655 \cdot 10^{-3}$	$-1.81891370 \cdot 10^{-3}$
δ_{5x} [m]	$5.45674090 \cdot 10^{-3}$	$5.45674098 \cdot 10^{-3}$	$5.45724232 \cdot 10^{-3}$	$5.45674144 \cdot 10^{-3}$
δ_{5y} [m]	$2.42521818 \cdot 10^{-3}$	$2.42521815 \cdot 10^{-3}$	$2.42517526 \cdot 10^{-3}$	$2.42521798 \cdot 10^{-3}$
δ_{6x} [m]	$5.45674090 \cdot 10^{-3}$	$5.45674070 \cdot 10^{-3}$	$5.45710518 \cdot 10^{-3}$	$5.45674065 \cdot 10^{-3}$
δ_{6y} [m]	$-3.63782727 \cdot 10^{-3}$	$-3.63782731 \cdot 10^{-3}$	$-3.63763418 \cdot 10^{-3}$	$-3.63782759 \cdot 10^{-3}$
δ_{7x} [m]	$1.61621718 \cdot 10^{-2}$	$1.61621718 \cdot 10^{-2}$	$1.61606005 \cdot 10^{-2}$	$1.61621722 \cdot 10^{-2}$
δ_{7y} [m]	$3.63782727 \cdot 10^{-3}$	$3.63782650 \cdot 10^{-3}$	$3.63730639 \cdot 10^{-3}$	$3.63782513 \cdot 10^{-3}$
δ_{8x} [m]	$1.737478098 \cdot 10^{-2}$	$1.737478055 \cdot 10^{-2}$	$1.73728087 \cdot 10^{-2}$	$1.73747803 \cdot 10^{-2}$
δ_{8y} [m]	$-4.24413181 \cdot 10^{-3}$	$-4.24413292 \cdot 10^{-3}$	$-4.24416919 \cdot 10^{-3}$	$-4.24413480 \cdot 10^{-3}$
δ_{9x} [m]	$4.24413181 \cdot 10^{-3}$	$4.24413010 \cdot 10^{-3}$	$4.24443048 \cdot 10^{-3}$	$4.24412783 \cdot 10^{-3}$
δ_{9y} [m]	$-1.908966920 \cdot 10^{-2}$	$-1.90896690 \cdot 10^{-2}$	$-1.90858478 \cdot 10^{-2}$	$-1.90896690 \cdot 10^{-2}$
δ_{10x} [m]	$1.798108552 \cdot 10^{-2}$	$1.79810834 \cdot 10^{-2}$	$1.79788210 \cdot 10^{-2}$	$1.79810809 \cdot 10^{-2}$
δ_{10y} [m]	$-1.969597374 \cdot 10^{-2}$	$-1.96959748 \cdot 10^{-2}$	$-1.96922747 \cdot 10^{-2}$	$-1.96959767 \cdot 10^{-2}$
δ_{11x} [m]	$3.63782727 \cdot 10^{-3}$	$3.63782379 \cdot 10^{-3}$	$3.63775137 \cdot 10^{-3}$	$3.63781893 \cdot 10^{-3}$
δ_{11y} [m]	$-3.575412022 \cdot 10^{-2}$	$-3.57541196 \cdot 10^{-2}$	$-3.57470982 \cdot 10^{-2}$	$-3.57541193 \cdot 10^{-2}$
δ_{12x} [m]	$1.798108552 \cdot 10^{-2}$	$1.79810817 \cdot 10^{-2}$	$1.797846122 \cdot 10^{-2}$	$1.79810766 \cdot 10^{-2}$
δ_{12y} [m]	$-3.575412022 \cdot 10^{-2}$	$-3.57541210 \cdot 10^{-2}$	$-3.574737653 \cdot 10^{-2}$	$-3.57541227 \cdot 10^{-2}$

[1] Verification of the loading force $\mathbf{F}_{Load}$ magnitude based on Node_11 equilibrium, according to Equations (3) and (4), used solely for optimizing the multiplication constant *con*.

Table 3. Example 3. A comparison of the theoretical results for forces N_1–N_{39} acting in the bars, as well as forces $\mathbf{F}_{1L}$, $\mathbf{F}_{20R}$ (not reactions) at the supports of nodes Node_1 and Node_20, and deflections δ_2–δ_{19} and δ_{21}, with the results from the MATLAB–Simscape Multibody model. The simulation time for the optimal value of *con* was 94 s. Hardware specifications: Processor i9–10900, 2.8 GHz, 32 GB RAM, Matlab 2024a, 64-bit. The same solver (ode15s) was used, with identical settings for absolute and relative tolerances (10^{-3}), and a maximum iteration step size of 0.01 within a variable step.

Quantity [Unit]	Theoretical Calculation	MATLAB Example 3	Quantity [Unit]	Theoretical Calculation	MATLAB Example 3
N_1 [N]	$1.11696551 \cdot 10^4$	$1.11696549 \cdot 10^4$	F_{20Ry} [N]	$-0.03267949 \cdot 10^4$	$0.03267948 \cdot 10^4$
N_2 [N]	$0.03773502 \cdot 10^4$	$0.03773502 \cdot 10^4$	F_{Loadx} [1] [N]	$0.2 \cdot 10^4$	$0.20000003 \cdot 10^4$
N_3 [N]	$-0.16980762 \cdot 10^4$	$-0.16980761 \cdot 10^4$	F_{Loady} [1] [N]	$1 \cdot 10^4$	$0.99999997 \cdot 10^4$
N_4 [N]	$0.33961524 \cdot 10^4$	$0.33961518 \cdot 10^4$	δ_{2x} [m]	$-0.10295513 \cdot 10^{-3}$	$-0.10295517 \cdot 10^{-3}$
N_5 [N]	$-0.03773502 \cdot 10^4$	$-0.03773503 \cdot 10^4$	δ_{2y} [m]	$1.63899678 \cdot 10^{-3}$	$1.63899669 \cdot 10^{-3}$
N_6 [N]	$0.03773502 \cdot 10^4$	$0.03773502 \cdot 10^4$	δ_{3x} [m]	$-0.81654806 \cdot 10^{-3}$	$-0.81654804 \cdot 10^{-3}$
N_7 [N]	$-0.13207259 \cdot 10^4$	$-0.13207258 \cdot 10^4$	δ_{3y} [m]	$1.25342200 \cdot 10^{-3}$	$1.25342194 \cdot 10^{-3}$
N_8 [N]	$0.30188021 \cdot 10^4$	$0.30188015 \cdot 10^4$	δ_{4x} [m]	$-0.18303134 \cdot 10^{-3}$	$-0.18303137 \cdot 10^{-3}$
N_9 [N]	$-0.03773502 \cdot 10^4$	$-0.03773502 \cdot 10^4$	δ_{4y} [m]	$2.12614960 \cdot 10^{-3}$	$2.12614946 \cdot 10^{-3}$
N_{10} [N]	$0.03773502 \cdot 10^4$	$0.03773502 \cdot 10^4$	δ_{5x} [m]	$-0.61063780 \cdot 10^{-3}$	$-0.61063780 \cdot 10^{-3}$
N_{11} [N]	$-0.09433756 \cdot 10^4$	$-0.09433756 \cdot 10^4$	δ_{5y} [m]	$1.90568920 \cdot 10^{-3}$	$1.90568908 \cdot 10^{-3}$
N_{12} [N]	$0.26414518 \cdot 10^4$	$0.26414514 \cdot 10^4$	δ_{6x} [m]	$-0.24022864 \cdot 10^{-3}$	$-0.24022866 \cdot 10^{-3}$
N_{13} [N]	$-0.03773502 \cdot 10^4$	$-0.03773502 \cdot 10^4$	δ_{6y} [m]	$2.32270114 \cdot 10^{-3}$	$2.32270096 \cdot 10^{-3}$
N_{14} [N]	$0.03773502 \cdot 10^4$	$0.03773502 \cdot 10^4$	δ_{7x} [m]	$-0.42760645 \cdot 10^{-3}$	$-0.42760645 \cdot 10^{-3}$
N_{15} [N]	$-0.05660254 \cdot 10^4$	$-0.05660253 \cdot 10^4$	δ_{7y} [m]	$2.24093681 \cdot 10^{-3}$	$2.24093665 \cdot 10^{-3}$
N_{16} [N]	$0.22641016 \cdot 10^4$	$0.22641010 \cdot 10^4$	δ_{8x} [m]	$-0.27454701 \cdot 10^{-3}$	$-0.27454702 \cdot 10^{-3}$

Table 3. *Cont.*

Quantity [Unit]	Theoretical Calculation	MATLAB Example 3	Quantity [Unit]	Theoretical Calculation	MATLAB Example 3
N_{17} [N]	$-0.03773502{\cdot}10^4$	$-0.03773501{\cdot}10^4$	δ_{8y} [m]	$2.28148798{\cdot}10^{-3}$	$2.28148782{\cdot}10^{-3}$
N_{18} [N]	$0.03773502{\cdot}10^4$	$0.03773501{\cdot}10^4$	δ_{9x} [m]	$-0.26745402{\cdot}10^{-3}$	$-0.26745405{\cdot}10^{-3}$
N_{19} [N]	$-0.01886751{\cdot}10^4$	$-0.01886751{\cdot}10^4$	δ_{9y} [m]	$2.31200143{\cdot}10^{-3}$	$2.31200128{\cdot}10^{-3}$
N_{20} [N]	$0.18867513{\cdot}10^4$	$0.18867509{\cdot}10^4$	δ_{10x} [m]	$-0.28598647{\cdot}10^{-3}$	$-0.28598646{\cdot}10^{-3}$
N_{21} [N]	$-0.03773502{\cdot}10^4$	$-0.03773502{\cdot}10^4$	δ_{10y} [m]	$2.05534673{\cdot}10^{-3}$	$2.05534661{\cdot}10^{-3}$
N_{22} [N]	$0.03773502{\cdot}10^4$	$0.03773502{\cdot}10^4$	δ_{11x} [m]	$-0.13018051{\cdot}10^{-3}$	$-0.13018053{\cdot}10^{-3}$
N_{23} [N]	$0.01886751{\cdot}10^4$	$0.01886750{\cdot}10^4$	δ_{11y} [m]	$2.17171965{\cdot}10^{-3}$	$2.17171953{\cdot}10^{-3}$
N_{24} [N]	$0.15094010{\cdot}10^4$	$0.15094006{\cdot}10^4$	δ_{12x} [m]	$-0.27454701{\cdot}10^{-3}$	$-0.27454697{\cdot}10^{-3}$
N_{25} [N]	$-0.03773502{\cdot}10^4$	$-0.03773502{\cdot}10^4$	δ_{12y} [m]	$1.69711398{\cdot}10^{-3}$	$1.69711389{\cdot}10^{-3}$
N_{26} [N]	$0.03773502{\cdot}10^4$	$0.03773501{\cdot}10^4$	δ_{13x} [m]	$-0.01578592{\cdot}10^{-3}$	$-0.01578590{\cdot}10^{-3}$
N_{27} [N]	$0.05660254{\cdot}10^4$	$0.05660252{\cdot}10^4$	δ_{13y} [m]	$1.87292807{\cdot}10^{-3}$	$1.87292798{\cdot}10^{-3}$
N_{28} [N]	$0.11320508{\cdot}10^4$	$0.11320505{\cdot}10^4$	δ_{14x} [m]	$-0.24022864{\cdot}10^{-3}$	$-0.24022858{\cdot}10^{-3}$
N_{29} [N]	$-0.03773502{\cdot}10^4$	$-0.03773501{\cdot}10^4$	δ_{14y} [m]	$1.25962634{\cdot}10^{-3}$	$1.25962628{\cdot}10^{-3}$
N_{30} [N]	$0.03773502{\cdot}10^4$	$0.03773503{\cdot}10^4$	δ_{15x} [m]	$0.07572974{\cdot}10^{-3}$	$0.07572976{\cdot}10^{-3}$
N_{31} [N]	$0.09433756{\cdot}10^4$	$0.09433753{\cdot}10^4$	δ_{15y} [m]	$1.46846330{\cdot}10^{-3}$	$1.46846324{\cdot}10^{-3}$
N_{32} [N]	$0.07547005{\cdot}10^4$	$0.07547003{\cdot}10^4$	δ_{16x} [m]	$-0.18303134{\cdot}10^{-3}$	$-0.18303129{\cdot}10^{-3}$
N_{33} [N]	$-0.03773502{\cdot}10^4$	$-0.03773503{\cdot}10^4$	δ_{16y} [m]	$0.79572039{\cdot}10^{-3}$	$0.79572037{\cdot}10^{-3}$
N_{34} [N]	$0.03773502{\cdot}10^4$	$0.03773501{\cdot}10^4$	δ_{17x} [m]	$0.14436650{\cdot}10^{-3}$	$0.14436654{\cdot}10^{-3}$
N_{35} [N]	$0.13207259{\cdot}10^4$	$0.13207256{\cdot}10^4$	δ_{17y} [m]	$1.01116193{\cdot}10^{-3}$	$1.01116192{\cdot}10^{-3}$
N_{36} [N]	$0.03773502{\cdot}10^4$	$0.03773501{\cdot}10^4$	δ_{18x} [m]	$-0.10295513{\cdot}10^{-3}$	$-0.10295508{\cdot}10^{-3}$
N_{37} [N]	$-0.03773502{\cdot}10^4$	$-0.03773501{\cdot}10^4$	δ_{18y} [m]	$0.35823274{\cdot}10^{-3}$	$0.35823275{\cdot}10^{-3}$
N_{38} [N]	$0.03773502{\cdot}10^4$	$0.03773502{\cdot}10^4$	δ_{19x} [m]	$0.19012433{\cdot}10^{-3}$	$0.19012439{\cdot}10^{-3}$
N_{39} [N]	$0.16980762{\cdot}10^4$	$0.16980758{\cdot}10^4$	δ_{19y} [m]	$0.55386056{\cdot}10^{-3}$	$0.55386056{\cdot}10^{-3}$
F_{1Lx} [N]	$-0.38867513{\cdot}10^4$	$0.38867509{\cdot}10^4$	δ_{21x} [m]	$0.21300325{\cdot}10^{-3}$	$0.21300326{\cdot}10^{-3}$
F_{1Ly} [N]	$-0.96732050{\cdot}10^4$	$0.96732051{\cdot}10^4$	δ_{21y} [m]	$0.14939578{\cdot}10^{-3}$	$0.14939579{\cdot}10^{-3}$
F_{20Rx} [N]	$0.18867513{\cdot}10^4$	$-0.18867508{\cdot}10^4$			

[1] Verification of the calculation of the load force magnitude $\mathbf{F}_{\mathrm{Load}}$ based on the equilibrium of Node_3 and Equations (51) and (52). This serves solely for the optimization of the multiplication constant *con*.

4.1. Results for Example 1

4.1.1. Theoretical Calculation for Example 1: Three Modeling Variants

Table 1 summarizes a comparison of the individual results for Models 1.1, 1.2, and 1.3 obtained from MATLAB–Multibody and from the theoretical calculation.

The following analytical expressions were obtained from the symbolic computation performed for Example 1. The force vectors, which do not represent reactions, at the supports of Node_1 and Node_3 are calculated as follows:

$$\mathbf{F}_1 = \begin{bmatrix} \frac{2\,F_x\,H+F_y\,L}{4\,H} \\ \frac{2\,F_x\,H+3\,F_y\,L}{4\,L} \end{bmatrix} \tag{5}$$

$$\mathbf{F}_3 = \begin{bmatrix} \frac{2\,F_x\,H-F_y\,L}{4\,H} \\ -\frac{2\,F_x\,H+F_y\,L}{4\,L} \end{bmatrix} \tag{6}$$

The forces in the bars are derived as follows:

$$N_1 = \frac{2\,F_x\,H - F_y\,L}{8\,H} \tag{7}$$

$$N_2 = -\frac{2\,F_x\,H - F_y\,L}{8\,H} \tag{8}$$

$$N_3 = -\frac{2\,F_x\,H - F_y\,L}{4\,H} \tag{9}$$

$$N_4 = \frac{\sqrt{4H^2 + L^2}\,(2F_x H + 3F_y L)}{8HL} \tag{10}$$

$$N_5 = -\frac{\sqrt{4H^2 + L^2}\,(2F_x H - F_y L)}{8HL} \tag{11}$$

$$N_6 = \frac{\sqrt{4H^2 + L^2}\,(2F_x H - F_y L)}{8HL} \tag{12}$$

$$N_7 = -\frac{\sqrt{4H^2 + L^2}\,(2F_x H - F_y L)}{8HL} \tag{13}$$

The displacement vectors $\boldsymbol{\delta}$ of the node displacements Node_2, Node_4, and Node_5 are calculated as follows:

$$\boldsymbol{\delta}_2 = \begin{bmatrix} \frac{L\,(2F_x H - F_y L)}{8ESH} \\ \frac{F_y\,(4H^2+L^2)^{3/2} + 2F_y L^3 - 4F_x H L^2}{16ESH^2} \end{bmatrix} \tag{14}$$

$$\boldsymbol{\delta}_4 = \begin{bmatrix} \frac{F_x\left(8\left(H^2+\frac{L^2}{4}\right)^{3/2} + 3L^3\right)}{8ESL^2} - \frac{3F_y L^2}{16ESH} \\ \frac{3F_y\,(4H^2+L^2)^{3/2} + 3F_y L^3 - 6F_x H L^2}{32ESH^2} \end{bmatrix} \tag{15}$$

$$\boldsymbol{\delta}_5 = \begin{bmatrix} \frac{F_x\left(8\left(H^2+\frac{L^2}{4}\right)^{3/2} - L^3\right)}{8ESL^2} + \frac{F_y L^2}{16ESH} \\ \frac{F_y\,(4H^2+L^2)^{3/2} + F_y L^3 - 2F_x H L^2}{32ESH^2} \end{bmatrix} \tag{16}$$

For all modeling variants—Model 1.1, Model 1.2, and Model 1.3—the force vector $\mathbf{F}_{load} = [F_x, F_y]^T$ has the numerical value $\mathbf{F}_{load} = [2, 10]^T$ kN. The lengths of the bars are determined by the distances between the nodes or by the dimensions of the system, with $L = 1$ m and $H = L{\cdot}cos(\pi/6)$ m. The bars have identical diameters of $d = 0.01$ m. The Young's modulus E is 210 GPa.

For Models 1.1 and 1.2, the shear modulus G is 79 GPa, and the material density (*Density*) is 7800 kg/m^3. Poisson's ratio is 0.28. The values for proportional (Rayleigh) damping are kept at their default settings for practical purposes, with $b_m = 0.01$ s^{-1} and $b_k = 0.01$ s. Essentially, these values can be set to nearly zero. The damping at the supports is $b_{rj} = 1$ N·m·s·rad^{-1}.

For Model 1.3, the damping at the supports is $b_{rj} = 6{\cdot}10^4$ N·m·s·rad^{-1}. The damping in the bars is $b = 10$ N·s·m^{-1}. The axial stiffness of the bars under tension/compression is defined as $k = E{\cdot}\pi{\cdot}d^2{\cdot}(4{\cdot}l)^{-1}$ N·s·m^{-1}.

4.1.2. Results for Example 1: Theoretical Calculation and MATLAB Models

For clarity, the results are presented in the Table 1, with the theoretical calculations listed first, followed by those from the individual models.

After optimization, Model 1.1 achieved guaranteed accuracy up to six decimal places compared to the theoretical model. Model 1.3 performed similarly. Model 1.2 attained guaranteed accuracy of approximately two to three decimal places, despite not fulfilling the assumptions of the theoretical model. In its nodes, there are no movable joints.

4.1.3. Results for Example 1: Visualization of the Deformed State of the Models

The visualization of the deformed state was performed for a specified norm of the displacement vector, $||\boldsymbol{\delta}_4|| = 0.1$ m, at Node_4; see, for example, Figure 1. This value is not critical, and serves solely to produce an exaggerated depiction of the deformation in theoretical models. For the material used in the physical model, copper (Cu) was selected due to its favorable ductility and plasticity. The bars had uniform diameters within each sample: $d_a = 1.2 \cdot 10^{-3}$ m in the first case and $d_b = 0.7 \cdot 10^{-3}$ m in the second. The dimensions were $L = 40 \cdot 10^{-3}$ m, $H = L \cdot cos(\pi/6)$ m. The model was constructed exclusively for the purpose of testing excessive deformation. In the theoretical models, the material was not changed, as it did not influence the resulting deformed shape.

For Model 1.3, the stiffness of the bars under uniaxial tension/compression needs to be reduced by approximately three orders of magnitude, calculated as $k = 10^{-3} \cdot E \cdot \pi \cdot d^2 \cdot (4 \cdot l)^{-1}$ N·s·m^{-1}.

The various models exhibit differences in the excessively deformed state when compared to the physical structure shown in Figure 17d. The best agreement with the physical structure was achieved by Model 1.1 in Figure 17a and Model 1.3 in Figure 17c. In the physical structure, the results are influenced by manufacturing precision and the method of applying support and load. Model 1.1 provided even better agreement when lower damping values b_{rj} were applied in the joints at the supports, combined with higher damping coefficients b_k in the bars. This allowed the force effect from the load point to be transmitted more effectively throughout the structure. However, this model entailed the disadvantage of increased computational time required for analysis.

4.2. Results for Example 2

4.2.1. Theoretical Calculation for Example 2: Three Modeling Variants

Table 2 above compares the theoretical calculation with the results obtained from Models 2.1, 2.2, and 2.3 developed in MATLAB–Multibody.

The following analytical expressions were obtained from the symbolic computation performed for Example 2. The force vectors, which do not represent reactions, at the supports of Node_1 and Node_2 are calculated as follows:

$$\mathbf{F}_1 = \begin{bmatrix} F_x \\ \frac{2\,F_x\,H + 2\,F_y\,L}{L} \end{bmatrix} \tag{17}$$

$$\mathbf{F}_2 = \begin{bmatrix} 0 \\ -\frac{2\,F_x\,H + 3\,F_y\,L}{L} \end{bmatrix} \tag{18}$$

The forces in the bars are determined as follows:

$$N_1 = F_x \tag{19}$$

$$N_2 = \frac{2\,F_x\,H + 2\,F_y\,L}{L} \tag{20}$$

$$N_3 = -\frac{F_x\,\sqrt{H^2 + L^2}}{L} \tag{21}$$

$$N_4 = -\frac{F_x\,H + 3\,F_y\,L}{L} \tag{22}$$

$$N_5 = F_x \tag{23}$$

$$N_6 = \frac{F_x\,H + 2\,F_y\,L}{L} \tag{24}$$

$$N_7 = -\frac{F_x\,\sqrt{H^2 + L^2}}{L} \tag{25}$$

$$N_8 = -3\,F_y \tag{26}$$

$$N_9 = F_x \tag{27}$$

$$N_{10} = 2\,F_y \tag{28}$$

$$N_{11} = -\frac{2\,F_y\,\sqrt{H^2+L^2}}{H} \tag{29}$$

$$N_{12} = \frac{2\,F_y\,L}{H} \tag{30}$$

$$N_{13} = \frac{F_x\,H - 2\,F_y\,L}{H} \tag{31}$$

$$N_{14} = -F_y \tag{32}$$

$$N_{15} = \frac{F_y\,\sqrt{H^2+L^2}}{H} \tag{33}$$

$$N_{16} = \frac{F_y\,L}{H} \tag{34}$$

$$N_{17} = \frac{F_x\,H - F_y\,L}{H} \tag{35}$$

$$N_{18} = -F_y \tag{36}$$

$$N_{19} = \frac{F_y\,\sqrt{H^2+L^2}}{H} \tag{37}$$

$$N_{20} = 0 \tag{38}$$

$$N_{21} = 0 \tag{39}$$

The displacement vectors $\boldsymbol{\delta}$ of the movable nodes, from Node_2 to Node_12, are determined as follows:

$$\boldsymbol{\delta}_2 = \begin{bmatrix} \frac{F_x\,L}{E\,S} \\ 0 \end{bmatrix} \tag{40}$$

$$\boldsymbol{\delta}_3 = \begin{bmatrix} \frac{F_x\left(\left(H^2+L^2\right)^{3/2}+2\,H^3+L^3\right)}{E\,S\,L^2} + \frac{2\,F_y\,H^2}{E\,S\,L} \\ \frac{2\,H\left(F_x\,H+F_y\,L\right)}{E\,S\,L} \end{bmatrix} \tag{41}$$

$$\boldsymbol{\delta}_4 = \begin{bmatrix} \frac{F_x\left(\left(H^2+L^2\right)^{3/2}+2\,H^3+2\,L^3\right)}{E\,S\,L^2} + \frac{2\,F_y\,H^2}{E\,S\,L} \\ -\frac{H\left(F_x\,H+3\,F_y\,L\right)}{E\,S\,L} \end{bmatrix} \tag{42}$$

$$\boldsymbol{\delta}_5 = \begin{bmatrix} \frac{2\,F_x\left(\left(H^2+L^2\right)^{3/2}+3\,H^3+L^3\right)}{E\,S\,L^2} + \frac{9\,F_y\,H^2}{E\,S\,L} \\ \frac{H\left(3\,F_x\,H+4\,F_y\,L\right)}{E\,S\,L} \end{bmatrix} \tag{43}$$

$$\boldsymbol{\delta}_6 = \begin{bmatrix} \frac{9\,F_y\,H^2}{E\,S\,L} + \frac{F_x\left(2\left(H^2+L^2\right)^{3/2}+6\,H^3+3\,L^3\right)}{E\,S\,L^2} \\ -\frac{H\left(F_x\,H+6\,F_y\,L\right)}{E\,S\,L} \end{bmatrix} \tag{44}$$

$$\boldsymbol{\delta}_7 = \begin{bmatrix} \frac{F_x\left(2\left(H^2+L^2\right)^{3/2}+10\,H^3+3\,L^3\right)}{E\,S\,L^2} + \frac{F_y\left(2\left(H^2+L^2\right)^{3/2}+21\,H^3\right)}{E\,S\,H\,L} \\ \frac{3\,H\left(F_x\,H+2\,F_y\,L\right)}{E\,S\,L} \end{bmatrix} \tag{45}$$

$$\boldsymbol{\delta}_8 = \begin{bmatrix} \frac{F_x\left(2\left(H^2+L^2\right)^{3/2}+10\,H^3+3\,L^3\right)}{E\,S\,L^2} + \frac{F_y\left(2\left(H^2+L^2\right)^{3/2}+21\,H^3+2\,L^3\right)}{E\,S\,H\,L} \\ -\frac{H\left(F_x\,H+7\,F_y\,L\right)}{E\,S\,L} \end{bmatrix} \tag{46}$$

$$\boldsymbol{\delta}_9 = \begin{bmatrix} \frac{2\,F_x\left(\left(H^2+L^2\right)^{3/2}+3\,H^3+2\,L^3\right)}{E\,S\,L^2} + \frac{F_y\left(9\,H^3-2\,L^3\right)}{E\,S\,H\,L} \\ -\frac{F_y\left(3\left(H^2+L^2\right)^{3/2}+19\,H^3+4\,L^3\right)}{E\,S\,H^2} - \frac{F_x\left(5\,H^3-L^3\right)}{E\,S\,H\,L} \end{bmatrix} \tag{47}$$

$$\delta_{10} = \begin{bmatrix} \frac{F_x\left(2\left(H^2+L^2\right)^{3/2}+10\,H^3+3\,L^3\right)}{E\;S\;L^2} + \frac{F_y\left(2\left(H^2+L^2\right)^{3/2}+21\,H^3+3\,L^3\right)}{E\;S\;H\,L} \\ -\frac{F_y\left(3\left(H^2+L^2\right)^{3/2}+20\,H^3+4\,L^3\right)}{E\;S\;H^2} - \frac{F_x\left(5\,H^3-L^3\right)}{E\;S\;H\,L} \end{bmatrix} \tag{48}$$

$$\delta_{11} = \begin{bmatrix} \frac{F_x\left(2\left(H^2+L^2\right)^{3/2}+6\,H^3+5\,L^3\right)}{E\;S\;L^2} + \frac{3\,F_y\,3\,H^3-L^3}{E\;S\;H\,L} \\ -\frac{2\,F_y\left(3\left(H^2+L^2\right)^{3/2}+16\,H^3+5\,L^3\right)}{E\;S\;H^2} - \frac{3\,F_x\,3\,H^3-L^3}{E\;S\;H\,L} \end{bmatrix} \tag{49}$$

$$\delta_{12} = \begin{bmatrix} \frac{F_x\left(2\left(H^2+L^2\right)^{3/2}+10\,H^3+3\,L^3\right)}{E\;S\;L^2} + \frac{F_y\left(2\left(H^2+L^2\right)^{3/2}+21\,H^3+3\,L^3\right)}{E\;S\;H\,L} \\ -\frac{2\,F_y\left(3\left(H^2+L^2\right)^{3/2}+16\,H^3+5\,L^3\right)}{E\;S\;H^2} - \frac{3\,F_x\left(3\,H^3-L^3\right)}{E\;S\;H\,L} \end{bmatrix} \tag{50}$$

For all modeling variants—Model 2.1, Model 2.2, and Model 2.3—the force vector $\mathbf{F}_{\text{load}} = [F_x, F_y]^{\text{T}}$ has the numerical value $\mathbf{F}_{\text{load}} = [0, 10]^{\text{T}}$ kN. The lengths of the bars are determined by the distances between the nodes or by the geometric dimensions of the system, with $L = 1$ m and $H = 1$ m. The bars all have the same diameter, $d = 0.01$ m. The Young's modulus E is 210 GPa.

For Models 2.1 and 2.2, the shear modulus G is 79 GPa, and the material density (*Density*) is 7800 kg/m^3. Poisson's ratio is 0.28. The values for Rayleigh damping are kept at their default settings for computational convenience, with $b_m = 0.01$ s^{-1}, $b_k = 0.01$ s. These values can essentially be considered nearly negligible. The damping in the supports is $b_{rj} = 1$ N·m·s·rad^{-1}.

In Model 2.3, the support damping is $b_{rj} = 10^3$ N·m·s·rad^{-1}, while the bar damping is $b = 10^2$ N·s·m^{-1}. The axial stiffness for tension/compression in the bars is $k = E{\cdot}10^9{\cdot}\pi{\cdot}d^2{\cdot}(4{\cdot}l)^{-1}$ N·s·m^{-1}.

4.2.2. Results for Example 2: Theoretical Calculation and MATLAB Models

For clarity, the results are presented in the Table 2, with the theoretical calculations listed first, followed by those from the individual models.

After optimization, Model 2.1 achieved consistent accuracy of up to five or six decimal places when compared with the theoretical model, except for values equaling zero. Model 2.3 exhibited similarly high accuracy. Model 2.2, despite not meeting the assumptions of the theoretical model, achieved an accuracy of up to approximately two decimal places This is because there are no movable joints in the nodes.

4.2.3. Results for Example 2: Visualization of the Deformed State of the Structures

Visualization of the deformed state was performed for a specified norm of the displacement vector, $||\delta_{11}|| = 0.1$ m, at Node_11 (see, e.g., Figure 9). This value is not critical and serves solely to exaggerate the deformation in the theoretical models for illustrative purposes. For the physical structure, copper (Cu) was selected as the material due to its favorable ductility and plasticity. The diameters of the bars were kept constant within each sample, with values of $d_a = 1.4{\cdot}10^{-3}$ m and $d_b = 1{\cdot}10^{-3}$ m. The dimensions were set as $H = 40{\cdot}10^{-3}$ m and $L = 40{\cdot}10^{-3}$ m. The model was constructed exclusively as a test for excessive deformation. In the theoretical models, the material was not changed, as this had no impact on the resulting deformed shape.

For Model 2.3, it was necessary to reduce the axial stiffness of the bars in tension/compression by approximately three orders of magnitude, calculated as $k = 10^{-3}{\cdot}E{\cdot}\pi{\cdot}d^2{\cdot}(4{\cdot}l)^{-1}$ N·s·m^{-1}.

The differences between the structural models and the physical structure depicted in Figure 18d are also evident in the state of excessive deformation. Model 2.3, shown in Figure 18c, offers a good geometric approximation of the deformation shape. For the physical structure, the results are influenced by manufacturing precision and the method of fixation, both at the load application point and at the supports. Models 2.1 in Figure 18a and 2.2 in Figure 18b produced a closer match when lower damping values were assigned to the joints at the supports b_{rj} and higher damping coefficients were used in the bars b_k. This enabled a more effective transmission of the load effect from the point of application throughout the rest of the structure. However, this came at the cost of increased computational time.

4.3. Results for Example 3: Test Calculation of a More Complex Structure Based on Example 2

Based on the positive results obtained in the previous examples, we selected a model in which the system consists of Euler–Bernoulli beams and joints. The aim was to test computational accuracy and, in particular, the required simulation time for a more complex structure in a theoretical context without a physical structure. This example will be available in [18]. And its sketch is in Figure 19.

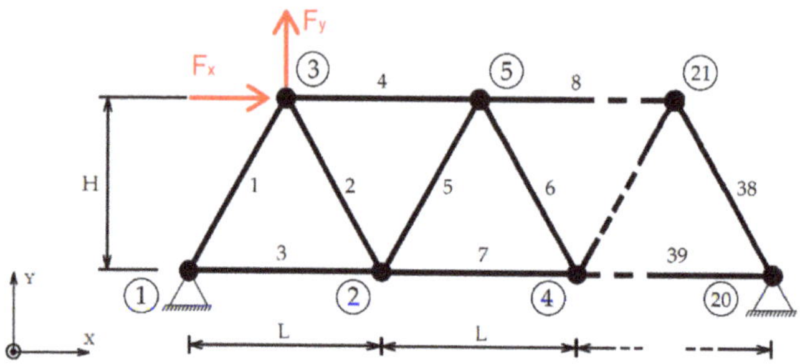

Figure 19. Example 3. Schematic of the truss system structure subjected to the force vector $\mathbf{F}_{\text{load}}$ at Node_3, comprising 39 bars and 21 nodes.

The system is essentially composed of a prototype subsystem that generally includes bars modeled as Euler–Bernoulli beams and joint nodes labeled 1, 2, 3, and 4. The last node has no further continuation, as it represents the end of the structure.

The force $F_{\text{load m}}$ serves as a verification check of the computed force F_{load}. The force $F_{\text{load m}}$ is derived from the forces within the system, specifically from the equilibrium at Node_3 in Subsystem 1; see Figure 20.

$$\sum F_x = 0;\ F_{\text{load }x} + N_4 - \frac{N_1}{2} + \frac{N_2}{2} = 0 \tag{51}$$

$$\sum F_y = 0;\ F_{\text{load }y} - \frac{\sqrt{3}\cdot N_1}{2} - \frac{\sqrt{3}\cdot N_2}{2} = 0 \tag{52}$$

The force $\mathbf{F}_{\text{load}}$ represents the external load acting on the system. In cases where multiple external forces are applied to the system, the maximum force is selected for the control measurement. For clarity, symbolic calculations are no longer included at this stage; instead, the individual results are organized in a table. First, the theoretical calculation is presented, followed by the results obtained from the model.

The model, after optimization, achieved an accuracy consistent with that of the theoretical model up to six decimal places. A visualization of the deformed state was performed for a defined norm of the displacement vector $||\delta_3|| = 0.1$ m at Node_3. This value is not critical, and is used solely for an exaggerated display of the deformation. See Figure 21.

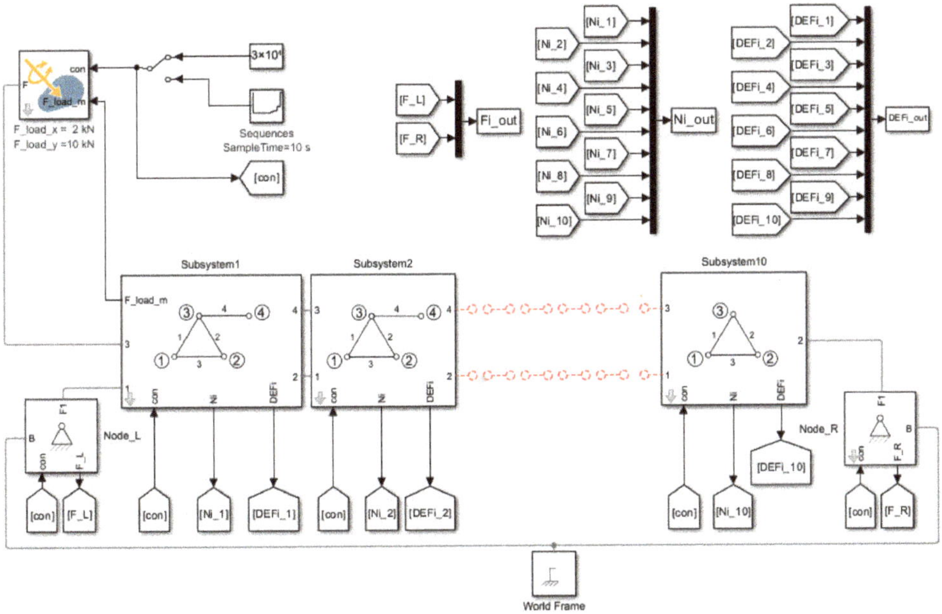

Figure 20. Example 3. Model comprising Euler–Bernoulli beams and joint nodes. Truss system structure subjected to the force vector $\mathbf{F}_{load}$ at Node_3, Subsystem 1. Created in MATLAB–Simscape Multibody based on the schematic in Figure 19.

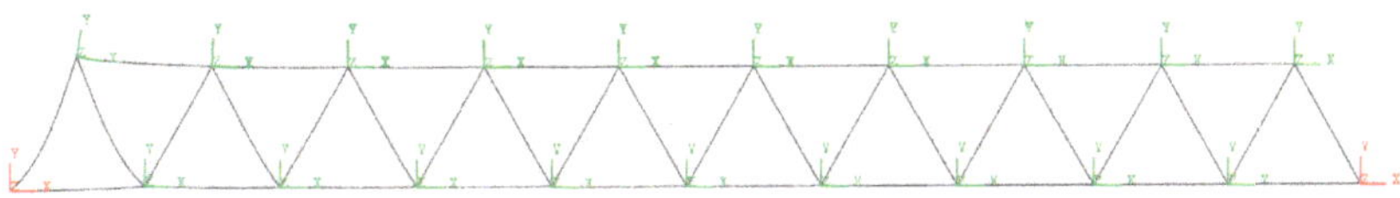

Figure 21. Visualization of the deformed state of the model for Example 3.

In this simulation, the computational time increased; however, the level of accuracy was maintained. The numerical computation conditions for all models in Example 1, Example 2, and Example 3 were identical. The same solver (ode15s) was used, along with consistent settings for the absolute tolerance (10^{-3}), relative tolerance (10^{-3}), maximum iteration step (0.01), within a variable step.

5. Discussion

During simulations in Simulink and Simscape models, certain differences in computational accuracy arise when the levels of input quantities are varied. These differences can naturally be addressed in "Model Settings–Solver" [19] (pp. 311–319) or the online aid [28] by imposing stricter requirements on parameters such as "Max step size", "Absolute tolerance", and "Relative tolerance", either with respect to zero or relative to previous iterations. For the computations involving the given truss systems, where a static problem was analyzed, the recommended solvers ode15s, ode23s, ode23t, and ode23tb were used for so-called stiff problems. These solvers exhibit significantly higher speed compared to methods designed for nonstiff problems, such as ode45, ode23, or ode113. Another interesting approach that can be taken to further improve computational accuracy involves adjusting the proportionality (scaling) of the input quantity in the system, whether the system under analysis is a static structure, as in our case, or another type of system. Based on this proportionality, it is sufficient to identify a known quantity within the system (in our case, $\mathbf{F}_{\text{load}}$) and redefine it based on the computational outputs (i.e., $\mathbf{F}_{\text{load m}}$). For the simplest evaluation of output quantities, we opted for implementing step changes in proportionality using the multiplier constant *con* within individual sequences. Data representing steady-state values were isolated using a time *Lag* and *Lead* to exclude transient responses, which would otherwise distort the output value of $\mathbf{F}_{\text{load m}}$. Figures 14–16 clearly demonstrate that the accuracy of the computation varies depending on the level of the load force $\mathbf{F}_{\text{load}}$. The optimum is achieved when the mean value of $\mathbf{F}_{\text{load m}}$ closely matches $\mathbf{F}_{\text{load}}$ while maintaining minimal swing. For example, if, for a given sequence k, the mean value of $\mathbf{F}_{\text{load m}}$ is closer to $\mathbf{F}_{\text{load}}$, but the swing is large enough to extend into the mean calculated in sequence $(k-1)$, then the probability of the calculation being accurate is not 100%. We observed that this proportionality cannot be easily generalized for all systems. It depends on numerous parameters of the system. If parameters such as the diameter of the bars or the Young's modulus are altered, the optimal setting for the computation also changes. Regarding damping, the coefficients were selected to be the minimum necessary. For bars subjected to Rayleigh damping, the default values of b_m and b_k were retained, as they can essentially be nearly negligible. This is because very small increments of deformation are considered, which are proportionally scaled. In the case of solid bars, the damping coefficient b and the damping of supports b_{rj} were only increased to the level necessary to suppress undesirable transient responses during the computation.

The examples are divided into models that differ in their types of nodes and bars. The aim was to assess their influence on static analysis and the visualization of deformation. The models constructed from Euler–Bernoulli beams or from solid bars with one degree of freedom (1 DOF) and joints delivered the best results when compared with the theoretical model, as shown in Tables 1–3. However, it should be noted that models with solid bars and one DOF are generally unsuitable for dynamic simulations due to the non-uniform distribution of mass elements along the length of the bar. Models using locked joints may, in some cases (such as welded joints), reflect physical reality, but they show lower agreement with the theoretical model. This discrepancy arises because the direction of internal forces in the bars changes during deformation, a behavior that conventional joints can accommodate. For visualizing deformation, physical structures composed of copper elements were deliberately created due to copper's excellent ductility (see Figures 17d and 18d). These structures were deformed between two holding plates to prevent undesired buckling along the z-axis. Models 1.3 and 2.3 (see Figures 17c and 18c) exhibited the fastest computations and provided a good approximation of the deformed shape. The time required for the calculation depends on the number of system elements. See Figure 22.

Models 1.1 and 2.1 (Figures 17a and 18a) also achieved accurate shape approximations when higher levels of Rayleigh damping were applied to the bars, albeit at the cost of significantly longer computational times. The final extended model depicted in Figure 19 was developed specifically to test the precision and computational speed for more complex structures.

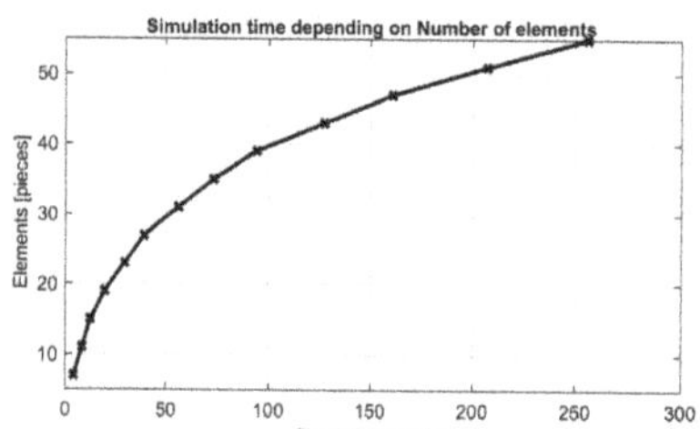

Figure 22. The calculation depends on the number of system elements.

6. Conclusions

This article provides a comprehensive explanation of a fundamentally simple approach to enhancing computational accuracy within the Simulink environment. It builds on previous work and explores the potential of utilizing built-in tools available as part of the software package employed in our research. Our interest lay in assessing the results in terms of accuracy and determining the extent to which the simulations in Simulink deviated from theoretical calculations. Although this may represent an extreme requirement in terms of precision, its scaling offers a viable means of improving computational accuracy.

As discussed in Section 5, the examples are categorized into models differing in their types of nodes and bars. The aim was to investigate their influence on computational accuracy and deformation visualization in order to identify the optimal model in terms of theoretical calculations and, where applicable, visualization. From a computational standpoint, the model composed of Euler–Bernoulli beams and joints proved to be the most suitable, and it is also appropriate for dynamic simulations. However, regarding the visualization of extreme deformations, the results were not as clear-cut as in the case of computational analyses. It is therefore not possible to definitively identify a single model as the most suitable in this context. It should also be noted that the physical structures themselves were not perfect either, considering aspects such as the precision of node placement within the system, their mechanical implementation, structural strength, material quality, and similar factors. Additionally, the ratio of the length of the bar to its cross-section played a significant role in the collapse behavior of the physical truss structure. Two physical structures were experimentally constructed, each using bars of different diameters. Under the specified dimensions, working with a smaller cross-section proved more advantageous during loading. For more complex structures, the required computational time increased even though the desired level of accuracy was maintained.

Author Contributions: Drafting of concept and theoretical analysis: Š.O.; model design and data processing in MATLAB: Š.O.; overview, supervision, and approval: J.S. (Jozef Svetlík); project administration and methodology: J.S. (Jozef Svetlík), R.J., J.S. (Ján Semjon), M.S., T.S., and P.M. All authors have read and agreed to the published version of the manuscript.

Funding: This paper was prepared with the support of the grant project KEGA 043TUKE-4/2024—Creation of prospective educational tools for the field of additive manufacturing with the implementation of progressive elements of virtual reality, and VEGA 1/0215/23—Research and development of robotic workplaces equipped with industrial and collaborative robots.

Data Availability Statement: The original contributions presented in this study are included in the article. Further inquiries can be directed to the corresponding author.

Conflicts of Interest: The authors declare no conflicts of interest. The funders had no role in the design of the study; in the collection, analyses, or interpretation of data; in the writing of the manuscript; or in the decision to publish the results.

Abbreviations

The following abbreviations are used in this manuscript:

FEA	Finite Element Analysis
FEM	Finite Element Method
DOF	Degree Of Freedom
$\mathbf{F}_{load}$	Load force vector
$\mathbf{F}_{load\ m}$	Control measurement of load force vector
$\mathbf{N}$, N	Vector, component of normal force acting on the bar
$\mathbf{F}$, F	Vector, component of the force (not a reaction force) acting on the support

l	Bar length, unless specified otherwise
L	Distance between nodes in the x–axis direction
H	Distance between nodes in the y–axis direction
d	Bar diameter
S	Bar cross-section
E	Young's modulus describing the bar's elasticity
G	Shear modulus
Density	Material density
μ	Poisson's ratio
$\boldsymbol{\delta}, \delta$	Vector, component of node deflection in plane
b_m	Coefficient proportional to the mass matrix in the Rayleigh damping model
b_k	Coefficient proportional to the stiffness matrix in the Rayleigh damping model
b_{rj}	Damping at the support
b	Damping in the joint of the solid bar with 1 DOF
k	Stiffness in the joint of the solid bar with 1 DOF, as well as sequence step size, sequence step
con	Multiplication (proportional) constant
x, y, z	Axes of the coordinate system, or components of the position vector
$u(m)$	mth element of input variable vector **u** for a function in Simulink
MATLAB®	(MathWorks, 1 Apple Hill Drive, Natick, MA 01760 USA, Founded in 1984)

References

1. Ondocko, S.; Svetlik, J.; Jánoš, R.; Semjon, J.; Dovica, M. Calculation of Trusses System in MATLAB—Multibody. *Appl. Sci.* **2024**, *14*, 9547. [CrossRef]
2. MathWorks, Inc. Simscape™ Multibody™ Getting Started Guide. *The MathWorks*. 2025. Available online: https://www.mathworks.com/help/pdf_doc/sm/sm_gs.pdf (accessed on 17 June 2025).
3. MathWorks, Inc. Simscape™ Multibody™ User's Guide. *The MathWorks*. 2025. Available online: https://www.mathworks.com/help/pdf_doc/sm/sm_ug.pdf (accessed on 17 June 2025).
4. Langtangen, H.P.; Pedersen, G.P. *Scaling of Differential Equations*; Springer Nature: Berlin/Heidelberg, Germany, 2016.
5. Bradley, D. What is mechatronics and why teach it? *Int. J. Electr. Eng. Educ.* **2004**, *41*, 275–291. [CrossRef]
6. Bishop, R.H. *Mechatronics: An Introduction*; CRC Press: Boca Raton, FL, USA, 2017.
7. MathWorks, Inc. *SimMechanics™ 3 User's Guide*; The MathWorks: Natick, MA, USA, 2009.
8. Hroncova, D.; Pastor, M. Mechanical system and SimMechanics simulation. *Am. J. Mech. Eng.* **2013**, *1*, 251–255.
9. Frankovský, P.; Hroncová, D.; Delyová, I.; Virgala, I. Modeling of dynamic systems in simulation environment MATLAB/Simulink–SimMechanics. *Am. J. Mech. Eng.* **2013**, *1*, 282–288.
10. MathWorks, Inc. System Scaling by Nominal Values. 2025. Available online: https://www.mathworks.com/help/releases/R2025b/simscape/ug/system-scaling-by-nominal-values.html?searchHighlight=command+scaling&s_tid=doc_srchtitle#mw_3cc30ae6-fec4-4118-a0d3-032dbf0901cc (accessed on 3 October 2025).
11. Miller, S.; Soares, T.; Weddingen, Y.V.; Wendlandt, J. Modeling flexible bodies with simscape multibody software. In *An Overview of Two Methods for Capturing the Effects of Small Elastic Deformations*; MathWorks: Natick, MA, USA, 2017.
12. Bauchau, O.A.; Craig, J.I. Euler–Bernoulli beam theory. In *Structural Analysis*; Springer: Dordrecht, The Netherlands, 2009; pp. 173–221.
13. Trebuňa, F.; Šimčák, F. *Odolnosť Prvkov Mechanických Sústav*; Technická Univerzita: Košice, Slovakia, 2004.
14. Lengvarský, P.; Pástor, M.; Bocko, J. Riveted Joints and Linear Buckling in the Steel Load–bearing Structure. *Am. J. Mech. Eng.* **2017**, *5*, 329–333.
15. Karnovsky, I.A.; Lebed, O. *Advanced Methods of Structural Analysis*; Springer Nature: Berlin/Heidelberg, Germany, 2021.
16. Plevris, V.; Ahmad, A. Deriving analytical solutions using symbolic matrix structural analysis for continuous beams. *Sci. Rep.* **2025**, *15*, 1–16. [CrossRef] [PubMed]
17. Florian, Z.; Pellant, K.; Suchánek, M. *Technická mechanika I–Statika*; Vysoké Učení Technické v Brně: Brno, Czech Republic, 2004.
18. Available online: https://github.com/stefanondocko/AddOptimInSimulink (accessed on 17 June 2025).
19. MathWorks, Inc. Using Simulink. In *Simulink–Model–Based and System–Based Design*; MathWorks Inc.: Natick, MA, USA, 2002.
20. MathWorks, Inc. Speed Up Simulation. 2023. Available online: https://www.mathworks.com/help//releases/R2021a/simulink/ug/speed-up-simulation.html (accessed on 22 June 2025).
21. Gavin, H.P. *Classical Damping, Non–Classical Damping and Complex Modes*; Duke University: Durham, NC, USA, 2018.

22. MathWorks, Inc. Choose an ODE Solver. Ordinary Differential Equations. 2025. Available online: https://uk.mathworks.com/help/matlab/math/choose-an-ode-solver.html (accessed on 17 June 2025).
23. Oršanský, P.; Ftorek, B. *Štatistické a Numerické Metódy*; EDIS–ŽU: Žilina, Slovakia, 2017.
24. Buša, J.; Pirč, V.; Schrotter, Š. *Numerické Metódy, Pravdepodobnosť a Matematická Štatistika*; FEI TU: Košice, Slovakia, 2006.
25. Available online: https://github.com/vplevris/SymbolicMSA-2DTrusses (accessed on 17 June 2025).
26. Murín, J.; Hrabovský, J.; Kutiš, V. *Metóda Konečných Prvkov—Vybrané Kapitoly pre Mechatronikov*; Slovenská technická univerzita v Bratislave: Bratislava, Slovakia, 2014.
27. Öchsner, A. Euler–Bernoulli beam theory. In *Classical Beam Theories of Structural Mechanics*; Springer: Berlin/Heidelberg, Germany, 2021; pp. 7–66.
28. MathWorks, Inc. Solver–Solver that Computes States and Outputs for Simulation. 2025. Available online: https://uk.mathworks.com/help/simulink/gui/solver.html (accessed on 17 June 2025).

applied sciences

MDPI

Article

Integration Approaches for Digital Twins in Dataspaces

Carlos Schmidt [1], Friedrich Volz [1,*], Ljiljana Stojanovic [1] and Holger Kett [2]

[1] Fraunhofer Institute of Optronics, System Technologies and Image Exploitation IOSB, 76131 Karlsruhe, Germany; carlos.schmidt@iosb.fraunhofer.de (C.S.); ljiljana.stojanovic@iosb.fraunhofer.de (L.S.)
[2] Fraunhofer Institute for Industrial Engineering IAO, 70569 Stuttgart, Germany; holger.kett@iao.fraunhofer.de
* Correspondence: friedrich.volz@iosb.fraunhofer.de

Abstract

Digital twins (DTs) based on the Asset Administration Shell (AAS) enable interoperable representations of industrial assets, while dataspaces like those defined by the International Data Spaces (IDS) facilitate sovereign, policy-governed data exchange across organizations. However, integrating AAS-based DTs into dataspaces remains challenging due to manual processes, synchronization issues, and varying implementation strategies. This paper addresses these gaps by first presenting a structured analysis of existing integration approaches and evaluating the user requirements to be met. We then introduce a flexible reference implementation extending the Eclipse Dataspace Connector (EDC), which automates AAS registration into dataspaces. Finally, we conduct a qualitative comparison with state-of-the-art prototypes from initiatives such as Catena-X and Factory-X and demonstrate the application of our integration approach in the MODAI project for sensor-driven manufacturing data sharing. Results indicate that the integration approach reduced manual effort and provides synchronization between digital twins and dataspace.

Keywords: dataspace; digital twin; Asset Administration Shell (AAS); Industry 4.0

1. Introduction

Digital twins (DTs) are digital representations of physical assets, systems, or processes that support real-time monitoring, simulation, and analysis throughout their life cycle [1,2]. Acting as a bridge between the physical and digital worlds, DTs enable data-driven decision-making across a wide range of domains, including manufacturing. As their adoption increases, the need for standardized models becomes critical to ensure interoperability, scalability, and reusability across heterogeneous systems and stakeholders.

In the industrial domain, the Asset Administration Shell (AAS) provides a standardized framework for DTs, offering a common, machine-interpretable information model and well-defined interfaces. The Industrial Digital Twin Association (IDTA) is now responsible for the ongoing development and standardization of the AAS [3], enabling machines and software systems to "speak the same language" regardless of manufacturer, technology, or life cycle phase.

While the AAS specification ensures syntactic and semantic interoperability within a production site, it previously lacked guidance on how data contained in an AAS can be securely shared across company borders. This gap is now addressed in the recently published Part 4 of the AAS [4], which introduces mechanisms for secure data exchange through attribute-based access control (ABAC). Additionally, it introduces the concept of dataspaces as the context in which usage contracts between companies can be applied, thereby fostering trusted and controlled cross-company collaboration [4].

https://doi.org/10.3390/app152111623

This vision of cross-organizational data sharing aligns closely with the work of the International Data Spaces Association (IDSA), which developed the International Data Spaces Reference Architecture Model (IDS-RAM) [5] to facilitate sovereign, policy-governed data exchange across organizational boundaries. The IDS architecture defines standardized building blocks (e.g., connectors, brokers, clearing houses) along with open-source reference implementations that support usage contracts, policy enforcement, and trust among participants.

Taken together, AAS and IDS offer complementary capabilities. The AAS enables plug-and-play interoperability on the shop floor—primarily within organizational boundaries. IDS-compliant dataspaces provide the governance, trust mechanisms, and contractual semantics required for data exchange between organizations. Recognizing this complementarity, several initiatives (e.g., Catena-X [6], Factory-X [7], CircularTwAIn [8]) have explored different patterns for integrating AAS with dataspaces. Several publications describe their integration of AAS into dataspaces [9–18]. However, these integration approaches vary significantly in terms of where the integration occurs, which extensions are introduced, and how much complexity is imposed on end users.

In [19], the authors identified nine articles regarding manufacturing networks with AAS and dataspaces based on IDS components. A common integration pattern involves mapping AAS submodels and metadata to IDS assets within an IDS connector. In this context, an IDS asset represents a data resource—such as an AAS submodel—along with its associated metadata and usage policies. The connector then publishes these assets as data offerings in a dataspace catalog, making them discoverable by other participants under specified usage policies.

In [20], the authors describe some limitations of this mapping approach. Namely, the unknown granularity for the AAS and the incompatibility between different dataspace implementations. Regarding the granularity, for each AAS instance, the specific parts intended for sharing (submodels, submodel-elements) must be registered in the catalog, with an associated usage policy that defines how the data can be accessed and under what conditions. Currently, this process is largely manual: data providers create IDS assets corresponding to selected AAS submodels or elements and explicitly define the associated metadata and contractual terms. In [21], the authors describe the manual creation of "resources in the supplier connector" for a manufacturing-as-a-service system. However, in their case, a resource is not created for a submodel or element but for a list of every AAS in the system, since their use case focuses on execution of manufacturing orders. The latter limitation of incompatibility has been mainly solved by the adoption of the Dataspace Protocol in IDS-based components [5].

Despite the high effort of manually describing the AAS and their contracts in the dataspace, this limitation is only occasionally addressed in publications (e.g., [12]). Especially in cases of high granularity on the element level, manual creation is a tedious workload. For example, the authors in [22] implemented an AAS Manager that creates the resources for each submodel to be shared.

Efforts to automate this process have led to several implementation strategies. One approach involves reading submodels from existing AAS services and dynamically generating IDS catalog entries. Another embeds AAS services directly within the connector, allowing for tighter integration and real-time synchronization. However, these strategies introduce trade-offs. In cases where companies already operate AAS services as part of their production infrastructure, automation may require re-hosting, redesign, or modification of existing systems—introducing additional complexity and increasing the overall burden of adoption.

In light of these challenges, this paper makes the following contributions: Firstly, a structured analysis of existing approaches for integrating AAS-based DTs into dataspaces, including an evaluation of how authorization, usage contracts, automation, etc., are addressed. Secondly, a flexible reference implementation built on the IDS open-source stack and the Eclipse Dataspace Connector (EDC) [23] is presented, which automates the integration of AAS instances into a dataspace with reduced manual effort. Finally, a qualitative comparison of our implementation against state-of-the-art prototypes (e.g., from Catena-X, Factory-X) is conducted.

The rest of this paper is organized as follows: Section 2 provides background on dataspaces and the AAS landscape for DTs. Section 3 describes the user requirements and architectural design for automated data integration into dataspaces. Section 4 presents our approach and implementation details. Section 5 compares our approach with existing solutions using the proposed comparison framework. Section 6 summarizes the findings and outlines future research directions, especially in the direction of "twin-native" connectors specifically designed for sharing DTs in dataspaces.

2. Background

2.1. Datspaces

Europe's data strategy has shifted from isolated data markets to federated dataspaces, i.e., interoperable infrastructures in which participants keep full control over the data they share. The Data Space Support Centre [24] defines the Data Space Blueprint, a technology-agnostic reference architecture. This blueprint is intended to give every dataspace project a common vocabulary, a canonical set of building blocks, and a minimal interoperability profile, while leaving enough freedom to choose concrete products or open-source implementations. In our work, we adopt the blueprint and implement most of its components with the Eclipse Dataspace Components [25].

The blueprint distinguishes two categories of dataspace building blocks:

- Technical building blocks: These cover all core capabilities—from secure data transfer in a "minimal dataspace" to advanced services such as marketplaces. Most are domain-agnostic, but the block dealing with data models and formats must be tailored to each sector (e.g., Industry 4.0) [26].
- Business and organizational building blocks: These capture the organizational, business, and operational agreements needed to realize and run a decentralized, multi-stakeholder architecture.

Within this ecosystem, dataspace participants are typically divided into data providers, who offer controlled access to data governed by usage policies, and data consumers, who access this data after accepting or negotiating usage terms.

A key technical component supporting these interactions is the EDC, part of the Eclipse Dataspace Components. The EDC is an open-source, Apache-2.0 licensed reference connector implementation that enables secure and controlled data sharing between companies, serving as the primary interface for data exchange within dataspaces.

The EDC consists of two logical planes:

- Control plane: Asset and policy registration, catalog API, contract negotiation, identity and trust, and management API.
- Data plane: Actual data transfer via protocols (usually HTTP pull/push).

The control plane is where data or a reference to the data is registered as assets in the internal AssetStore. Contracts are attached to the assets to describe who may access the data and which usage policies need to be accepted by the data consumers. Assets and contracts are together offered as datasets in the provider's catalog. Data consumers also have the option to make counteroffers based on the provider's contract. Figure 1 shows the interaction between data providers and data consumers with the different planes of the EDC.

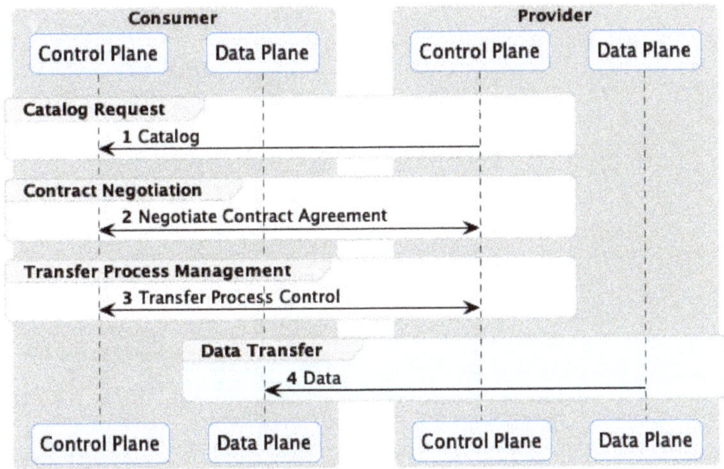

Figure 1. Control and data plane between consumer and provider with arrows indicating data flow, reproduced with permission from [23].

The consumer requests the catalog from the provider and selects an asset to negotiate on. In many cases the contract attached to the asset is accepted as-is by the consumer, but in every case a contract offer is sent. The transfer process is then initiated either by provider push or consumer pull so that the data transfer between the data planes starts.

2.2. Asset Administration Shell as Digital Twin

The IDTA assumed responsibility for further developing and maintaining the AAS specification, keeping it up to date with industry needs and international standards. It describes the AAS as a standardized DT of a physical asset that enables different manufacturing applications to work together. It clearly points to the asset it represents, contains digital submodels that represent many aspects of the asset, and lists the technical services offered by the asset [3].

Ideally, an AAS covers most information and functions related to an asset—its features, parameters, sensor data, and operating abilities. All these data are arranged in submodels, which describe one aspect of the asset, for example, the technical data or the carbon footprint. So-called submodel templates predefine the submodel and are managed by the IDTA to cover common aspects of the asset in a standardized way. Figure 2 shows the structure of the AAS.

Each submodel consists of submodel elements, which can be properties, operations, files, etc., of the asset. They can be grouped in SubmodelElementCollections and SubmodelElementLists. Different communication protocols can connect the AAS to the real-world asset to fill the properties and execute the operations.

The AAS standard is split into several parts. Part 1 explains the meta-model and how data is serialized, while Part 2 defines the API for communication with AAS instances during operation. Further parts are still being worked on. Part 4 was recently published and deals with security in the form of ABAC. It also mentions a dataspace with AAS where the AAS can "support Identity Providers of dataspaces or any other multilateral company use cases" [4].

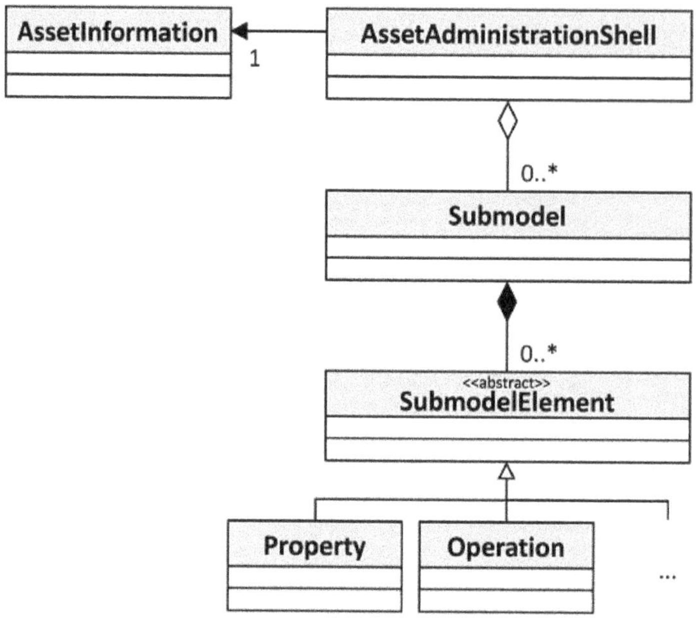

Figure 2. UML class diagram of simplified AAS meta-model, reproduced with permission from [27]. 0,*,1 are standard UML notation for min, max, multiplicity.

Part 2 of the AAS specifies different AAS APIs, such as the submodel repository API to create and delete submodels or the submodel registry API to create and access descriptors of submodels. In this context, a submodel registry is a software component that contains a list of descriptors that describe the submodels and where to find them. Analogously, there is an AAS registry that describes the AAS. So-called AAS services or AAS servers usually combine the AAS repository API with the AAS service API to manage and grant access to the AAS.

2.3. Why to Use Dataspaces with Digital Twins

While data sharing can be conducted solely with DTs, e.g., AAS communication with AAS API, most DT specifications do not deal with data-sovereign sharing to the same extent as the IDS. For example, the topic of data usage contracts is not addressed by the AAS. The AAS defines ABAC, an attribute-driven authorization model concerned with granting access, while Usage Control is a superset that governs the entire usage phase, extends the decision with obligations/conditions, and can continuously enforce or revoke permissions during and after use. Table 1 compares an AAS-only approach with a dataspace integration approach.

Table 1. Comparison between a data integration approach with only AAS or dataspace.

Aspect	AAS-Only Approach	Dataspace Approach
Identity	Entity-specific	Company-specific
Access control	ABAC only	Usage control policies and usage contracts
API	AAS API	Dataspace protocol
Datasets	AAS metamodel and data	any

In a dataspace, the identity of a provider or consumer is usually company-specific, while the AAS identifies the asset itself. The Dataspace Protocol is the API that all connectors utilize [5].

While dataspaces and DTs are two distinct concepts, they can be combined to enable interoperable and standardized data exchange within industrial ecosystems [28]. Several research projects have defined data integrations between DTs and dataspace, which will be discussed in Section 5.

2.4. Business Use Case of MODAI Project

Our use case consists of multiple press shops with several press machines and sensors, continuously monitoring elastic deformations that occur during each press stroke. The resulting force curves serve as digital fingerprints of the forming operation, enabling real-time detection of anomalies, wear, or manual interventions. This highly localized, high-frequency sensor data is preprocessed by embedded AI and enriched with contextual information close to the machine—creating a valuable data asset at the machine edge. Describing this data asset with AAS submodel templates allows easier use outside the press itself, for example, by the press manufacturer or the customers of the press shop. Figure 3 shows the part manufacturer setup in the use case.

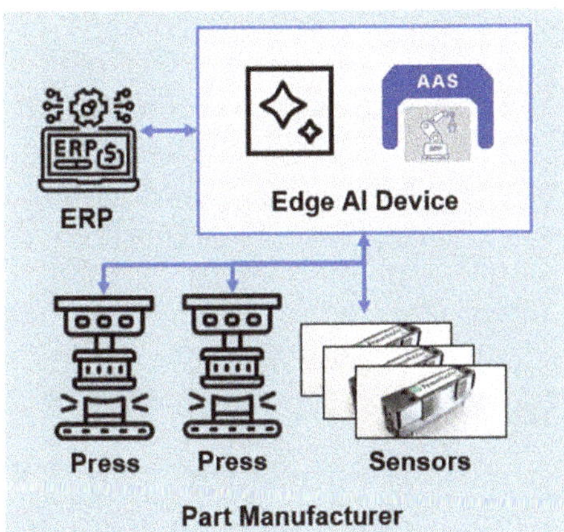

Figure 3. Part manufacturer in MODAI use case with arrows indicating data flow.

The potential of such sensor data extends well beyond the press itself. In quality management, deviations in press signatures can be linked to downstream part defects or machine faults or traced back to material inconsistencies, thereby enhancing root-cause analysis. Finally, procurement and supplier management gain access to live evidence about material and machine behavior during forming—opening the door to collaborative quality assurance with upstream partners.

Sharing this data across the supply chain is useful but not yet common since the way to share it is not harmonized, and data usage terms must be defined in actual contracts. By utilizing a dataspace, the press shop would be able to share this data with the customer in a harmonized way. However, if the press shop must actively manage the data to be shared and update metadata, onboarding friction and additional effort occur. The software landscape of the part manufacturers also heavily varies case by case with different software implementations to be utilized or already in use. Initial workforce orientation efforts with the DT and dataspace software, as well as contracts and policies, also limit broad adoption. The challenge that the press shops are faced with when trying to integrate an AAS—or any kind of data—into the dataspace is the need of expert DT and dataspace knowledge in the field. Additionally, the frequency of the data makes it necessary to expand on the structure of the AAS, e.g., creating new time-series data segments. The press shops are a highly dynamic environment with frequent tool, configuration, and setup changes.

What enables this broad organizational impact is not just the sensor itself, but the structured exposure of its data through an interoperable framework. By wrapping the sensor, its metadata, and its AI-generated features within a standardized AAS, the press becomes semantically visible and machine-readable within IT systems across the factory.

It is also vital to prepare machines and sensors in the factory for a bidirectional data and control flow across the value chain, where decisions are made not only for one factory location.

When integrated into an industrial dataspace, these AAS-based digital twins can be securely discovered, accessed, and governed across department and company boundaries. This allows, for example, quality managers to retrieve anonymized press performance profiles from suppliers, or planners to orchestrate forming lines based on live status data.

In essence, AAS and dataspaces together act as the bridge that turns isolated sensor events into enterprise-wide intelligence. While the sensor provides the raw capability, it is the structured, sovereign, and policy-controlled integration of its data into dataspaces that truly unlocks its cross-functional value. This example illustrates how a seemingly narrow sensor application can catalyze transformation across organizational layers—when embedded in a federated, interoperable data architecture.

3. User Requirements and Architecture for AAS Dataspace Integration

3.1. Requirements

Table 2 shows an overview of functional and non-functional user requirements for a component ("system") that helps to automate the process of integrating digital twins into the dataspace.

The base requirement is to support providers to automatically share the AAS from a service or registry at the property level or submodel level (FR1). The property level means that data consumers can negotiate for access on a single property, while at the submodel level, consumers will negotiate for the full submodel.

Every time there is a structural change in the AAS, users would have to update or delete the EDC asset for it while creating a new one. This is why a synchronization between AAS structure and published dataspace metadata is required (FR2).

Some data providers prefer to only share a subset of the AAS [21] and grant access at the submodel level, which means that after successful negotiation on a certain submodel, all properties of this submodel are accessible (FR3).

Different projects utilize dataspace connectors like the EDC and modify it for their use case, for example, the Tractus-X EDC [29] or MX-Port [30]. Likewise, there are AAS implementations with a focus on different features [27]. A successful integration should thus support different connectors and AAS implementations (FR4).

Additionally, managing and attaching data contracts to the data results in high manual effort for the user. An integration component should assist the user in attaching contracts to certain elements while also allowing attachment of a pre-configured contract to all shared elements (FR5).

Use cases with high data throughput require a high amount of scaling in the cloud, which means the integration approach should not tightly couple any software components (FR6).

As the authors of [12] also point out, another important requirement is offering data consumers a discovery mechanism for a published AAS in the dataspace (FR7). In dataspaces, a catalog is typically used to offer any published data to all partners. A limitation of this is the flat structure in which datasets are presented within the catalog. To account for structured entities like an AAS, an AAS-aligned discovery mechanism is needed.

Table 2. Requirements for an integration component between digital twins and dataspace.

#	Requirement Name	Description
FR1	Assisted AAS Registration	The system shall assist in the registration of existing AAS repositories/registries as datasets so that their metadata and endpoints are known to the dataspace. In case an AAS is available only as a file, the system shall instantiate the corresponding AAS repository.
FR2	Structure and Metadata Synchronization	The system shall reflect changes to the AAS structure and any published metadata in the dataspace. This includes, for example, new properties or changing a schema version.
FR3	Selective Sharing of AAS	The system shall provide an option to register only specific parts of an AAS. This includes sharing a single entity, an entity with all its child elements, or a child element by itself.
FR4	Support of Different Connectors and AAS Implementations	The system shall support different implementations of dataspace connectors and AAS repository implementations.
FR5	Assisted Contract Management	The system shall provide per-entity selection of access and usage policies for sharing the AAS in the dataspace as well as supply default policies for an assisted onboarding process.
FR6	Scaling Distributed in Cloud	The implementation of the system shall not interfere with the modularity of the overall dataspace connector, i.e., not require a direct connection to both the administrative and the data management parts of the connector such that both can be individually scaled.
FR7	AAS Discovery in Dataspace	The system shall provide an AAS-aligned discovery mechanism, showing the published AAS entities structured within the AAS environment.
FR8	Zero-Downtime Integration	Integration of AAS into the dataspace shall be available with running AAS services and dataspace connectors, requiring no reconfiguration to both.
FR9	AAS API Support	The system shall support not only secure data transfer of the AAS submodel data (or their elements) but also other AAS functionalities like invoking operations and monitoring events.
NFR1	Security	The system should be secure and not tamper with the security configured in the connector. Additionally, it should support integrating access-protected AAS.
NFR2	Standard Compliance	Extensions or modifications to either DTs or dataspace specifications should be minimal, as they may require modifying implementations when onboarding or transferring data.
NFR3	Synchronization Latency and Consistency	After an update is acknowledged by an AAS, any subsequent catalog request by a dataspace participant shall contain that update.

The next requirement is about the support of existing and running DT or dataspace services by the integration components (FR8). In brownfield use cases where dataspace or AAS infrastructure is already up and running, it is important to provide a solution where deployments are unaffected by the integration component.

Lastly, to offer the full feature set of an AAS over dataspaces, an integration solution must provide functionality beyond reading provider AAS data (FR9). This includes supporting manipulation of AAS data or invoking an AAS operation. Invoking an operation may, for example, require input variables by the data consumer, which must be transmitted, typically over the data plane.

A non-functional requirement is the security of the connector that should not be compromised by the integration components, which have access to the connector API (NFR1). Additionally, an AAS may be access-protected via authentication or access rules. Supporting such an AAS in a way where the access is only granted to authorized users can be beneficial.

Some integration approaches choose to extend parts of either DT or dataspace standards to simplify the onboarding process or reduce user workload. For example, the

EDC asset could be extended to store the AAS metadata. However, any modification to a standard might result in failures when utilizing reference implementations and reduces compatibility with tools following the specifications (NFR2).

Lastly, some manufacturing use cases require low latency when new data from the DT is published to the dataspace (NFR3), for example, for fault detection.

3.2. Comparison of Approaches for AAS–Dataspace Integration

Based on the availability of existing approaches [6–8,22,26], we identified three software architectures for a component that automates the integration of digital twins into the dataspace, which are conceptually shown in Figure 4.

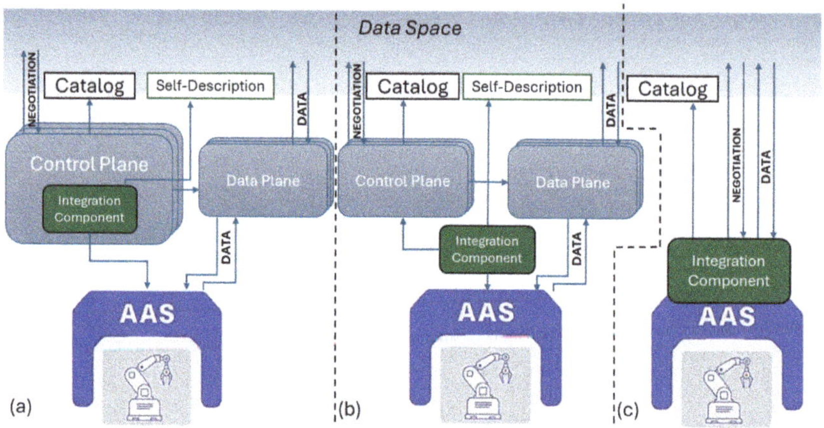

Figure 4. Three possible architectures for a component to integrate AAS into the dataspace separated by dotted lines: (**a**) implement an extension of the EDC control plane; (**b**) the integration component is independent of AAS and dataspace components; (**c**) integrate the dataspace connector into the AAS repository.

EDC Extension. In the first approach, the integration functionality is incorporated into the EDC by leveraging its extensibility, specifically creating a connector extension that contains integration logic. This allows for direct access to connector components, even if they are not accessible via an outward-facing API, as well as simplifying the installation and deployment process and thereby improving usability. However, this tight integration of the extension into the connector comes with the limitation of existing connector deployments being difficult to extend the integration component, because an extension needs to be present during the connector's startup process (FR8). Also, compatibility of the extension is restricted to the connector it was implemented for, possibly but not necessarily including its derivatives like the Tractus-X EDC (FR4).

Standalone Component. The second alternative is to keep the integration component independent from the connector's control plane, handling registration and synchronization of AAS only with the connector's outward-facing APIs like the EDC management API [23]. This approach allows for easier integration into existing systems (FR8), as no build configurations must be changed. Also, a standalone integration component is compatible with any dataspace connector implementations exposing the same API, with the potential to extend compatibility to connectors with different APIs (FR4). Another benefit of deploying the integration component separately is enhanced robustness to connector code changes, as it remains unaffected by modifications to the connector's internal components, which may change more frequently than its API. However, the independence from the connector results in greater overhead in terms of configuration, deployment, and administration by operating the extension separately (FR6). Also, in comparison to the EDC extension approach, configuring external connector endpoints to provide the complete structure of the shared AAS as a discovery mechanism is not possible. So, a data consumer is limited to the dataspace catalog or needs the API endpoint of the integration component for AAS-aligned discovery (FR7). The authors of [22] have implemented the standalone approach to connect an AAS server to the EDC. Although their component, the "AAS Manager", can register submodels to the EDC, it cannot synchronize changes and does not register submodel elements as separate entities.

AAS Repository Extension. In the third approach, the necessary logic to publish AAS data in the dataspace is injected into the AAS repository in the form of a software library, as depicted in Figure 4. This approach was introduced in the Factory-X project [7] as the "Dataspace Protocol Lib" [31] and implements cataloging, negotiating, and data transfer via the Dataspace Protocol, while access control must be implemented specific to each AAS repository. Combining the AAS repository with the dataspace component results on the one hand in an easier deployment and maintenance, but on the other hand creates a monolithic software component with no possibility of scaling as well as no separation of concern. The component is agnostic to the type of data it can provide in a dataspace; therefore, the AAS repository API must be connected manually in the form of additional code. This also means that to integrate the component into the repository, they need to share the same programming language—in the case of the Dataspace Protocol Lib, this would be Java—restricting the component's overall compatibility to possible AAS repositories.

Table 3 summarizes the comparison of different integration approaches regarding the above requirements.

Table 3. Requirements met by the three alternatives to integrating AAS into dataspaces based on current implementations. [a]: Possible with additional implementation effort per AAS repository.

	EDC Extension	Standalone	AAS Repository Extension
		Requirement Fulfillment	
FR1	Yes	Yes	No [a]
FR2	Yes	No [a]	Yes
FR3	Partial	Partial	Yes
FR4	Yes	Yes	No [a]
FR5	Partial	Partial	No
FR6	Yes	Yes	No
FR7	Yes	No	No
FR8	No	Yes	No
FR9	Partial	Partial	No [a]
NFR1	Yes	Yes	Yes
NFR2	Yes	Yes	Yes
NFR3	No	No	Yes

4. Our AAS Integration Approach

Based on the analysis given in the previous chapter, in this chapter we provide detailed information on how we have realized the first approach and how to realize the second approach to meet further requirements. The third approach requires tight integration into any AAS repository implementation with additional source code. As such, it lacks interoperability with respect to AAS repository and dataspace connector implementations. Therefore, we did not consider this approach for the development of our integration component. In Section 4.1, we present the existing solution for integrating an AAS into dataspaces, analyze its architecture, and discuss which requirements from Section 3 it currently fulfills. In Section 4.2, we propose a set of changes to the architecture to better accommodate the stated requirements, which leads to the second approach for integration. In Section 4.3, we present our approach in the MODAI use case.

4.1. Current Architecture of EDC Extension

The EDC extension for AAS [32] is a suite of software components assisting in the registration and use of Asset Administration Shells (AASs) in dataspaces via the EDC. Figure 5 shows the current architecture of the EDC extension for AAS, consisting of three control plane extensions as well as one data-plane extension:

- The AAS extension registers and synchronizes AAS elements at the EDC.
- The client extension provides functionality for assisted contract negotiation and data transfer.
- The public management API extension provides a public endpoint for the extension's discovery mechanism, the AAS self-description.
- The AAS data plane is an extension of the EDC's default HTTP data plane and offers additional communication possibilities with AAS repositories for a data consumer, such as invoking AAS operations.

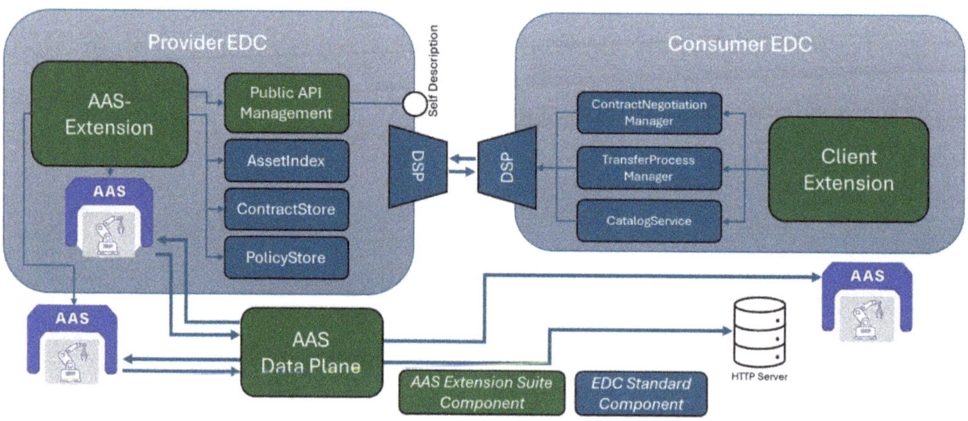

Figure 5. Current state of the EDC extension suite for AAS (green) within the context of the EDC (blue) with arrows indicating interaction flows.

The AAS extension can register any referable entity of an AAS environment into the dataspace by automatically registering them as EDC assets. However, only a reference to an AAS element along with its metadata is stored inside the EDC, not the actual data (e.g., the value of an AAS property). Along with each AAS element, the extension also registers a contract containing access policies making the AAS elements accessible through the dataspace as well as contract (or usage) policies defining the usage rules of the data. A default contract can be specified, and modification of each contract is possible during runtime. Another key feature of the extension is the continuous synchronization of the EDC's AssetStore with the AAS repository. This means that whenever an AAS element is added, removed, or altered, the extension automatically updates the EDC's internal state to reflect these changes.

From the data consumer perspective, the client extension allows for automatic contract negotiation and data transfer using the Dataspace Protocol. For this, a data consumer needs to register a set of policies, which, when offered alongside a trusted data provider's asset, will be automatically accepted to then obtain the asset data. With the client extension, AAS data from a provider can also be received directly at the consumer EDC to immediately view the transferred data.

The AAS data plane serves as a specialized communication channel to AAS repositories and is based on the EDC's HTTP data plane. On the one hand, it provides an abstraction to register AAS elements by supporting AAS references to them instead of URL paths. On the other hand, it provides data consumers the ability to invoke AAS operations on provider AASs. This is carried out by forwarding the necessary input variables from a consumer's data transfer request to the AAS repository of the provider. An example data transfer request containing operation variables could look like this (Listing 1):

Listing 1. Example TransferRequest of a data consumer containing operation variables.

```
{
"@type": "TransferRequest",
"protocol": "dataspace-protocol-http",
"counterPartyAddress": "https://provider-connector",
"contractId": "agreement-id",
"transferType": "HttpData-PUSH",
"dataDestination": {
"@type": "DataAddress",
"type": "HttpData",
"properties": {
"baseUrl": "https://my-http-receiver/receiveEndpoint",
"method": "PATCH",
"operation": {
"inputArguments": [
{
"value": 4,
"idShort": "in"
}
],
"inoutputArguments": {
"value": "original value",
"idShort": "note"
}
}
}
}
}
```

Lastly, with the public API management extension, the AAS discovery mechanism through self-descriptions is published, offering a structured view of the AAS in the dataspace. Additionally, temporary and authenticated endpoints can be created as a data transfer sink for EDC-to-EDC data transfers.

Having outlined the functionality of each extension making up the AAS extension suite, we now evaluate to what degree they satisfy the requirements established in Section 3.

FR1 is satisfied: Users can publish any number of AAS repositories or registries to the dataspace.

FR2 concerns the synchronization of structural and metadata changes from an AAS repository to the dataspace. This requirement is also satisfied, as the AAS extension periodically checks the AAS repository and compares it to the currently published assets.

FR3, which addresses selective sharing of an AAS, is only partly met: While the extension allows users to choose to only share the AAS submodels or all AAS elements, a user cannot select specific AAS entities to share.

FR4, which demands compatibility with different connectors and AAS implementations, is fully met: The AAS extension supports both the mainline EDC as well as the Tractus-X EDC. It is also compatible with all AAS repositories and registries that implement the AAS specification Part 2, including FA^{3}ST Service and Registry as well as their BaSyx counterparts.

FR5, which concerns assisted contract management, is partly fulfilled: While default policies are provided to support onboarding, the extension does not support per-entity selection of access and usage policies.

FR6 is fulfilled: The extension preserves the modularity of the EDC, allowing independent scaling of the control and data planes. Additionally, new AAS repositories can be created and registered with the extension.

FR7 is also satisfied: The AAS extension exposes a self-description endpoint showing all registered AAS environments' structures.

FR8, however, is not fulfilled: Since the extension must be deployed within the control plane, any running control plane needs to be redeployed when adopting the extension.

FR9 is partially fulfilled: Reading AAS data and invoking operations are supported, whereas data manipulation and event reading/subscription are not supported (note that the latter is not yet defined in AAS Specification Part 2).

NFR1 is related to security and is fulfilled: The extension leverages the same security mechanisms as the EDC control plane.

NFR2 is about compliance and met: Neither the AAS nor EDC specifications are modified or extended. AAS metadata is represented using EDC asset properties.

NFR3 is not fulfilled: The synchronization interval is configurable both at deployment time and at runtime, with a minimum interval of one synchronization per second. However, if a catalog request by a dataspace participant is received before a synchronization run completes, an inconsistent state of the AAS could potentially be sent in response. Thus, there is no guarantee that the catalog response always reflects the latest AAS state, due to the nature of periodic polling.

4.2. Redesign of the Current Architecture

To meet requirement FR8 (Zero-Downtime Integration) and make our integration component compatible with more brownfield use cases, the AAS extension needs to be deployable next to running EDC control plane and data plane instances. Consequently, all internal connections of the AAS extension to either of them must be substituted by remote connections. There are multiple ways of realizing this change, including extracting the AAS extension's functionality to a standalone runtime, for example, with the Spring Boot framework. The most straightforward and least invasive solution, though, is to create an EDC extension handling the communication with the remote control plane, as shown in Figure 6. This ensures continuing compliance with the EDC ecosystem and the ability to use EDC components such as the base runtime, extension handling, or the EDC's HTTP client implementation, while allowing for keeping much of the AAS extension in the current state. In fact, only borderline cases where the communication to the control plane fails would have to be handled. This would also mean that the AAS extension would still be usable in the same way as before, by deploying it as part of the EDC control plane.

The new extension introduces two primary capabilities: First, the control plane interfaces that are used by the AAS extension are implemented to allow for seamless transition between local and remote control plane communication, specifically the control plane's persistence layer components pertaining to EDC assets, policies, and contracts. Second, because communication with the control plane will take place over its management API via HTTP, all data is transferred in JSON format. This requires the serialization and deserialization of data objects, which is supported by EDC transformer libraries.

Additionally, to remove data plane dependencies, a new AAS library will consolidate common elements and eliminate dependency points (e.g., AASDataAddress).

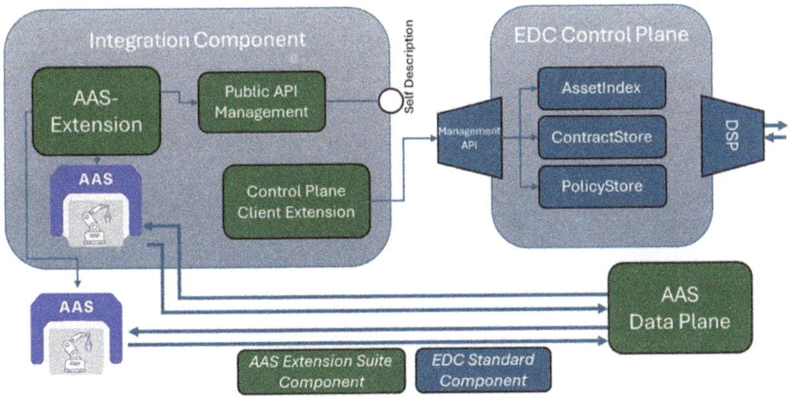

Figure 6. Revised architecture meeting requirement FR8: The integration component (marked green) is extracted from the EDC control plane (marked blue), and connections to the entity stores of the control plane are handled by a new component. Arrows indicate interaction flows between the components.

4.3. Integrating Factory Data in MODAI Project

In this section, we apply our approach in a press shop exchanging data with the customer who receives the finished parts. The data preprocessing of the sensors, press machine, and relevant MES/ERP systems takes place on an edge device. For the AAS service on the edge device, the Eclipse FA^{3}ST Service is set up via a Docker image. We utilized the current extension implementation, which is not a standalone integration approach. Figure 7 shows the implemented integration approach.

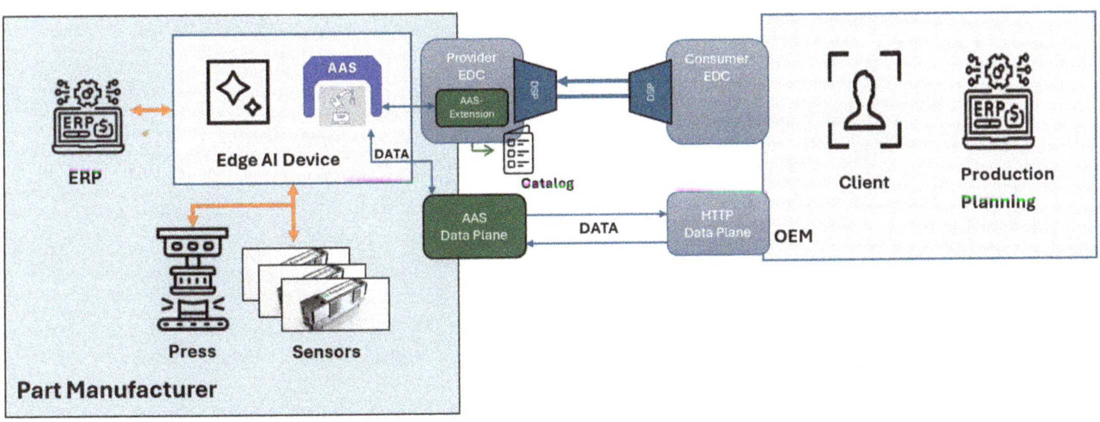

Figure 7. Press machine data integration with AAS and EDC extensions for AAS. Arrows indicate data and control flow.

The Eclipse FA^{3}ST Service contains several submodels for which EDC assets are created by the extension and kept synchronized. The onboarding takes place by sending the URL of the Eclipse FA^{3}ST Service to the extension. FR1 and FR2 are fulfilled by helping the MODAI part manufacturers with initial AAS onboarding and reducing management effort when new submodels or properties arrive in the AAS. Currently, all submodels inside the Eclipse FA^{3}ST Service are automatically chosen to be shared; therefore, FR3 is not fulfilled, but the implementation could be changed to allow for selective sharing, e.g., share only submodels "Time-Series" and "OperationalData". The extension implementation supports

EDC variants such as the Tractus-X EDC (FR4) and attaches the pre-configured default contract to each submodel (FR5). We opted to deploy the mainline EDC in MODAI since a registry-based discovery is not utilized. However, the regular EDC API must be used by the part manufacturer if he wishes to change the contract. This can be improved by a graphical interface for contract management that configures each asset. The synchronization delay is also a configuration parameter that specifies the time interval to wait before the Eclipse FA^{3}ST Service is checked for new changes. A low value increases network traffic but keeps the dataspace metadata neatly synchronized with the AAS. For a production forecast based on quality assessment of the produced parts, setting synchronization delay to sixty seconds is more than sufficient and does not overload the network.

Regarding FR8, the extension implementation already supports running AAS services such as the Eclipse FA^{3}ST Service on the edge device, but the EDC connector must be started in conjunction with the extension. This is a key drawback of the extension approach, because some of the part manufacturers deploy and utilize the connector standalone. While our extension approach also supports FR9 by extending the HTTP data plane with specific AAS calls, the AAS data plane must also be started with the connector. The advantage of the AAS data plane is that the OEM has more AAS interactions available compared to just pulling data via the HTTP data plane.

NFR1 security is partially fulfilled since the extension implementation requires the same API key as any other EDC API call. However, we currently only support AAS services/registries protected by basic authentication and not by the newly released Part 4 ABAC rules. The extension approach does not make any changes to the AAS or dataspace specifications. However, we added a new API endpoint in the connector to make discovery for the OEM easier so that it does not need to browse the catalog, which is presented as a flat list of EDC assets. This is especially useful if the part manufacturers have many different AAS services running and adds some privacy because the available AAS are only shared after negotiation for DTR access. However, all of these discovery mechanisms are project-specific and currently not part of the DT or dataspace specifications.

Because the use case is an ongoing research project, we can only make qualitative statements regarding the evaluation so far. Before the project, the press shop testbed was not equipped with any sensors, edge devices, or DTs. Since there are no running connectors or DT services, our extension approach is suitable. However, the scenario might be different in other press shops. There are also no DT or dataspace experts present for prolonged durations in the press shop. The systems are deployed initially and then maintained at intervals. If the DT structure grows because of new batches or new pressed product types, the dataspace catalog would not include these changes. A workaround would be to share the complete AAS as one entity, but the customer would have to request the complete AAS with each data transfer request. Within the AAS, we opted to structure the data in batches, implying that new elements are created for a batch. The customer can then request specific batch data by negotiating on the specific batch dataspace entry. To reflect new batches in the dataspace, an integration component with synchronization is indispensable, and a lack thereof would mean that the press shop and customer would not consider this kind of dataspace data sharing.

5. Comparison with State-of-the-Art

In this chapter we will elaborate on the integration approaches applied in different dataspace projects that utilize AAS. After that, we compare it with our approach and discuss whether switching to another integration component would fulfill the requirements better.

In other related dataspace projects, the AAS is also used for providing data to the dataspace by mapping AAS metadata to IDS assets. The main differences are in the

IDS connectors and AAS implementations used, their integration patterns, and whether integration components are used or not. This section compares our EDC extension for AAS against state-of-the-art projects and evaluates how well each approach meets the user requirements defined in Section 3.

The prevailing pattern is to map the AAS entities (typically submodels and their descriptors) to dataspace assets and contracts in the connector. The concrete choices of AAS stack (e.g., registry/service), connector, and discovery mechanism vary, and these differences determine how much of the integration effort is manual, how reliably the mapping stays in sync over time, and how approachable the setup is for non-experts.

An AAS registry can contain descriptors of many AAS and their submodels. Submodel descriptors carry endpoint information pointing to the concrete service offering the submodel. Our extension implementation supports both AAS registries and services. It reads AAS descriptors from either an AAS service or an AAS registry and then maintains the corresponding EDC assets, data addresses, and contract definitions.

Catena-X uses a Digital Twin Registry (DTR) at the provider side to make the AAS and their submodels discoverable. It is based on the AAS registry specifications, but their submodel descriptors reference EDC endpoints instead of AAS endpoints. This DTR is used to discover available AAS and submodels. An EDC asset is created for this DTR so that consumers need to negotiate on DTR access first. In this approach, providers still need to create and maintain EDC assets that correspond to the AAS and submodels described in the DTR. The connector used is the Tractus-X EDC, a modified variant based on the mainline EDC. We are not aware of an integration component provided by Catena-X to automate the AAS integration. In short, the providers are required to manually create the EDC assets, and therefore the requirements for an integration component do not apply. Their DTR approach could be supplemented by any integration approach specified in Section 3.2. While Catena-X adds EDC endpoints to submodel descriptors in the DTR, the Tractus-X EDC still needs EDC asset and contract definitions for the AAS and submodels. Our extension implementation could automate this end-to-end mapping from AAS/DTR descriptors since support for the Tractus-X EDC was added. This would be useful for providers that operate outside or alongside Catena-X components or that need a uniform process across multiple ecosystems.

Factory-X is a project based on the Catena-X results but applies the results in manufacturing use cases. They are both part of the Manufacturing-X initiative. It provides hardened connector deployments, often referred to as MX-Ports, in different architecture configurations such as Hercules, Leo, and Orion [33]. AAS plays a central role in the Hercules and Leo approaches of Factory-X, but only the Hercules configuration utilizes the EDC and the DTR. The MX-Port is currently based on the Tractus-X EDC developed in Catena-X. Like in Catena-X, our extension approach can be adopted for the MX-Port Hercules configuration to create dataspace assets for AAS servers and registries. We recently added support for the Tractus-X EDC variant in the extension implementation, which the current MX-Port is based on. However, with time, the implementation might undergo critical changes requiring specific changes in the extension implementation.

In Factory-X, the third integration approach mentioned in Section 3.2 has also been developed. For this, a Dataspace Protocol library is available here [31] that was integrated into the Eclipse BaSyx Server, an AAS service implementation, as a proof of concept. Limitations are that the current library is implemented for Eclipse BaSyx [34] (FR4). We are currently evaluating which steps are necessary to integrate the DSP library into the Eclipse FA^{3}ST Service [35]. Our current extension implementation is an abstraction of common AAS 3.0.1 APIs and is therefore designed to accommodate any AAS implementation. The DSP library integration approach currently dynamically creates the data assets for the catalog when a catalog request is sent to it. It is implementation-specific as to which elements are shared and which contracts are attached (FR2, FR3). Currently, data assets are created for the AAS elements to which the consumer has access, defined in the BaSyx access control policy. The approach is limited to assisting the user in creating data assets in the catalog for all the AAS and submodels inside the Eclipse BaSyx Server equipped with the DSP library.

To scale the DSP library approach in the cloud, the AAS services must be scaled, effectively providing their own control and data plane. However, it currently does not yet support multiple control and data planes. Regarding the AAS Discovery (FR7), the library implementation approach does not offer any additional mechanisms, and it does not support integrating with running connectors or running AAS services/registries (FR8). The DSP library provides a common HTTP data plane as found in the EDC (FR9). Regarding security, the mapping between AAS security and the dataspace catalog is not implemented, as is the case in all other integration approaches. However, Eclipse BaSyx Server provides its own access control implementation currently not based on AAS Part 4. Except for additional discovery mechanisms, all integration approaches are fully compliant with the AAS and dataspace specifications. Regarding the synchronization latency and consistency (NFR2), this approach offers the least synchronization delay by coupling the catalog and the AAS in one component. Compared to the other integration approaches, high-frequency synchronization does not cause high network load but requires some processing and memory bandwidth. The current library implementation synchronizes on user catalog requests but could be modified to synchronize at given time intervals.

In [36], the authors describe the System of Records integration into the AAS dataspace, but the work focuses on data integration into the AAS, for example, data from ERP systems. In [22], the authors implement a standalone approach called AAS Manager. This tool is located between the AAS servers and the EDC. Their current implementation fulfills many requirements of a standalone approach, like FR1, FR3, FR4, FR5, FR6, and FR8. We could not identify a synchronization mechanism (FR2), so the user must repeat the request to the AAS Manager to update the resources. The authors introduce a novel approach to manage the contracts in the AAS, an ODRL submodel template. In this submodel, the policies are described, and the AAS Manager maps them into EDC contracts later. A variant of this was also introduced in [10], where the authors add the concept of usage policies inside a shared AAS, complementing the AAS access control [4]. A data consumer then must evaluate those policies when it shares the AAS with yet another party. In [26], the authors propose an EDC extension publishing AAS elements. Their focus lies on the selective sharing (FR3) and per-element usage policies (FR5), while also fulfilling requirements FR1 and NFR3. Drawbacks of their approach are the lack of synchronization (FR2) and the need to modify each AAS element regarding if it should be published and which policies apply by adding "HasExtension" attributes.

In summary, the tooling for the integration components for AAS into a dataspace is still at an early stage, with many developments and improvements yet to come. The focus of current dataspace projects is the specification and development of connector reference implementations while fulfilling specific use case requirements outside onboarding and integration tooling.

6. Conclusions and Future Work

In this paper, we discussed the integration of digital twins in dataspaces, more specifically the automated integration of the AAS into the IDS with an integration software component. The recurrent pattern to share AAS is the creation of IDS assets pointing to AAS data in the dataspace catalog.

The dataspace projects have converged on reusing AAS as the semantic anchor for dataspace publication, for example, in Catena-X and Factory-X. While their focus is currently not on simplified onboarding and integration tooling, some approaches have been implemented, namely the DSP-Lib and our EDC extension for AAS. We followed the extension approach but decided to switch to a standalone approach because the requirements are better covered, especially in supporting already running connectors.

The main pain points of dataspace integration are repetitive asset modeling inside connectors and the steep learning curve for non-experts. These are all addressed by integration tools following the three integration approaches discussed in Section 3.2. Our EDC extension for AAS is one example that closes these gaps by automatically deriving EDC assets and attaching contracts to AAS metadata and descriptors. As such, the connectors are extended to "twin-native" connectors, which specialize in handling DTs like the AAS.

We contributed (i) a structured requirements framework for automated AAS–dataspace integration, (ii) an analysis of three architectural approaches (connector extension, standalone component, repository extension), (iii) a description of our current implementation of an EDC extension with an AAS-aware data plane, (iv) a redesign to enable zero-downtime integration against running connectors, and (v) a comparative perspective on Catena-X and Factory-X practices and our application in the MODAI use case.

Both the connector extension and standalone approaches are viable, but they optimize for different constraints. The extension provides deep integration and feature completeness (e.g., specialized AAS operations via a custom data plane) and strong coupling with the connector. The standalone approach is superior for brownfield deployments, avoiding connector redeployment and supporting the EDC connector landscape. The repository extension approach (as seen with the Dataspace Protocol Lib) simplifies deployment in homogeneous stacks by creating monolithic coupling with the AAS services.

Requirements coverage with our current EDC extension is strong for automated onboarding (FR1), synchronization (FR2, FR7), scaling compatibility (FR6), multi-connector/ AAS support (FR4), and AAS-specific data operations (FR9). It partially fulfills requirements for selective sharing (FR3) and assisted contract management (FR5), but does not yet meet requirements for zero-downtime of running connectors (FR8). Security and standard compliance are partially preserved (NFR1, NFR2), with support for AAS ABAC being planned for the future

An AAS-aware data plane enables operation invocation with parameters, paving the way for richer interactions between business partners. Discovery remains a critical issue because flat connector catalogs hinder usability. It is therefore often complemented with AAS registries (e.g., DTR) or publishing structured self-descriptions.

Our results in the MODAI use case indicate that integration automation in connectors significantly lowers onboarding friction, reduces manual asset modeling, and brings contract management closer to operational and supply-chain data. However, we did not evaluate the benefits in a quantitative manner. The comparison was focused on whether the requirements of Section 3.1 were met. Moving from a connector extension to a standalone approach appears to be the most promising path for broad adoption, especially in brownfield contexts.

Regarding future work, we are planning to adapt our implementation to be a more modular tool so that users with a different requirement focus have more options in integrating DTs. Some users might not require a brownfield integration and want to focus on AAS-specific data transfers. For example, the exchange of AAS machine events is a research focus in the Factory-X project. We are also active in the development of frontend user interfaces to make interaction with the integration tool easier. A particular point of interest is the ongoing implementation of AAS Part 4 Security, which allows users to

33. Friedrich, C.; Vogt, S.; Rudolph, F.; Patolla, P.; Grützmann, J.M.; Hohmeier, O.; Richter, M.; Wenzel, K.; Reichelt, D.; Ihlenfeldt, S. Enabling Federated Learning Services Using OPC UA, Linked Data and GAIA-X in Cognitive Production. *J. Mach. Eng.* **2024**, *24*, 18–33. [CrossRef]
34. Eclipse Foundation. Eclipse BaSyx. Available online: https://eclipse.dev/basyx/ (accessed on 26 September 2025).
35. Eclipse Foundation. Eclipse FA^{3}ST. Available online: https://projects.eclipse.org/projects/dt.fa3st (accessed on 26 September 2025).
36. Danish, M.G.A.; Schnicke, F. Integrating Systems of Record (SOR) into the Asset Administration Shell (AAS) Dataspace. In Proceedings of the 2024 IEEE 29th International Conference on Emerging Technologies and Factory Automation (ETFA), Padova, Italy, 10–13 September 2024; IEEE: New York, NY, USA, 2024; pp. 1–8. [CrossRef]

applied sciences

MDPI

Article

Investigation of Key Process Parameters Affecting Product Quality in Robotic Milling: A Comprehensive Analysis

Lukáš Hanko [1], Ondrej Chlebo [1], Peter Križan [1,*], Stanislav Strigáč [1], Miloš Matúš [1] and Lenka Matejáková [2]

[1] Institute of Production Engineering and Quality, Faculty of Mechanical Engineering, Slovak University of Technology in Bratislava, 812 31 Bratislava, Slovakia; lukas.hanko@stuba.sk (L.H.); ondrej.chlebo@stuba.sk (O.C.); stanislav.strigac@stuba.sk (S.S.); milos.matus@stuba.sk (M.M.)

[2] Institute of Hygiene, Medical Faculty, Comenius University, 813 72 Bratislava, Slovakia; lenka.matejakova@fmed.uniba.sk

* Correspondence: peter.krizan@stuba.sk; Tel.: +421-257-296-539

Abstract

Industrial robots are a widely used technology in various industries, capable of performing multiple tasks, particularly machining operations. However, compared to CNC machines, industrial robots have certain drawbacks, such as lower rigidity, precision, and repeatability. These limitations are significant obstacles to fully integrating industrial robots into manufacturing processes. The deficiencies in robots directly impact the quality of machined workpieces, leading to reduced surface quality and geometric accuracy. Researchers have been exploring ways to improve production accuracy and quality using robotic arms for decades. Despite technological advancements, the full implementation of robotic arms in machining processes remains in its early stages. This article focuses on robotic milling and the effect of technological parameters on the final quality of the machined material, aiming to determine their importance for designing the main experimental plan. In our experiment, we used an ABB IRB 1100 robotic arm with a payload of 4 kg and a maximum reach of 0.58 m. The milling process was programmed in RobotStudio, which also supports milling complex shapes. Our statistical analysis showed that feed rate had the greatest impact on quality, while depth of field had the least influence. This information will guide further research in developing a mathematical model for robotic milling.

Keywords: robotic milling; vibrations; surface roughness; technological parameters

1. Introduction

Industrial robotics is generally considered one of the key sectors of industrial automation, most commonly used in engineering but also in fields such as medicine, the military, biomechanics, and others [1]. Industrial robots are widely applied in various tasks, often replacing human labour. Over the past decade, the number of industrial robot applications has increased dramatically, indicating the potential for robotics, with implementations expected to continue growing in the future [2]. Robotic arms have undergone numerous changes and modifications since their inception, improving their manoeuvrability and efficiency in performing manipulation tasks in the engineering industry. Initially, robotic arms were used for simpler tasks, relieving workers from strenuous and repetitive work. Over time, they have been applied to more complex tasks in industries such as engineering and automotive manufacturing [3].

Robotic arms quickly found their place in the automotive industry, especially in specific applications such as assembling various components and more precise tasks like welding individual car body parts [4], followed by spraying the welded body [5]. With the increasing complexity of these applications, collaborative robotics emerged, involving the cooperation of multiple robots to achieve common goals. This developing field has significant potential to transform industrial processes [6]. In addition to using multiple robots in applications, robotics has expanded to areas where robots are equipped with external devices for environmental sensing through industrial cameras [7]. These cameras help robots recognise and determine their position in space, detect obstacles or loads to be moved, and correctly position parts into other components.

In the engineering industry, industrial robotics was initially used for basic tasks and pick-and-place applications, where six-axis industrial robots provided wide manoeuvrability within their extensive workspaces [8]. Over time, robotics has evolved, particularly in handling tasks, utilising its potential for moving and positioning workpieces into machining tools, such as clamping and removing workpieces from CNC machine spindles.

Today, robotics has advanced to a higher level, with researchers striving to integrate robots directly into machining processes such as milling, drilling, and grinding [4,9].

In most cases, robotic arms were initially implemented for machining processes with low loads, generating minimal cutting forces. Processes like grinding, polishing, and drilling produce low cutting forces, as they typically remove small amounts of material and apply minimal abrasive force. The main goal of polishing operations is not to alter the shape of a structure but to improve its surface, meaning these operations do not require high positional accuracy. Later, robotic arms began to be used for more complex and demanding applications, such as removing larger volumes of material. Robotic machining is a newer application of industrial robots, offering a larger working area at lower costs, though with reduced precision. Robotic arms are commonly used in milling machining processes that do not involve heavy machining loads or require high tool path accuracy. Consequently, robotic machining is often applied to softer, lighter materials that do not generate significant machining forces [10,11]. The biggest obstacle to the full implementation of robotic arms in machining is their limited rigidity, which affects both their precision and repeatability. These factors are significantly lower compared to conventional machining tools [12]. The rigidity of an industrial articulated robot is approximately 1 N/μm, which is several times lower than that of a standard CNC machine, typically around 50 N/μm. This lack of rigidity, combined with cutting forces, leads to deformations in the end-effector, causing vibrations, poor surface quality, and reduced geometric accuracy during machining. In some cases, end-effectors experience deflections of up to 10 mm when working on parts. Manufacturers are focusing on reducing these end-effector deviations during machining by improving the robot's rigidity or applying special control algorithms. To enhance the manipulator's rigidity, engineers either increase the cross-section of the robot's joints or use modern composite materials. However, the first approach increases the mass of the moving parts, which reduces the robot's dynamic properties. This low rigidity significantly impacts the quality of the machined surface due to increased vibrations in the entire robotic structure during milling [13,14]. To some extent, vibrations occur in all machining processes. If not properly controlled, they can lead to poor surface quality, low productivity, tool wear, and damage. The natural frequency of robotic arms typically ranges from 10 to 20 Hz, which is lower than that of CNC machines. Given that cutting forces in machining are periodic and can sometimes change unpredictably, the occurrence of vibrations or chatter is not surprising.

Machining vibrations can be categorised into two types: forced vibrations and self-excited chatter. Forced vibrations are primarily caused by time-varying external disturbances, such as the regular interaction of the cutter's teeth with the workpiece. Once the main source of these vibrations is identified, they can often be easily mitigated or even

https://doi.org/10.3390/app16041832

eliminated. Chatter, on the other hand, causes instability in the machining process and is more undesirable and harder to control compared to forced vibrations. In their study, Tlustý and Poláček [10] identified two major sources of self-excited vibrations: regenerative chatter and mode-coupling chatter.

These parameters of the robotic arm are closely interconnected and determined by the specific kinematics of the arm. In our experiment, we used a six-axis industrial robot. These robots, also known as articulated robots, are defined by their rotary joints, often referred to as axes. Each axis is powered by servomotors, and every joint provides one degree of freedom, corresponding to an independent motion of the robot. The axes are typically arranged in a chain so that each axis supports the next along the robot's structure. The structure of an articulated robot starts with a base, which is perpendicular to the ground and contains the first rotary joint. Through this joint, the robot's main body is connected to the base. The next rotary joint is located at the end of the main body, linking it to the robot's arm. Further joints or axes are found at the end of the arm, connecting the robot's wrist to the end-effector. This robotic design closely mimics the movement of a human arm. Articulated robots offer more degrees of freedom than any other type of robot, with a wide range of motion that closely follows human movements, making them ideal for many applications [15]. Their adaptability to changes in the production process or workpieces makes them highly versatile. The enhanced motion allows robots to operate in a larger workspace, enabling them to handle various workpieces, from small to large. Additionally, they are capable of performing a wide range of tasks, such as arc welding, material handling, assembly, part transfer, pick and place, packaging, machine loading, and palletising, among others.

The milling process is a growing area of interest, attracting more and more researchers. Although robotic arms cannot compete with CNC machines in terms of precision, repeatability, and, most notably, rigidity, their key advantages lie in their larger working space, manoeuvrability, and, above all, lower acquisition costs. The working space and versatility of robotic arms make them particularly useful in manufacturing more complex part geometries [16]. While CNC machines would produce higher-quality parts, achieving the necessary kinematics for five-axis milling would require modifications, increasing production costs. In robotic milling, the main advantage is the ability to create topologically optimised parts for prototype production [16]. The main objective of this article is to investigate the influence of key technological parameters on the resulting surface quality of the machined Corian material using an industrial robotic arm. Corian exhibits a homogeneous and isotropic structure, ensuring stable behaviour during chip-forming machining. Its material properties minimise process variability, enabling a more reliable assessment of the effects of technological parameters without the presence of confounding factors. Corian is also commonly used in robotic milling, particularly in the production of design surfaces, prototype parts, and geometrically complex components, making it a suitable and representative material for this type of research. This article presents the results of preliminary screening experiments that examine the relationship between various technological parameters and surface quality. Additionally, this study investigates the impact of tool vibration on the machined surface. This article concludes by identifying the technological parameters that will be influential in developing the main experimental plan for robotic milling.

https://doi.org/10.3390/app16041832

2. Materials and Methods

2.1. Industrial Robotic Arm IRB 1100

The robotic industrial arm used in the screening experimental research was from ABB, s.r.o. (Bratislava, Slovakia) specifically the industrial robot designated as IRB 1100, as shown in Figure 1.

Figure 1. Robotic arm IRB 1100.

The working range of the robotic arm (working envelope) at the maximum extension of the Tool Centre Point (TCP) of the last sixth axis is 0.58 m, as illustrated in the following Figure 2. In Table 1, we can see the maximum values for the positions of the wrist's midpoint in the X and Z axes, as well as the maximum rotation of each axis.

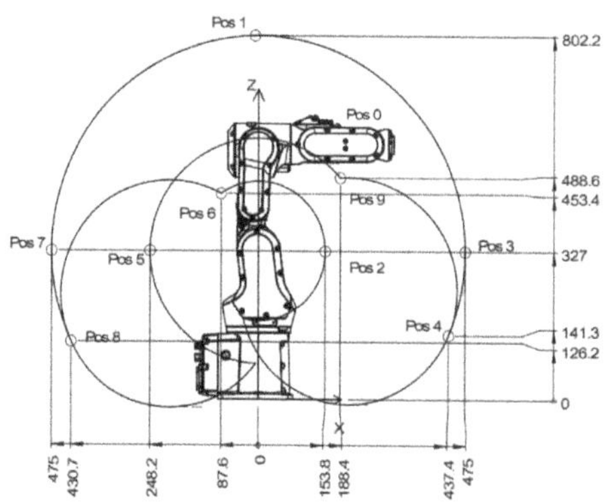

Figure 2. Workspace of robotic arm IRB 1100.

The robot has a payload capacity of 4 kg at the end effector. The payload capacity was considered a significant parameter when selecting the experimental spindle and designing the mounting joint that allows for the attachment of the spindle to the robotic arm. The robotic arm features mounting holes for additional equipment and the attachment of extra payloads. The maximum allowable load on the arm depends on the centre of gravity, the load on the arm, and the robot's payload capacity. The value of the maximum external load is 0.5 kg. The attachment of the additional payload was used for routing cables, whether from the experimental spindle or from sensors.

https://doi.org/10.3390/app16041832

Table 1. Position of the wrist centre of the robotic arm and angle of the axes 2 and 3.

The Individual Positions Corresponding to Figure 2	Position of the Wrist Centre (mm)		Angle (°)	
	X	Z	Axis 2	Axis 3
Pos0	364	627	0	0
Pos1	0	907.2	0	−88
Pos2	184.6	327	12.5	55
Pos3	580	327	90	−88
Pos4	534	100.3	113	−88
Pos5	−304	237	−28.3	−205
Pos6	−112.4	473.5	−115	55
Pos7	−580	327	−90	−88
Pos8	−525.8	81.8	−115	−88
Pos9	237.3	517.1	113	−205

2.2. Experimental Milling Spindle

In the screening experiments, a spindle with a power of 500 W, air-cooled (Figure 3), was used. The spindle speed is variable and can be regulated through a voltage converter in the range of 1000 to 12,000 RPM using a potentiometer. The spindle also includes a set of collets, allowing for the attachment of various diameters of cutting tools. The individual technical parameters of the spindle can be seen in Table 2.

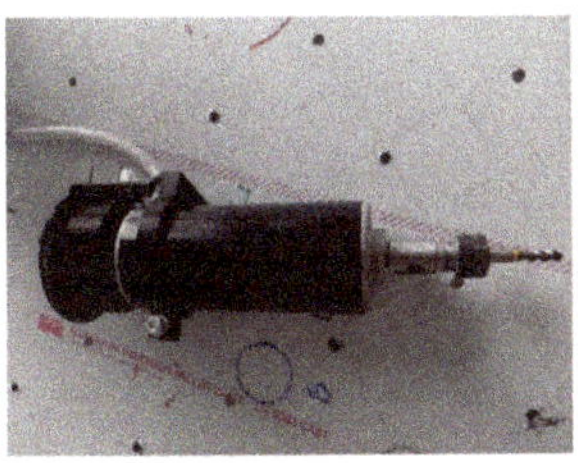

(a)

(b)

Figure 3. The assembly of the experimental spindle used in the robotic milling screening experiment: (a) experimental spindle with a power of 500 W; (b) power supply and voltage converter.

Table 2. Technical parameters of experimental spindle.

Parameter	Value
Power	0.5 kW
Maximal revolutions	12,000 rpm
Voltage	110–230V DC
Current	3.9 A
Weight	0.85 kg
Dimensions	50 × 200 mm
Collets	ER 11

2.3. Material of Milled Workpiece

The chosen material for the experiment was Corian. Also known as artificial stone, it has a uniform appearance that can mimic the look of natural stone or other materials. It consists of one-third transparent acrylic resin (polymethyl methacrylate), (PMMA) and two-thirds natural minerals. The main mineral is aluminium trihydrate (ATH), a white powder derived from bauxite, the mineral from which aluminium is extracted. In Figure 4, we can see a plate of Corian material used in the experimental research. Detailed technical properties of Corian material are shown in the following Table 3.

https://doi.org/10.3390/app16041832

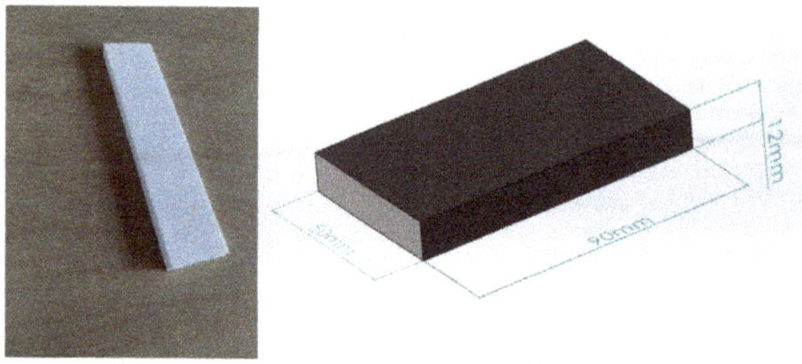

(a) (b)

Figure 4. Milling material Corian (**a**) test plate of Corian material (**b**) geometric dimensions of the test plate (50 mm × 12 mm × 90 mm).

Table 3. Mechanic properties of Corian material for 13 mm thickness.

Properties	Value	Unit	Test Method
Specific density	1.7	g/cm^3	ISO 1183 [17]
Tensile strength	34.9–39.5	N/mm^2	DIN 53 455 [18]
Weight	23	Kg/m^2	
Tensile strength upon rupture	0.36–0.49	%	DIN 53 445 [19]
Bending strength	60	N/mm^2	DIN 53 445
Modulus of elasticity	10.000 ± 400	N/mm^2	DIN 53 457-Z-B 3 [20]
Ball impact hardness	337.3 ± 8,4	N/mm^2	DIN 53 456 [21]
Impact resistance	4.7	KJ/m^2	DIN 53 453 [22]
Notch resistance	1.9	KJ/m^2	DIN 53 453
Compressive strength	119	MPa	ISO 604 [23]
Thermal expansion	3.9×10^{-5}	Mm/mm °C	ASTM D 696 [24]
Hardness Rokwell	>85		
Hardness Barcol	56		ISO 19712-2 [25]
Surface hardness (Mosh index)	2–3		DIN EN 101 [26]
Flexural modulus	8.800	MPa	ISO 178 [27]

2.4. Cutting Tool

The most important component for the proper execution of the milling process, whether by a robotic arm or conventional machines, is the cutting tool. The proposed cutting tool is a three-blade cylindrical end mill with a cutting diameter of 5 mm and a shank diameter of 6 mm (Figure 5). More detailed parameters of the three-blade mill can be seen in Table 4. The choice of cutting tool was made based on professional consultation with Hoffman.

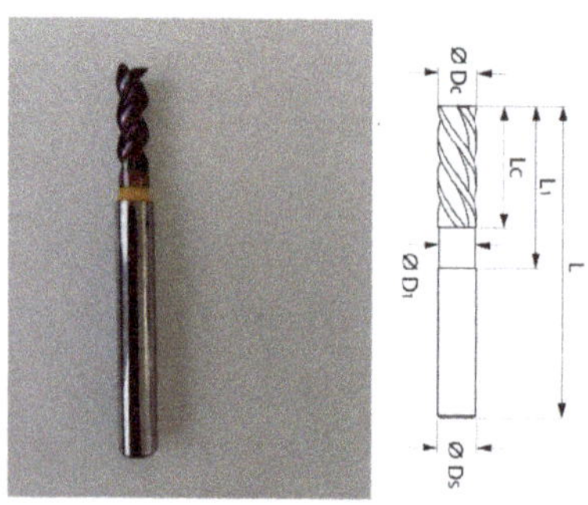

Figure 5. Three-blade cylindrical end mill with a diameter of 5 mm.

https://doi.org/10.3390/app16041832

Table 4. Parameters of tree-blade mill.

Parameter	Value
∅ cutting edge	5 mm
Surface finish	DLC
Cutting material	VHM
Number of teeth Z	3
Angle of the inclination of the teeth	45
Total length L	57
∅ shank	6 mm
Shank shape	HA
Length of the cutting blade	13 mm

2.5. Preparation of the Experimental Environment for Experimental Milling with a Robotic Arm

Experiments were conducted using the IRB 1100 robotic arm. This robotic arm is part of a workbench (Figure 6), to which it is securely mounted. The workbench also includes a work surface where the Corian material was attached.

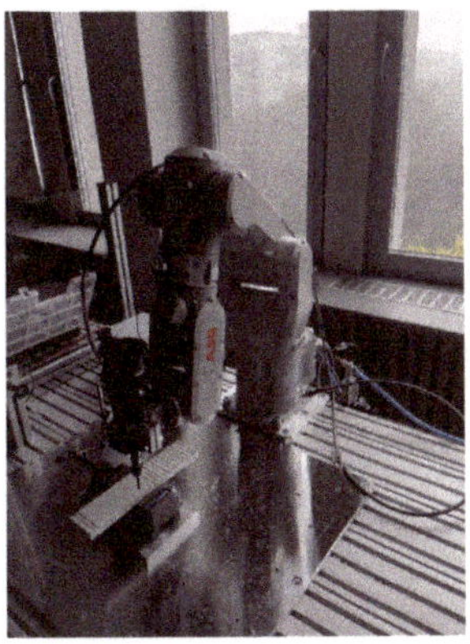

Figure 6. Robotic milling cell.

Since this surface contains several holes, there was no need to use complex clamping fixtures, and a precise mechanic vice was utilised for securing the material (Figure 7).

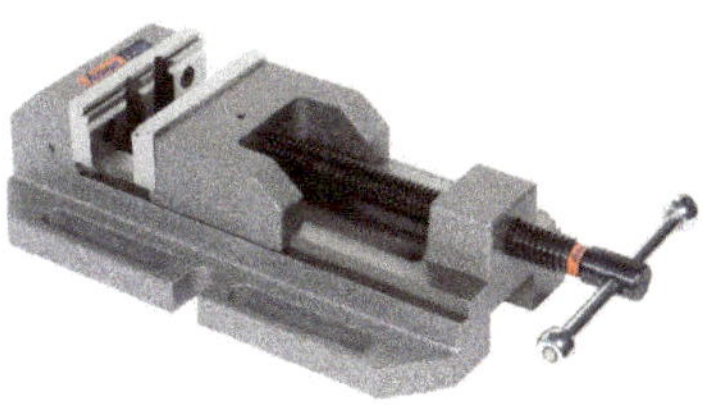

Figure 7. Mechanic vice.

However, more attention was given to the design of the clamping joint, which was used to firmly attach the experimental spindle together with the robotic arm. The spindle attachment was designed to allow for the secure mounting of accelerometers for vibration measurement while keeping the weight as low as possible due to the low payload capacity of the robotic arm. In Figure 8, we can see the design of the spindle attachment along with the spindle itself, created in SolidWorks.

https://doi.org/10.3390/app16041832

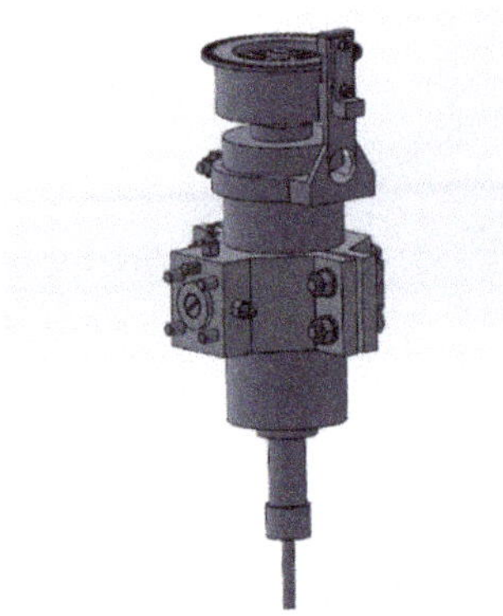

Figure 8. The model of the clamping joint together with the spindle created in SolidWorks 2024 (Dassault Systèmes, Vélizy-Villacoublay, France).

The attachment consisted of two parts, where one part was specially modified to allow for the attachment of an accelerometer directly inside the clamping joint, as shown in Figure 9. This clamping joint design was created for measuring accelerations along the X axis. The other accelerometers were mounted on the outer surfaces of the clamping joint. These accelerometers measured accelerations in the Y and Z axes.

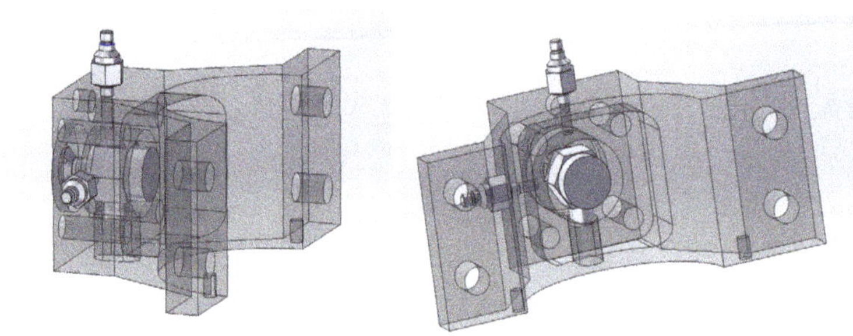

(**a**) (**b**)

Figure 9. Model of the part of the clamping joint: (**a**) location of accelerometers on the outer surfaces; (**b**) location of the accelerometer in the inner part of the clamping joint.

The second part of the clamping joint was used solely to securely grip the spindle between the two parts of the clamping joint. In the following Figure 10, one part of the clamping joint is shown along with the securely mounted accelerometers.

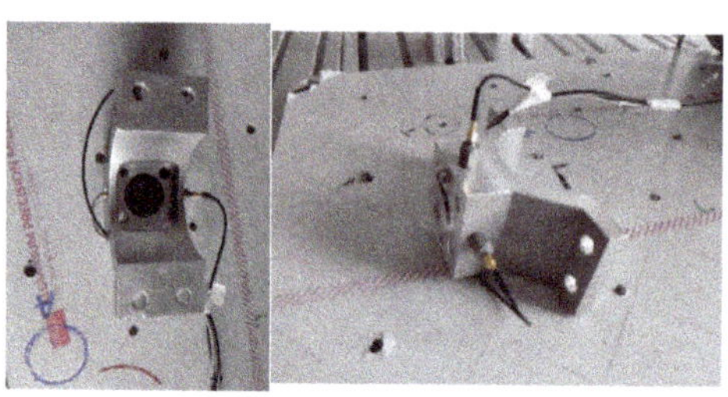

(**a**) (**b**)

Figure 10. Real clamping joint: (**a**) placement of accelerometers on the outer surfaces; (**b**) placement of accelerometers in the inner part of the clamping joint.

https://doi.org/10.3390/app16041832

Acceleration sensors were fixed in three places on the clamping joint so that we measured accelerations in each axis, X, Y, and Z. Accelerometers CCLD ACCELEROMETER TYPE TYPE 4534-b (Brüel & Kjær, Nærum, Denmark) were used to measure accelerations, which can also be seen in Figure 11.

Figure 11. Acceleration sensor CCLD ACCELEROMETER TYP TYPE 4534-b.

The universal CCLD accelerometer with TEDS has a coaxial connector mounted on top and a low-impedance output for long cable runs. Type 4534-b provides a wide frequency range from 0.2 Hz to 12.8 kHz and low noise levels for low-level measurements. Additionally, the unit offers low sensitivity to environmental factors, making it ideal for use in changing or demanding environments. The parameters of the accelerometer are listed in Table 5.

Table 5. Technical parameters of accelerometer TYPE 4534-b.

Parameter	Value
Frequency range	0.2–12.8 Hz
Working temperature	−55–125 °C
Weight	8.6 g
Sensitivity	1 mV/ms^{-2}
Resonance frequency	38 kHz
Maximum shock level	10.000 g

To record and store vibration data, a measuring module was used that allows for monitoring vibration patterns and subsequently recording, storing, and further processing the measured data. The measuring module is shown in Figure 12, and as can be seen, it offers up to 12 channels for connecting accelerometers and a microphone.

Figure 12. Measuring module for collecting data from accelerometers.

https://doi.org/10.3390/app16041832

2.6. Programming the Robot's Movement Instructions in the RobotStudio Program

The robotic arm from ABB provides users with the programming software RobotStudio (Version 25.3.11371.0), which, among many other applications, includes packages for the milling process. Therefore, the entire simulation and generated codes for the real robot were created in this program.

RobotStudio allowed for modelling the trajectory paths along which the cutting tool (milling cutter) would move. Additionally, the program set the values for the tool's feed (TCP tool) and the depth of material removal. Programming motion instructions was handled in a virtual environment, where a copy of the robotic workstation was created that corresponded to the real workstation, as seen in Figure 13.

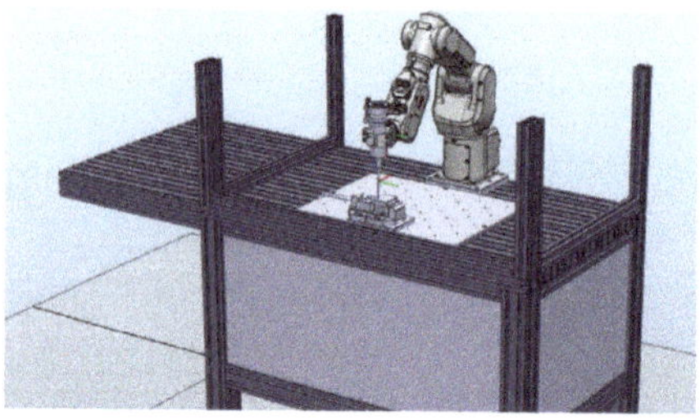

Figure 13. Virtual workspace of robotic milling station.

In the virtual environment, all tools and materials needed to simulate the milling process were positioned. The created motion instructions were pre-simulated, allowing for the prediction of undesirable states or collisions. The RAPID program itself was modified to the level of parameterisation of individual changing parameters for the sake of simplicity and fluidity of the program.

The main part of the program was divided into two sections (Figure 14), where one section was programmed for the cutting tool to traverse the surface of the material being machined to level its surface and achieve a plane. In the second part of the program, the value for the depth of material removal was read from this plane.

```
PROC Zrovnanie_drazky_2()

    MoveL odsadenie_2,vneutral,fine,Vreteno_freza_5mm\WObj:=Frezovanie;

    odsadenie:=Offs(vstup_2,0,0,5);
    MoveL odsadenie,vneutral,fine,Vreteno_freza_5mm,\WObj:=Frezovanie;
    WaitTime\InPos,0.5;
    MoveL vstup_2,vodsadenie,fine,Vreteno_freza_5mm\WObj:=Frezovanie;
    WaitTime\InPos,2;
    MoveL vystup_2,vzarovnanie,fine,Vreteno_freza_5mm\WObj:=Frezovanie;
    WaitTime\InPos,2;
    odsadenie:=Offs(vystup_2,0,0,5);
    MoveL odsadenie,vodsadenie,fine,Vreteno_freza_5mm,\WObj:=Frezovanie;
    WaitTime\InPos,0.5;

    MoveL odsadenie_2,vneutral,fine,Vreteno_freza_5mm\WObj:=Frezovanie;

ENDPROC
```

(**a**)

```
PROC Drazka_2()

    MoveL odsadenie_2,vneutral,fine,Vreteno_freza_5mm\WObj:=Frezovanie;

    odsadenie:=Offs(vstup_2,0,0,5);
    MoveL odsadenie,vneutral,fine,Vreteno_freza_5mm,\WObj:=Frezovanie;
    WaitTime\InPos,0.5;
    odsadenie:=Offs(vstup_2,0,0,-Hfreza);
    MoveL odsadenie,vodsadenie,fine,Vreteno_freza_5mm,\WObj:=Frezovanie;
    WaitTime\InPos,2;
    odsadenie:=Offs(vystup_2,0,0,-Hfreza);
    MoveL odsadenie,vnasa,fine,Vreteno_freza_5mm,\WObj:=Frezovanie;
    WaitTime\InPos,2;
    odsadenie:=Offs(vystup_2,0,0,5);
    MoveL odsadenie,vodsadenie,fine,Vreteno_freza_5mm,\WObj:=Frezovanie;
    WaitTime\InPos,0.5;

    MoveL odsadenie_2,vneutral,fine,Vreteno_freza_5mm\WObj:=Frezovanie;

ENDPROC
```

(**b**)

Figure 14. RAPID code for robotic grooves milling: (**a**) movement instructions for levelling the surface of the plane; (**b**) movement instructions for groove milling with specific parameters.

2.7. Design of a Screening Experiment

The screening experiment consisted of several individual tests focused on milling grooves with gradually changing technological parameters. These tests allowed us to determine the range of technological parameter values for the design and implementation of the main experiment, which aimed to investigate the impact of technological parameters

https://doi.org/10.3390/app16041832

in the milling process using a robotic arm on the final quality of the machined material (surface roughness).

The grooves were milled at a specifically defined position that did not change during the screening test. This position corresponded to the maximum extension of the robotic arm. The material being machined was secured in a mechanic vice (Figure 15a), which achieved a constant extension position of the robotic arm and, consequently, a constant tool path (Figure 15b).

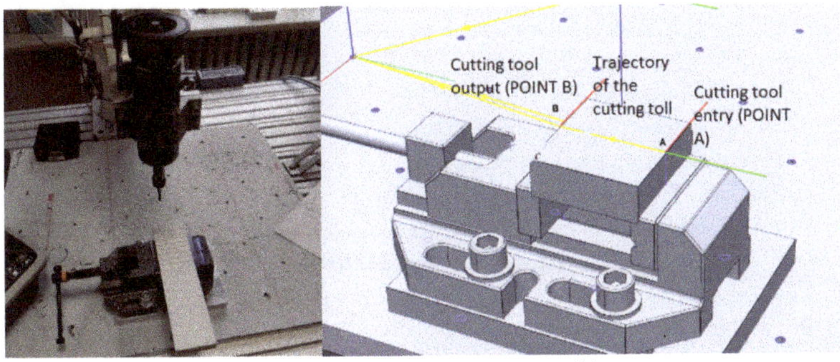

(a) (b)

Figure 15. Positioning of the mechanic vice at the maximum position: (**a**) clamping of the machined material in the mechanic vice; (**b**) path generation in the virtual environment in RobotStudio.

The technological process for milling the groove involved milling a specific groove using predetermined technological parameters. After the groove was milled, the workpiece was released from the mechanic vice and repositioned so that a new groove could be milled at the same extension position of the robotic arm using a different set of technological parameters.

The robot's motion instructions were programmed so that, upon the cutting tool's entry into the workpiece, there was a time delay that ensured the entire system was stabilised. This prevented other vibrations, such as those occurring during the transition of the cutting tool from the home position to the groove milling position, from affecting the test results.

In this test, we repeatedly performed the milling process with all gradually changing technological parameters, ranging from minimum to maximum values. Simultaneously, we recorded data from the accelerometers, which we analysed along with the measured roughness in each groove. The roughness was measured on the bottom surface of all the grooves.

The test began by setting the minimum spindle speed, which was set at 6000 RPM. We then selected the minimum value for the depth of material removal, set at 0.5 mm. The feed rate of the cutting tool was also set to the minimum value of 5 mm/s. This parameter was varied from the minimum value to the maximum value of 90 mm/s, while the spindle depth of field remained constant. After milling the last groove at a speed of 90 mm/s, we changed the depth of field to 1 mm and repeated the test with varying feed rates. The same approach was applied for a depth of 1.5 mm. After milling all combinations, we increased the spindle speed to its maximum and repeated the milling process for all combinations of feed rates and depths of field. Through this testing, we achieved a total of 66 settings, which provided us with 220 surface roughness values and 66 vibration values that were evaluated through statistical analysis.

From the acquired vibration data, a calculation was performed that yielded RMS values representing values directly related to the energy content of the vibration profile.

https://doi.org/10.3390/app16041832

In other words, it is the effective value of the signal calculated based on the entire sample according to the following formula:

$$\text{RMS} = \sqrt{\frac{1}{T}\int_0^T v(t)^2 dt} \quad (1)$$

As the formula indicates, the RMS (Root Mean Square) is the square root of the average of the squared values of the waveform. In the case of a sinusoidal waveform, the RMS value is 0.707 times the peak value (as shown in Figure 16), but this only applies to sine waves. The RMS value is proportional to the area under the curve—if there are negative values, the peaks are rectified, meaning they become positive, and the area under the resulting curve is averaged to a constant level, which will be proportional to the effective value. The RMS value of a vibration signal is an important measure of its amplitude. As previously mentioned, it numerically equals the square root of the mean of the squared amplitude values. To calculate this value, the instantaneous amplitude values of the waveform must be squared, and these squared values must be averaged over a specific time interval. This time interval must cover at least one period of the wave to achieve an accurate value. Since all squared values are positive, their average will also be positive. The square root of this average value is then taken to obtain the effective mean value.

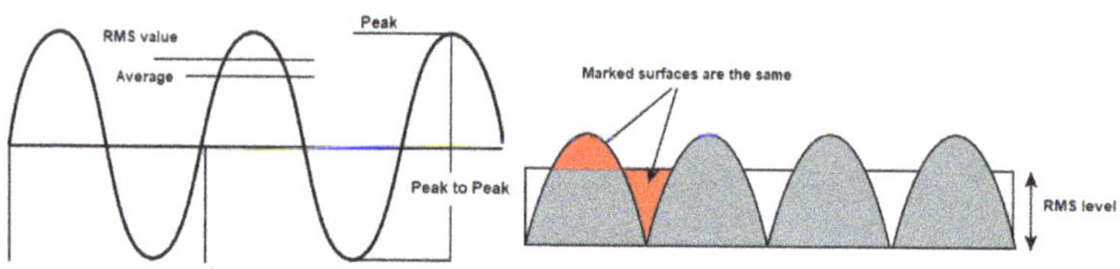

Figure 16. RMS value obtained from a sinusoidal signal.

Surface roughness was measured using the Mitutoyo S-210 roughness tester (Mitutoyo, Kawasaki, Japan.), as shown in Figure 17. The surface roughness of the inner surface of the groove was measured at five locations (Figure 18). This means that the total length of the groove was divided into five segments, and multiple measurements of the surface roughness value were taken in each segment.

Figure 17. Measuring the roughness of grooves on a Mitutoyo S-210 roughness tester.

The total length of the groove was 50 mm, which means that each segment measured 10 mm (Figure 18). Specifically, each segment underwent four roughness measurements, and from the obtained values, we calculated the average roughness for each segment. Subsequently, we expressed the overall average value for the entire groove based on the average values of the individual segments. In the statistical evaluation, the measured values of roughness and vibrations in the segments 0–10 mm and 40–50 mm were not considered.

https://doi.org/10.3390/app16041832

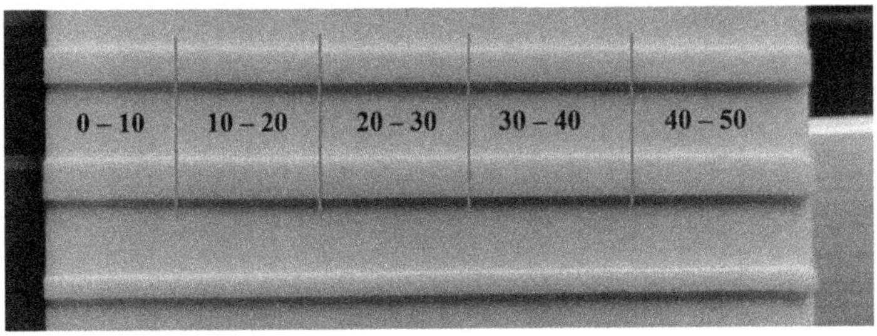

Figure 18. A milled groove divided into five parts with a length of 10 mm.

The reason for evaluating the measured surface roughness values in the sections from 10–20 mm to 30–40 mm is that the RMS value was calculated from the same segment of the groove, while the entry of the cutting tool into the material (0–10 mm) and the exit from the material (40–50 mm) were not considered. The entry (0–10 mm) and exit (40–50 mm) segments of the groove were intentionally excluded from the statistical evaluation because these regions represent non-steady-state cutting conditions, in which the tool is either engaging with or disengaging from the material. These transition phases generate higher transient vibration peaks, irregular chip formation, and fluctuations in cutting forces, which do not represent the stable machining regime intended for comparison across different parameter combinations.

To ensure repeatability and to avoid distortion of the results due to transient effects, only the central section of the groove (10–40 mm), where the machining process is dynamically stable, was used for RMS vibration analysis and surface roughness evaluation.

Vibrations were recorded using three accelerometers, each measuring vibrations along the individual X, Y, and Z axes. In the statistical analysis, the overall RMS vector value (RMS V) was considered according to the following formula:

$$\text{RMS V} = \sqrt{(RMS \quad X^2) + (RMS \quad Y^2) + (RMS \quad Z^2)} \tag{2}$$

3. Results

In our experiment, we conducted 220 roughness measurements at various technological settings in different sections of the groove, where we performed four measurements in each section and averaged them. We also calculated the average roughness value (Ra average) in the middle section of the groove, specifically in the segments 10–20 mm to 30–40 mm. During the milling of the groove in this section, vibration values were measured, from which the RMS was calculated according to Formula (1), and the RMS vector (RMS V) was calculated using Formula (2). In the following Tables 6–11, we can see the average roughness values for individual technological settings, as well as the average values for each section of the groove and the resulting roughness of the entire groove. The tables also display the RMS values in all axes processed from the obtained vibrations.

In the statistical analysis, the values of average roughness Ra average and the RMS vector vibration values were taken into account. The measured roughness and vibration values were statistically evaluated using IBM SPSS Statistics version 26. We tested the normality of the data using the Kolmogorov–Smirnov test. The result of the normality test yielded a p-value of 0.007 for the average roughness Ra average, and for the RMS vector value, we obtained a p-value of 0.003. The results of the normality test can be seen in the following images, where Image 19-a shows unevenly distributed values for Ra average, and Image 19-b graphically represents the unevenly distributed values for the RMS vector values.

https://doi.org/10.3390/app16041832

Table 6. Surface roughness values Ra and RMS from the first setting of technological parameters.

Parameter	Values										
n [rpm]	6000	6000	6000	6000	6000	6000	6000	6000	6000	6000	6000
h [mm]	0.5	0.5	0.5	0.5	0.5	0.5	0.5	0.5	0.5	0.5	0.5
v [mm · s^{-1}]	5	10	15	20	30	40	50	60	70	80	90
0–10 [µm]	0.85	1.72	1.60	2.27	2.64	3.43	3.73	3.86	3.91	4.03	4.76
10–20 [µm]	1.04	1.54	2.13	1.98	2.88	3.39	3.95	3.98	4.47	5.06	5.66
20–30 [µm]	1.28	2.11	2.58	2.22	2.71	3.51	4.05	4.44	4.80	5.09	5.76
30–40 [µm]	1.29	2.06	2.41	2.16	3.05	3.41	4.59	5.16	4.73	3.53	4.52
40–50 [µm]	1.71	2.10	2.40	2.21	3.07	3.36	4.03	4.45	4.38	4.67	5.08
Ra average [µm]	1.20	1.90	2.37	2.12	2.88	3.43	4.20	4.53	4.67	4.56	5.32
RMS Z	4.67	6.24	7.20	8.07	8.48	9.57	10.22	11.19	12.60	14.04	14.25
RMS X	2.35	3.84	4.90	5.58	6.02	7.17	7.64	8.84	9.39	10.18	10.42
RMS Y	2.99	4.71	5.40	6.08	6.57	7.30	7.75	8.85	9.59	10.47	10.94
RMS V	6.02	8.71	10.24	11.55	12.30	14.01	14.93	16.78	18.41	20.26	20.77

Table 7. Surface roughness values Ra and RMS from the second setting of technological parameters.

Parameter	Values										
n [rpm]	6000	6000	6000	6000	6000	6000	6000	6000	6000	6000	6000
h [mm]	1	1	1	1	1	1	1	1	1	1	1
v [mm · s^{-1}]	5	10	15	20	30	40	50	60	70	80	90
0–10 [µm]	0.81	1.32	1.88	2.31	3.15	4.22	4.62	4.34	5.13	4.51	5.25
10–20 [µm]	0.78	1.22	1.65	1.99	3.07	3.51	4.06	5.08	5.60	4.99	5.52
20–30 [µm]	0.83	1.16	1.93	2.20	3.33	3.71	3.94	4.87	5.74	4.51	4.78
30–40 [µm]	0.79	1.31	2.15	2.23	3.58	3.77	4.24	6.47	4.44	8.11	4.31
40–50 [µm]	2.36	1.68	2.06	2.00	3.19	3.03	4.04	4.82	5.31	9.16	7.63
Ra average [µm]	0.80	1.23	1.91	2.14	3.33	3.66	4.08	5.47	5.26	5.87	4.87
RMS Z	8.20	11.2	13.1	15.3	17.1	18.7	21.2	23.5	25.2	26.6	28.8
RMS X	2.60	5.12	6.61	7.81	9.10	10.9	12.3	14.2	14.9	15.9	17.0
RMS Y	4.89	6.98	8.39	9.56	10.8	11.7	13.1	14.3	15.2	17.0	17.7
RMS V	9.89	14.1	16.93	19.66	22.22	24.70	27.89	31.04	33.06	35.44	37.87

Table 8. Surface roughness values Ra and RMS from the third setting of technological parameters.

Parameter	Values										
n [rpm]	6000	6000	6000	6000	6000	6000	6000	6000	6000	6000	6000
h [mm]	1.5	1.5	1.5	1.5	1.5	1.5	1.5	1.5	1.5	1.5	1.5
v [mm · s^{-1}]	5	10	15	20	30	40	50	60	70	80	90
0–10 [µm]	0.88	1.96	2.90	3.14	4.99	4.29	5.49	6.06	6.82	8.28	6.56
10–20 [µm]	0.68	1.39	2.92	4.51	4.43	4.11	6.08	8.24	6.66	6.52	5.88
20–30 [µm]	0.73	1.72	4.52	3.57	4.21	5.62	8.08	8.30	6.53	7.80	8.91
30–40 [µm]	0.81	2.29	3.40	3.66	6.29	8.99	5.91	8.91	5.41	6.01	5.87
40–50 [µm]	1.74	4.09	4.07	3.39	3.44	4.74	6.94	3.05	6.03	6.71	9.32
Ra average [µm]	0.74	1.80	3.62	3.91	4.97	6.24	6.69	8.48	6.20	6.78	6.89
RMS Z	9.63	11.3	13.34	15.41	19.52	23.84	27.66	31.30	35.80	37.97	41.42
RMS X	3.04	5.40	7.01	8.51	11.0	13.42	16.25	18.39	20.49	21.50	23.85
RMS Y	4.65	5.90	7.43	9.00	11.3	13.97	17.00	19.78	22.66	24.23	26.56
RMS V	11.1	13.9	16.81	19.77	25.14	30.72	36.31	41.34	47.07	49.91	54.68

https://doi.org/10.3390/app16041832

Table 9. Surface roughness values Ra and RMS from the fourth setting of technological parameters.

Parameter	Values										
n [rpm]	10,500	10,500	10,500	10,500	10,500	10,500	10,500	10,500	10,500	10,500	10,500
h [mm]	0.5	0.5	0.5	0.5	0.5	0.5	0.5	0.5	0.5	0.5	0.5
v [$mm \cdot s^{-1}$]	5	10	15	20	30	40	50	60	70	80	90
0–10 [μm]	0.50	0.73	0.93	0.83	1.23	1.60	1.64	1.89	1.96	2.31	2.49
10–20 [μm]	0.47	0.67	0.88	0.98	1.35	1.69	1.65	1.85	1.93	2.16	2.28
20–30 [μm]	0.52	0.71	0.89	0.99	1.13	1.68	1.60	1.85	2.16	2.11	2.39
30–40 [μm]	0.68	1.32	1.15	1.01	1.30	1.48	1.68	1.87	2.05	2.20	2.44
40–50 [μm]	1.79	1.47	1.39	1.26	1.34	1.58	1.73	2.44	2.32	2.68	2.60
Ra average [μm]	0.56	0.90	0.97	0.99	1.26	1.61	1.64	1.86	2.05	2.16	2.37
RMS Z	4.05	4.88	5.68	6.58	8.25	9.44	10.07	10.13	10.71	10.45	10.62
RMS X	1.75	2.62	3.40	4.28	5.84	6.56	6.68	6.90	7.57	7.88	8.63
RMS Y	2.88	3.93	4.54	5.96	7.34	6.83	7.15	7.40	8.02	8.49	8.91
RMS V	5.27	6.79	8.02	9.86	12.49	13.37	14.04	14.31	15.38	15.60	16.33

Table 10. Surface roughness values Ra and RMS from the fifth setting of technological parameters.

Parameter	Values										
n [rpm]	10,500	10,500	10,500	10,500	10,500	10,500	10,500	10,500	10,500	10,500	10,500
h [mm]	1	1	1	1	1	1	1	1	1	1	1
v [$mm \cdot s^{-1}$]	5	10	15	20	30	40	50	60	70	80	90
0–10 [μm]	1.19	1.00	1.15	2.08	2.49	3.02	3.45	3.58	3.40	3.59	3.55
10–20 [μm]	1.07	1.25	1.30	1.53	2.43	3.19	3.52	3.82	3.67	3.32	4.02
20–30 [μm]	1.02	0.97	1.25	1.67	2.39	3.60	3.22	3.98	3.84	3.56	3.64
30–40 [μm]	0.90	0.86	1.29	1.62	2.34	3.14	4.45	4.01	3.80	3.96	3.23
40–50 [μm]	1.07	1.12	1.47	1.50	2.24	2.53	3.42	3.48	3.65	4.17	3.80
Ra average [μm]	0.99	1.02	1.28	1.61	2.39	3.31	3.73	3.94	3.77	3.61	3.63
RMS Z	5.25	7.31	8.74	10.10	13.12	16.16	18.22	20.31	20.65	22.00	21.71
RMS X	2.07	3.99	5.12	6.53	8.29	10.14	11.15	12.49	12.95	14.03	14.72
RMS Y	3.18	6.97	7.20	8.13	9.51	10.76	11.64	12.54	13.17	13.78	14.36
RMS V	6.48	10.8	12.43	14.52	18.20	21.91	24.33	26.94	27.71	29.50	29.91

Table 11. Surface roughness values Ra and RMS from the sixth setting of technological parameters.

Parameter	Values										
n [rpm]	10,500	10,500	10,500	10,500	10,500	10,500	10,500	10,500	10,500	10,500	10,500
h [mm]	1.5	1.5	1.5	1.5	1.5	1.5	1.5	1.5	1.5	1.5	1.5
v [$mm \cdot s^{-1}$]	5	10	15	20	30	40	50	60	70	80	90
0–10 [μm]	0.98	0.74	1.04	1.44	2.47	3.52	3.75	4.06	4.11	4.15	4.40
10–20 [μm]	0.87	0.73	1.05	1.43	2.42	3.39	3.86	4.36	3.65	4.06	4.81
20–30 [μm]	1.16	0.80	1.04	1.35	2.35	4.13	4.24	4.19	4.29	3.84	3.85
30–40 [μm]	0.96	0.73	1.12	1.45	2.14	3.37	4.01	3.91	3.89	4.14	4.64
40–50 [μm]	2.22	2.55	3.94	3.16	4.09	3.64	4.27	4.04	4.46	5.29	4.29
Ra average [μm]	1.00	0.75	1.07	1.41	2.30	3.63	4.04	4.15	3.94	4.01	4.43
RMS Z	11.99	14.47	13.81	14.42	16.82	19.61	23.26	25.41	29.20	31.89	33.59
RMS X	4.47	4.28	3.86	4.70	7.20	9.89	12.87	14.20	16.32	17.14	19.28
RMS Y	7.79	7.72	8.86	9.93	11.12	12.89	15.14	16.50	18.23	19.04	21.24
RMS V	14.98	16.95	16.85	18.13	21.42	25.47	30.59	33.46	38.09	40.90	44.17

Since the data were unevenly distributed based on descriptive statistics, we opted for non-parametric tests (Figure 19). In preparing to compare the individual values of all segments, we used the Independent Samples Kruskal–Wallis Test. When observing the differences in roughness measured in the various segments across all changing technological

https://doi.org/10.3390/app16041832

parameter settings, no statistically significant differences in the average values were found (Figure 20), with $p = 0.588$.

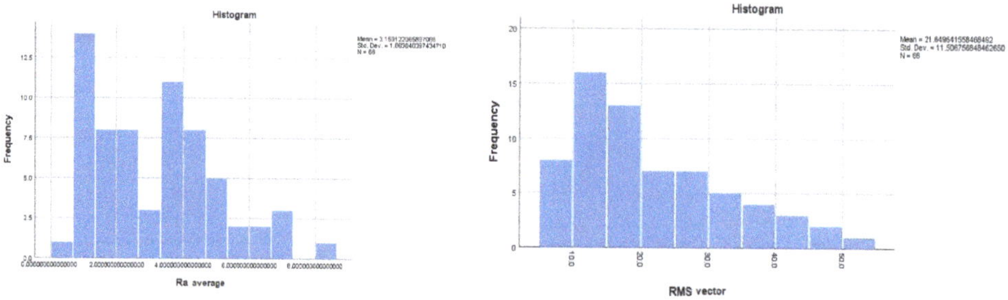

(**a**) (**b**)

Figure 19. The results of the statistical test of normality: (**a**) unevenly distributed data of Ra average roughness values; (**b**) unevenly distributed data of RMS values.

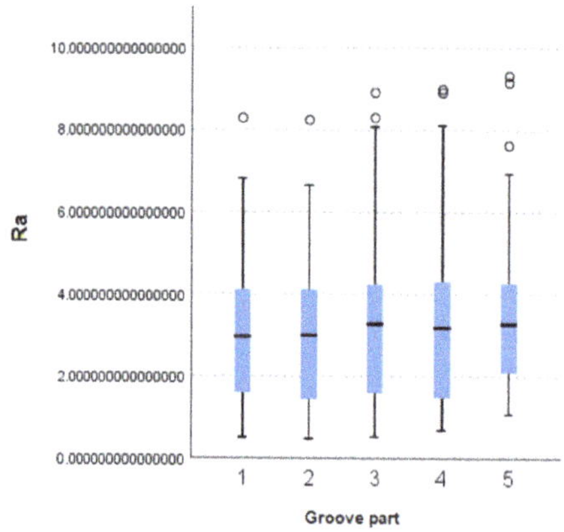

Figure 20. Comparison of roughness values measured in all sections of the groove was performed using the Kruskal–Wallis independent test.

Based on the results, we can determine that there is no difference in the measured values of Ra among the individual segments (10–20 to 30–40) in the given grooves. This means that the roughness measurement in the main experiment will only take place in the middle part of the groove (20–30), which will significantly reduce the number of repeated measurements and decrease the time required for conducting the main experiment.

Another aspect of our statistical analysis was to examine the correlations between selected technological parameters and roughness. Since the previous results showed that we were working with unevenly distributed data, we chose the non-parametric equivalent for correlation analysis (Spearman's correlation test; Table 12). In Table 12, the significance level is indicated with asterisks, showing the strength of statistical significance, where *—$p < 0.05$, **—$p < 0.01$, and ***—$p < 0.001$.

The values of the correlation coefficient r ranging from 0.8 to 1 (or −0.8 to −1) are considered particularly strong, indicating a very strong mutual dependence between the variables. Values from 0.4 to 0.8 (or −0.4 to −0.8) are considered moderately strong, while values from 0 to 0.4 (or −0.4 to 0) are considered weak. From the correlation results, it is clear that the r values for the spindle speed parameter take on negative values, indicating an inverse relationship. In our case, this explains the fact that as we set the spindle speed parameter to a higher value, the resulting roughness values tend to be lower. The

https://doi.org/10.3390/app16041832

correlation values for the spindle speed parameter fall within the moderate range of 0.4 to 0.8, suggesting that spindle speed has a moderately strong impact on the resulting surface quality.

Table 12. Results of correlation test.

		rpm	h	v	0–10	10–20	20–30	30–40	40–50	Ra	Ra Average	RMS Z	RMS X	RMS Y	RMS V
rpm	r	1.000	0.000	0.000	−0.40 **	−0.40 **	−0.448 **	−0.435 **	−0.354 **	−0.429 **	−0.438 **	−0.138	−0.174	−0.079	−0.126
	p		1.000	1.000	0.001	0.001	0.000	0.000	0.004	0.000	0.000	0.271	0.162	0.530	0.312
h	r	0.000	1.000	0.000	0.327 **	0.276 *	0.285 *	0.265 *	0.420 **	0.326 **	0.282 *	0.675 **	0.386 **	0.552 **	0.599 **
	p	1.000		1.000	0.007	0.025	0.020	0.032	0.000	0.007	0.022	0.000	0.001	0.000	0.000
v	r	0.000	0.000	1.000	0.781 **	0.792 **	0.735 **	0.727 **	0.695 **	0.776 **	0.760 **	0.667 **	0.854 **	0.768 **	0.733 **
	p	1.000	1.000		0.000	0.000	0.000	0.000	0.000	0.000	0.000	0.000	0.000	0.000	0.000

In the following figure (Figure 21), we can observe the dependence of roughness on spindle speed, where we compared the change in roughness at spindle speeds of 6000 rpm and 10,500 rpm. This comparison was conducted with a constant depth of field parameter of 0.5, 1, and 1.5 mm. As shown in Figure 21, roughness at 6000 rpm reaches higher values than at 10,500 rpm for all values of material removal and varying feed rates of the tool.

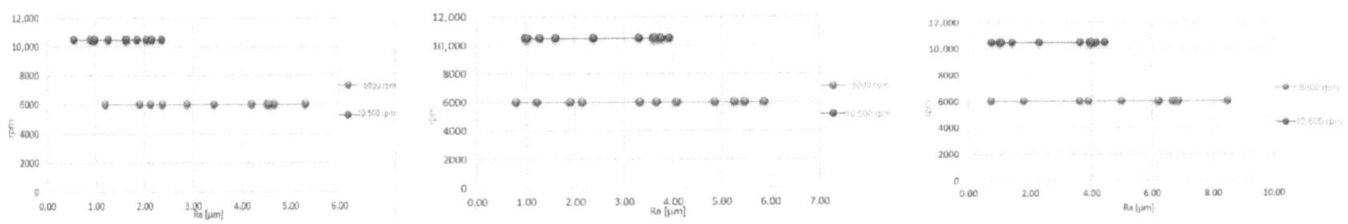

(a) (b) (c)

Figure 21. Dependence of roughness on spindle speed: (**a**) depth of field 0.5 mm; (**b**) depth of field 1 mm; (**c**) depth of field 1.5 mm.

As seen, the parameter representing spindle speed has an effect on the resulting roughness of the groove, and with increasing spindle speeds, the roughness values decrease. This fact is also reflected in the following comparisons of vibration results, where we compared the actual vibration profiles between the maximum and minimum spindle speeds. Figure 22 illustrates the different amplitudes between the set spindle speeds of 6000 rpm and 10,500 rpm, with a depth of field of 0.5 mm and a tool feed rate of 10 mm/s. The upper part of the figure displays the accelerations at 6000 rpm, while the lower part represents the vibration profile at 10,500 rpm. This comparison further emphasises the relationship between spindle speed and vibration behaviour during the machining process, highlighting the impact of varying speeds on the system's dynamic response.

The second technological parameter examined, affecting the surface roughness of the groove, was the depth of field. For this parameter, we recorded values indicating a weak correlation (0 to 0.4), which means that the depth of field has the least influence on the final surface quality of the machined material. When comparing the vibration profiles with the same spindle speeds and tool feed rates but with varying depths of field, we can see that the amplitude differences are minimal. To illustrate this slight amplitude difference, we compared the vibration profiles (Figure 23) during groove milling at a spindle speed of 6000 rpm and a tool feed rate of 10 mm/s. The upper part of the figure represents the vibration profile for a depth of field of 0.5 mm, the middle part shows the profile for a depth of 1 mm, and the bottom part represents the profile for a depth of 1.5 mm.

https://doi.org/10.3390/app16041832

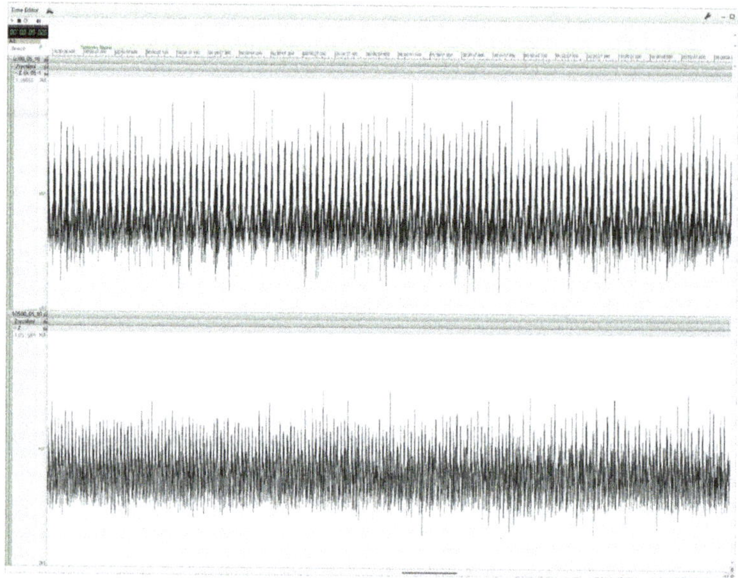

Figure 22. Vibration values measured when changing the technological parameter of the spindle speed.

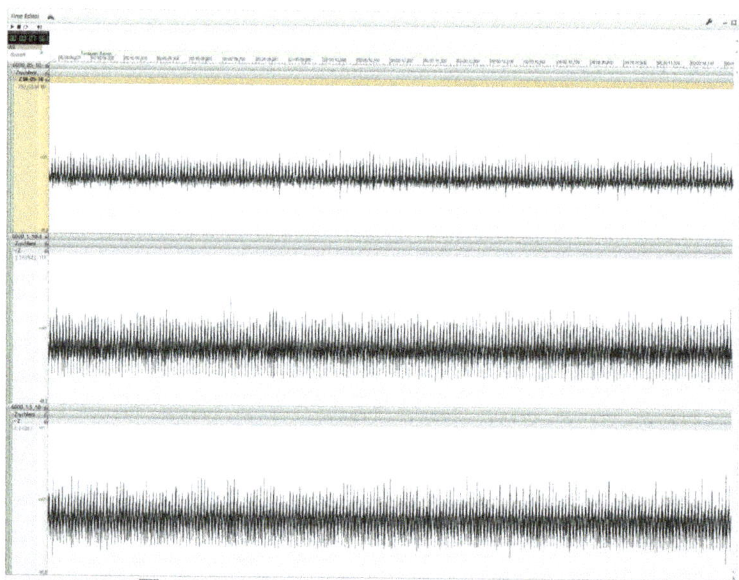

Figure 23. Comparison of the vibration value when changing the technological parameter of the depth of field.

It is worth noting that the tool feed rate parameter shows particularly strong correlation values (0.8 to 1), indicating that this parameter has the greatest influence on the resulting surface quality. This can also be illustrated by Figure 24, where we observe an increase in surface roughness as the tool feed rate increases. Figure 24 depicts the dependence of surface roughness on tool feed rate at a depth of field of 1 mm and spindle speed of 10,500 rpm. Significant differences were also observed in the vibrations, as shown in Fig-

https://doi.org/10.3390/app16041832

ure 25. The vibration samples were displayed for corresponding technological parameters with feed rates of 5 mm/s, 10 mm/s, 30 mm/s, and 80 mm/s. The higher feed rates led to increased surface roughness and vibration amplitudes, highlighting the importance of this parameter in achieving the desired surface finish.

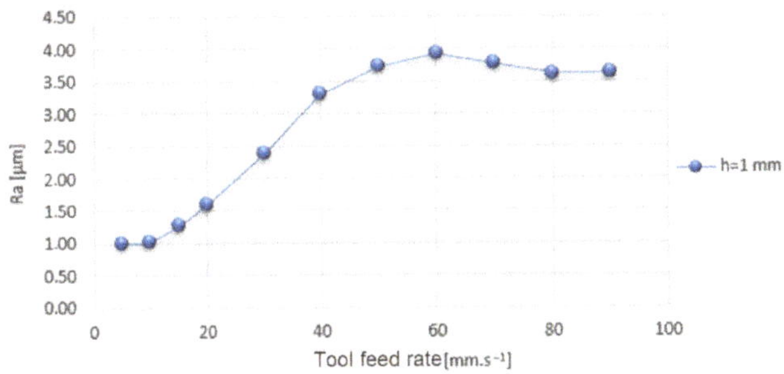

Figure 24. Dependence of surface roughness on tool feed rate at spindle speed n = 10,500 rpm, depth of field 1 mm.

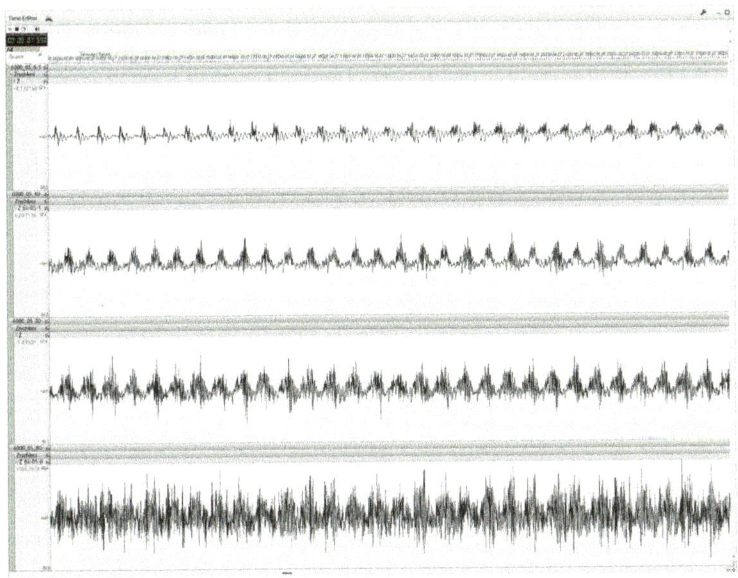

Figure 25. Change in the vibration amplitude with a change in the speed of the cutting tool.

In Figure 26, we can see an overall comparison of the feed rate of the tool in relation to the resulting surface quality of the machined material, considering changes in both the depth of field and spindle speed. From the figure, we can clearly observe the comparison of the significance and level of influence of the technological parameters (spindle speed and depth of field) on the resulting surface roughness. If we take into account the Ra average value (purple cross mark) at a speed of 90 mm $\cdot$ s^{-1} (point A1), which corresponds to technological parameters such as a spindle speed of 10,500 rpm and a depth of field of 0.5 mm, and compare it with point A2 (marked with a blue star), which has identical technological parameters except for a depth of field of 1 mm, we can observe a minimal difference in the Ra roughness value (ΔA1–A2). However, if we apply such a comparison while maintaining the depth of field of 0.5 mm and include point B2 (marked with a blue diamond), where the technological parameter spindle speed is changed to 6000 rpm and

https://doi.org/10.3390/app16041832

compared with point A1, we obtain a larger difference in Ra roughness values (ΔA1–B1). This graphical representation allows us to highlight the results of the statistical analysis, which suggests the significance of individual technological parameters on the resulting surface roughness. This means that spindle speed has a more significant impact on surface roughness than changes in depth of field, especially as the tool feed rate increases.

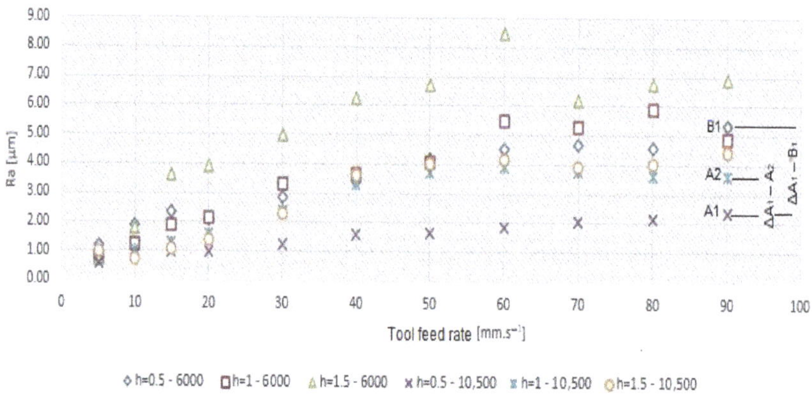

Figure 26. Dependence of Ra average on tool feed speed with changing depth of field and spindle speed.

The final parameter investigated through statistical analysis was the RMS V value. The RMS V value showed the highest correlation with the Ra mean roughness value (r = 0.793). From previous results, we know that the most significant influence on surface roughness comes from the tool feed rate parameter. The results from the statistical analysis indicate that the RMS V value significantly affects the surface roughness of the machined material. From the combined results, we can conclude that the RMS V value significantly influences the final quality of the machined material, Ra, meaning that the surface roughness is significantly dependent on vibrations. The results can be visually presented in Figure 27, where the left side shows a graph of the dependence of roughness on tool speed, and the right side illustrates the graph of roughness dependence on the RMS value. In the first graph, we observe an increase in the Ra value with increasing speed, which correlates with the increasing roughness from the RMS V value.

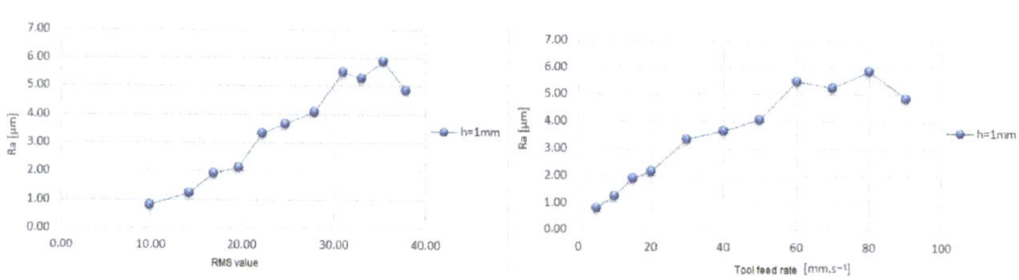

(a) (b)

Figure 27. The dependence of the roughness value on the vibration value: (**a**) the course of increasing Ra values from RMS V values; (**b**) the course of increasing Ra values from the tool feed rate.

In addition to the significant dependence on the Ra value, RMS V values also show a high correlation coefficient with all technological parameters, except for spindle speed. A strong correlation was noted between the tool feed rate and the depth of field, where these parameters exhibited approximately the same correlation value. The significance and

https://doi.org/10.3390/app16041832

level of influence of the technological parameters on the RMS V values is also illustrated in Figure 28.

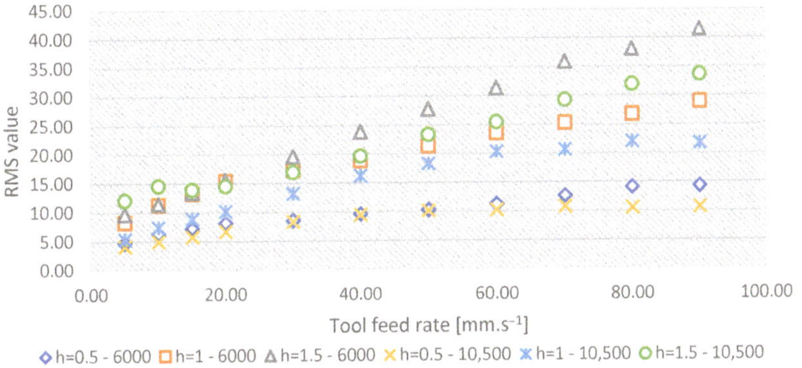

Figure 28. Dependence of RMS on tool feed rate with changing material removal depth and spindle speed.

If we compare the RMS V values at technological parameters such as a spindle speed of 10,500 rpm with a depth of field of 0.5 mm across the entire range of tool feed rates (yellow cross) to those at a spindle speed of 6000 rpm with the same depth of field of 0.5 mm across the entire range of tool feed rates (blue diamond), we observe minimal differences in the RMS V values. Furthermore, if we compare the RMS V values at a spindle speed of 10,500 rpm and a depth of field of 0.5 mm across the entire range of tool feed rates with the RMS V values corresponding to a spindle speed of 10,500 rpm and a depth of field of 1 mm across the entire range of tool feed rates (blue cross), we can see a significant difference in the RMS V values. If we were to conduct all comparisons of spindle speeds and depth of field in relation to the RMS V value, we would find consistent results showing larger differences in values between the depth of field and the RMS V and minimal differences between spindle speeds and the RMS V value. This aligns with the results of the statistical analysis.

4. Discussion

Machining is one of the most complex processes in the manufacturing field, which presents a significant challenge for a robotic manipulator with low stiffness. In numerous discussions on robotic machining, technological parameters and their influence on the milling process have been emphasised. The generation of the tool path is the main distinction when comparing the mechanism between a robot and a CNC machine tool [28]. Robots used in machining processes have many advantages over traditional machines, but the phenomenon of vibrations in robotic milling prevents its widespread adoption. As a result, discussions have arisen about the issue of vibrations during planar milling with robots and the method of generating the milling tool path, as well as the positioning of the workpiece clamping, with the aim of providing guidance on how to suppress vibrations and improve production quality [29]. In most cases, researchers have focused on the position and placement of the machining material relative to the base of the robotic arm, as well as the tool movement strategy.

Authors Jing Li and Biao Li [30], in their study, focused on the vibrations of the robotic arm and examined the surface quality of the machined material in relation to the change in the cutting tool movement along the X and Y axes under identical technological parameters. The results indicate that there is a difference in which direction the cutting tool, attached

https://doi.org/10.3390/app16041832

to the robotic arm, moves, due to the robot's stiffness characteristics. This is because the stiffness of the robot varies across the X, Y, and Z axes. The experimental findings also demonstrate that the clamping position of the workpiece relative to the robot significantly impacts the stability of vibrations in the robotic arm. The surface quality of the workpiece is better in areas where the robotic arm is more stable. At the optimal position, the surface quality improved by 57.3%, and vibrations were reduced by 50.3%. The second examined factor was the tool path, where the authors revealed that the highest stiffness of the robotic arm is along the X axis of the global coordinate system, while the lowest stiffness is along the Y axis. Additionally, the further the cutting tool is positioned from the robot's global coordinate system, the lower the stiffness of the robot becomes. These results highlight the importance of positioning the workpiece relative to the base of the robotic arm.

In the study by Xu et al. [31], the authors investigated the influence of two key parameters (robot performance dependent on position and the type and geometry of the cutting tool) on cutting force and surface quality improvement. A robotic machining system was developed for this research, consisting of a six-axis industrial robot, a two-axis positioning device, and a linear track. The combined effect of milling process parameters and the robot's position on machining outcomes was experimentally examined. A multi-objective optimisation was conducted based on Taguchi's grey relational analysis to achieve lower cutting forces and improved surface quality. The optimal combination of process parameters and robot positions was identified, resulting in the best overall milling performance. The study's results demonstrate that the cutting force is heavily influenced by the choice of process parameters, highlighting their importance in affecting the cutting force. Additionally, the position of the robot was found to have a significant impact, which cannot be overlooked. Overall, the depth of field was identified as the most critical factor affecting cutting force in both the Y and Z directions. A deeper field intuitively generates a higher cutting force, which in turn leads to increased vibrations throughout the entire assembly. The study further revealed that the main factors influencing the surface quality of the machined material are the milling process parameters, where the key components were spindle speed, tool path, and the orientation of the positioning table (worktable). It was found that vibrations are influenced not only by the process parameters but also by the configuration of the robot. This is likely because the robot's dynamic behaviour changes within its workspace. The results suggest that the robot exhibits higher stiffness in certain positions, which are more favourable for robotic machining. The statistical method ANOVA showed that the depth of field, orientation of the positioning table, and spindle speed were the three most significant factors affecting the cutting force in the Y direction (Fy), with contributions of 27.18%, 19.44%, and 15.83%, respectively. The analysis revealed similar results when examining the parameters' influence on the cutting force in the Z direction (Fz), where the depth of cut contributed 31.69%, the orientation of the positioning Table 18.4%, and the spindle speed 18.29%. Regarding surface quality, the spindle speed had the most significant influence, accounting for 40.12%, while the orientation of the positioning table represented the primary factor with a contribution of 11.89%.

In the analysed literature, Pillai et al. [32] also addressed the issue of machining using a robotic arm, where the goal of their study was to derive sets of optimal combinations of process parameters for face milling on the aforementioned six-axis robotic machining centre. The study also examined the influence of process parameters such as tool path, spindle speed, and feed rate on performance characteristics, including machining time and surface roughness. The study utilised Taguchi's Grey relational analysis (GRA) and ANOVA to explore the effects of machining parameters (spindle speed, tool path, and feed rate). The optimal combination of process parameters estimated through Taguchi's grey relational analysis was as follows: the optimal tool path was the raster path, spindle speed

https://doi.org/10.3390/app16041832

was 14,000 rpm, and feed rate was 800 mm/min. Upon closer examination, the analysis revealed that the tool path strategy was one of the parameters that most significantly influenced performance characteristics like machining time and surface roughness. The tool path strategy accounted for up to 70.58% of the contribution to surface quality, followed by cutting speed with a share of 28.47% and feed rate with a contribution of only 0.95%.

Studies focused on machining using robotic arms indicate that robotic manipulators exhibit low structural stiffness, which, in combination with technological parameters, significantly affects the resulting surface quality. The most influential factor with respect to stiffness is the position of the workpiece relative to the base of the robotic arm. In our study, the maximum extension of the robotic arm limited by the workspace was used, as this position was considered the most critical in terms of arm stiffness. Future research will focus on this issue by mapping the entire workspace in order to determine the optimal workpiece positioning, which will serve as the basis for the main experimental investigation. The influence of technological parameters reported in the literature highlights the significant importance of individual parameters. Our study focused on investigating the effect of selected technological parameters on the resulting surface roughness of the material. In the study by Xu et al. [31], the depth of cut was reported as the most influential parameter; however, that work investigated cutting forces acting on the cutting tool in relation to changes in technological parameters. In that context, the depth of cut had the greatest influence. In contrast, in our study this parameter exhibited the lowest significance, as cutting forces were not monitored; instead, surface quality was evaluated, which is primarily influenced by the tool feed rate and spindle speed. Consequently, the depth of cut showed the least effect in terms of forces acting on the surface of the workpiece. Many studies primarily focus on cutting forces, whereas in our research vibration levels (RMS) and surface roughness were evaluated. It was also noted that technological parameters influence cutting forces differently than they affect surface roughness, which naturally leads to differing conclusions. While depth of cut is often a dominant factor in force-orientated studies, our results indicate that the tool feed rate is significantly more influential, which can be attributed to the lower stiffness of the robotic arm structure and its higher sensitivity to vibrations. Differences in the evaluated response variables (forces vs. roughness), as well as differences in material, robot type, and structural stiffness, logically result in differing outcomes. All of these insights will be taken into account in future research on robotic milling, contributing to the accurate design of the main experiment and the development of a mathematical model describing the milling process using a robotic arm.

5. Conclusions

In the screening experiment, we monitored the quality of the machined surface based on various changing technological parameters of milling. The goal of the entire experiment was to establish the limits of robotic milling and to determine the settings for individual technological parameters that will be used in the main experiment, as well as to propose the main experiment itself. The screening experiment was conducted in a specific sequence. In the first phase of the experiment, information was gathered about the equipment and systems on which the experiment took place, as well as information about the milled material and measurement devices. The core of the experiment was to establish the correct procedure for the experimental plan, where independent variables such as technological parameters and the positioning of the workpiece were selected. The essence of milling involved measuring vibrations and the change in surface roughness of the material as influenced by the varying aforementioned technological parameters.

Based on the measured values of roughness and vibrations, we used statistical analysis to determine the significance of individual technological parameters affecting the final

https://doi.org/10.3390/app16041832

quality of the machined material. The results of the statistical analysis are crucial for further measurements concerning research in the field of robotic milling and serve as the foundation for designing the main experiment. In this main experiment, we will focus more deeply on robotic milling and the impact of technological parameters on the final quality of the machined material, represented in the form of a mathematical model.

Author Contributions: Conceptualization, L.H. and P.K.; Data curation, L.M.; Formal analysis, O.C., S.S., M.M. and L.M.; Funding acquisition, P.K.; Investigation, L.H. and O.C.; Methodology, L.H. and P.K.; Project administration, P.K. and M.M.; Resources, P.K.; Software, O.C.; Supervision, P.K. and M.M.; Validation, S.S., M.M. and L.M.; Writing—original draft, L.H. and S.S.; Writing—review and editing, P.K. and M.M. All authors have read and agreed to the published version of the manuscript.

Funding: This research received no external funding.

Institutional Review Board Statement: Not applicable.

Informed Consent Statement: Not applicable.

Data Availability Statement: Data available on request from the corresponding author.

Acknowledgments: This paper is a part of the research carried out within the project APVV-23-0084 "Robotic-Based Hybrid Manufacturing of Workpiece for the Concept of Smart Production" funded by the Slovak Research and Development Agency, and the Grant STU "Research on path generation a motion planning of industrial robot and its use in hybrid manufacturing with MEX technology and subtractive processes" funded by STU in Bratislava.

Conflicts of Interest: The authors declare no conflicts of interest.

References

1. Hanko, L.; Križan, P.; Bábics, J. Design of experiment (DOE) for robotic machining operation. *Int. J. Adv. Eng. Manag.* **2022**, *4*, 267–277.
2. Zbiss, K.; Kacem, A.; Santillo, M.; Mohammadi, A. Automatic Optimal Robotic Base Placement for Collaborative Industrial Robotic Car Painting. *Appl. Sci.* **2024**, *14*, 8614. [CrossRef]
3. Urrea, C.; Sari, P.; Kern, J.; Torres, H. Enhancing Adaptability and Autonomy in Cooperative Selective Compliance Assembly Robot Arm Robots: Implementation of Coordination and Rapidly Exploring Random Tree Algorithms for Safe and Efficient Manipulation Tasks. *Appl. Sci.* **2024**, *14*, 6804. [CrossRef]
4. Yang, M.; Liu, J. Research on Six-Degree-of-Freedom Refueling Robotic Arm Positioning and Docking Based on RGB-D Visual Guidance. *Appl. Sci.* **2024**, *14*, 4904. [CrossRef]
5. Wu, N.; Jia, D.; Li, Z.; He, Z. Trajectory Planning of Robotic Arm Based on Particle Swarm Optimization Algorithm. *Appl. Sci.* **2024**, *14*, 8234. [CrossRef]
6. Hanko, L.; Križan, P.; Kijovský, J.; Strigáč, S. Kinematic and dynamic analysis of a robotic arm. In *Zborník Vedeckých Článkov z Konferencie Vývoj Výrobných Systémov a Technológií*; Publishing House of the Slovak University of Technology: Bratislava, Slovakia, 2022.
7. Borys, S.; Kaczmarek, W.; Laskowski, D.; Polak, R. Experimental Study of the Vibration of the Spot Welding Gun at a Robotic Station. *Appl. Sci.* **2022**, *12*, 12209. [CrossRef]
8. Liu, J.; Zhao, Y.; Niu, Y.; Cao, J.; Zhang, L.; Zhao, Y. Optimization of Redundant Degrees of Freedom in Robotic Flat-End Milling Based on Dynamic Response. *Appl. Sci.* **2024**, *14*, 1877. [CrossRef]
9. Gierlak, P.; Warmiński, J. Analysis of Bifurcation Vibrations of an Industrial Robot Arm System with Joints Compliance. *Appl. Sci.* **2023**, *13*, 11941. [CrossRef]
10. Tlusty, J.; Polacek, M. *Beispiele der Behandlung der Selbsterregten Schwingung der Werkzeugmaschinen FoKoMa*; Hanser Verlag: Munich, Germany, 1957.
11. Lin, J.-L.; Huang, H.-P.; Lin, C.-Y. Iterative Learning Control for Vibration Suppression of a Robotic Arm. *Appl. Sci.* **2023**, *13*, 828. [CrossRef]
12. Iglesias Sánchez, I.; Ares, J.E.; González Gaya, C.; Rosales Prieto, V. A New Approach to the Consideration and Analysis of Critical Factors in Robotic Machining. *Appl. Sci.* **2020**, *10*, 8885. [CrossRef]
13. Wu, K.; Kuhlenkoetter, B. Experimental Analysis of the Dynamic Stiffness in Industrial Robots. *Appl. Sci.* **2020**, *10*, 8332. [CrossRef]

https://doi.org/10.3390/app16041832

14. Zhang, L.; Dhupia, J.S.; Wu, M.; Huang, H. A Robotic Drilling End-Effector and Its Sliding Mode Control for the Normal Adjustment. *Appl. Sci.* **2018**, *8*, 1892. [CrossRef]
15. Wolniakowski, A.; Trochimczuk, R.; Moulianitis, V.; Miatliuk, K. Kinematic Modeling of a Trepanation Surgical Robot System. *Appl. Sci.* **2023**, *13*, 9110. [CrossRef]
16. Hanko, L.; Križan, P.; Kijovský, J.; Strigáč, S. Application of topology optimization in the modification of a robotic gripper. *Int. J. Res. Eng. Sci.* **2022**, *10*, 578–587.
17. *ISO 1183*; Plastics—Methods for Determining the Density of Non-Cellular Plastics—Part 1: Immersion Method, Liquid Pycnometer Method and Titration Method. International Organization for Standardization (ISO): Geneva, Switzerland, 2019.
18. *DIN 53455*; Testing of Plastics—Tensile Test. Deutsches Institut für Normung (DIN): Berlin, Germany, 1981.
19. *DIN 53445*; Testing of Plastics—Bending Test. Deutsches Institut für Normung (DIN): Berlin, Germany, 1975.
20. *DIN 53457-Z-B 3*; Testing of Plastics—Determination of Modulus of Elasticity. Deutsches Institut für Normung (DIN): Berlin, Germany, 1987.
21. *DIN 53456*; Testing of Plastics—Ball Indentation Hardness. Deutsches Institut für Normung (DIN): Berlin, Germany, 1984.
22. *DIN 53453*; Testing of Plastics—Impact Strength. Deutsches Institut für Normung (DIN): Berlin, Germany, 1983.
23. *ISO 604*; Plastics—Determination of Compressive Properties. International Organization for Standardization (ISO): Geneva, Switzerland, 2002.
24. *ASTM D 696*; Standard Test Method for Coefficient of Linear Thermal Expansion of Plastics Between −30 °C and 30 °C with a Vitreous Silica Dilatometer. ASTM International: West Conshohocken, PA, USA, 2016.
25. *ISO 19712-2*; Plastics—Determination of Hardness—Part 2: Rockwell Hardness. International Organization for Standardization (ISO): Geneva, Switzerland, 2000.
26. *DIN EN 101*; Testing of Plastics—Determination of Scratch Hardness According to Mohs. Deutsches Institut für Normung (DIN): Berlin, Germany, 1994.
27. *ISO 178*; Plastics—Determination of Flexural Properties. International Organization for Standardization (ISO): Geneva, Switzerland, 2019.
28. Owen, W.S.; Croft, E.A.; Benhabib, B. A multi-arm robotic system for optimal sculpting. *Robot Comput.-Integr. Manuf.* **2008**, *24*, 92–104. [CrossRef]
29. Wei, L.; Wang, L. Industrial robotic machining: A review. *Int. J. Adv. Manuf. Technol.* **2019**, *103*, 1239–1255. [CrossRef]
30. Li, J.; Li, B.; Shen, N.; Qian, H.; Guo, Z. Effect of the cutter path and the workpiece clamping position on the stability of the robotic milling system. *Int. J. Adv. Manuf. Technol.* **2017**, *89*, 2919–2933. [CrossRef]
31. Xu, P.; Gao, Y.; Yao, X.; Ng, Y.H.; Liu, K.; Bi, G. Influence of process parameters and robot postures on surface quality in robotic machining. *Int. J. Adv. Manuf. Technol.* **2022**, *124*, 2545–2561. [CrossRef]
32. Unnikrishna, P.J.; Sanghrajka, I.; Shunmugavel, M.; Muthuramalingam, T.; Goldberg, M.; Littlefair, G. Optimisation of multiple response characteristics on end milling of aluminium alloy using Taguchi-grey relational approach. *Measurement* **2018**, *124*, 291–298. [CrossRef]

https://doi.org/10.3390/app16041832

MDPI AG
Grosspeteranlage 5
4052 Basel
Switzerland
Tel.: +41 61 683 77 34

Applied Sciences Editorial Office
E-mail: applsci@mdpi.com
www.mdpi.com/journal/applsci

www.ingramcontent.com/pod-product-compliance
Lightning Source LLC
Chambersburg PA
CBHW040809250926
51850CB00007BA/1118
* 9 7 8 3 7 2 5 8 7 8 0 7 9 *